T0275825

LONDON MATHEMATICAL SOCIETY LECTURE NOTE SERIES

119 Triangulated categories in the representation theory of finite-dimensional algebras, D. HAPPEL
121 Proceedings of *Groups - St Andrews 1985*, E. ROBERTSON & C. CAMPBELL (eds)
122 Non-classical continuum mechanics, R.J. KNOPS & A.A. LACEY (eds)
124 Lie groupoids and Lie algebroids in differential geometry, K. MACKENZIE
125 Commutator theory for congruence modular varieties, R. FREESE & R. MCKENZIE
126 Van der Corput's method of exponential sums, S.W. GRAHAM & G. KOLESNIK
127 New directions in dynamical systems, T.J. BEDFORD & J.W. SWIFT (eds)
128 Descriptive set theory and the structure of sets of uniqueness, A.S. KECHRIS & A. LOUVEAU
129 The subgroup structure of the finite classical groups, P.B. KLEIDMAN & M.W.LIEBECK
130 Model theory and modules, M. PREST
131 Algebraic, extremal & metric combinatorics, M-M. DEZA, P. FRANKL & I.G. ROSENBERG (eds
132 Whitehead groups of finite groups, ROBERT OLIVER
133 Linear algebraic monoids, MOHAN S. PUTCHA
134 Number theory and dynamical systems, M. DODSON & J. VICKERS (eds)
135 Operator algebras and applications, 1, D. EVANS & M. TAKESAKI (eds)
136 Operator algebras and applications, 2, D. EVANS & M. TAKESAKI (eds)
137 Analysis at Urbana, I, E. BERKSON, T. PECK, & J. UHL (eds)
138 Analysis at Urbana, II, E. BERKSON, T. PECK, & J. UHL (eds)
139 Advances in homotopy theory, S. SALAMON, B. STEER & W. SUTHERLAND (eds)
140 Geometric aspects of Banach spaces, E.M. PEINADOR and A. RODES (eds)
141 Surveys in combinatorics 1989, J. SIEMONS (ed)
142 The geometry of jet bundles, D.J. SAUNDERS
143 The ergodic theory of discrete groups, PETER J. NICHOLLS
144 Introduction to uniform spaces, I.M. JAMES
145 Homological questions in local algebra, JAN R. STROOKER
146 Cohen-Macaulay modules over Cohen-Macaulay rings, Y. YOSHINO
147 Continuous and discrete modules, S.H. MOHAMED & B.J. MÜLLER
148 Helices and vector bundles, A.N. RUDAKOV et al
149 Solitons, nonlinear evolution equations and inverse scattering, M.J. ABLOWITZ & P.A. CLARKSON
150 Geometry of low-dimensional manifolds 1, S. DONALDSON & C.B. THOMAS (eds)
151 Geometry of low-dimensional manifolds 2, S. DONALDSON & C.B. THOMAS (eds)
152 Oligomorphic permutation groups, P. CAMERON
153 L-functions and arithmetic, J. COATES & M.J. TAYLOR (eds)
154 Number theory and cryptography, J. LOXTON (ed)
155 Classification theories of polarized varieties, TAKAO FUJITA
156 Twistors in mathematics and physics, T.N. BAILEY & R.J. BASTON (eds)
157 Analytic pro-*p* groups, J.D. DIXON, M.P.F. DU SAUTOY, A. MANN & D. SEGAL
158 Geometry of Banach spaces, P.F.X. MÜLLER & W. SCHACHERMAYER (eds)
159 Groups St Andrews 1989 volume 1, C.M. CAMPBELL & E.F. ROBERTSON (eds)
160 Groups St Andrews 1989 volume 2, C.M. CAMPBELL & E.F. ROBERTSON (eds)
161 Lectures on block theory, BURKHARD KÜLSHAMMER
162 Harmonic analysis and representation theory for groups acting on homogeneous trees, A. FIGA-TALAMANCA & C. NEBBIA
163 Topics in varieties of group representations, S.M. VOVSI
164 Quasi-symmetric designs, M.S. SHRIKANDE & S.S. SANE
165 Groups, combinatorics & geometry, M.W. LIEBECK & J. SAXL (eds)
166 Surveys in combinatorics, 1991, A.D. KEEDWELL (ed)
167 Stochastic analysis, M.T. BARLOW & N.H. BINGHAM (eds)
168 Representations of algebras, H. TACHIKAWA & S. BRENNER (eds)
170 Manifolds with singularities and the Adams-Novikov spectral sequence, B. BOTVINNIK
173 Discrete groups and geometry, W.J. HARVEY & C. MACLACHLAN (eds)
174 Lectures on mechanics, J.E. MARSDEN
175 Adams memorial symposium on algebraic topology 1, N. RAY & G. WALKER (eds)
176 Adams memorial symposium on algebraic topology 2, N. RAY & G. WALKER (eds)
177 Applications of categories in computer science, M.P. FOURMAN, P.T. JOHNSTONE, & A.M. PITTS (eds)
178 Lower K- and L-theory, A. RANICKI
179 Complex projective geometry, G. ELLINGSRUD, C. PESKINE, G. SACCHIERO & S.A. STRØMME (eds)

London Mathematical Society Lecture Note Series. 173

Discrete Groups and Geometry

Papers dedicated to A. M. Macbeath

Proceedings of a conference at Birmingham University

Edited by

W. J. Harvey

King's College London

in collaboration with

C. Maclachlan

University of Aberdeen

CAMBRIDGE UNIVERSITY PRESS
Cambridge, New York, Melbourne, Madrid, Cape Town, Singapore, São Paulo

Cambridge University Press
The Edinburgh Building, Cambridge CB2 8RU, UK

Published in the United States of America by Cambridge University Press, New York

www.cambridge.org
Information on this title: www.cambridge.org/9780521429320

First published 1992

A catalogue record for this publication is available from the British Library

ISBN 978-0-521-42932-0 paperback

Transferred to digital printing 2008

Contents

Contents

Preface

This book is a collection of articles addressing a range of topics in the theory of discrete groups; for the most part the papers represent talks delivered at a conference held at the University of Birmingham in January 1991 to mark the retirement of A. M. Macbeath from his chair at the University of Pittsburgh.

The central theme of the volume is the study of groups from a geometric point of view. Of course, the geometric aspect takes many forms; thus one may find a group which operates on Euclidean or hyperbolic space rubbing shoulders with free groups or some generalisation studied with the help of graphs or homological algebra. Groups with presentation also relate to abstract or formal geometry: in recent years the study of groups which act on trees has brought algebraic structures back into the geometric fold, and the seminal idea that a countable group itself carries a geometric essence has reinforced this return to geometry within rather than through group theory, giving Klein's Programme a fresh cutting edge.

A major part of group theory today relates directly to explicit algebraic or geometric objects — permutation groups, Coxeter groups and discrete subgroups of Lie groups are prominent — and one of the strengths of this field lies in the wealth of fascinating interactions with complex analysis and low dimensional topology. The serious study of discrete groups via combinatorial techniques began in hyperbolic space with Poincaré and Dehn, and the reader will find here many echoes of their original ideas and interests. In particular, applications of the structural properties of Fuchsian and crystallographic groups, the representation of finite groups as automorphisms of surfaces, orientable or not, and the continuing drive to understand the classification of manifolds in low dimension by way of numerical invariants, such as Euler characteristic or invariant volume; for instance, a topic of great interest recently has been the attempt to estimate the minimum volume attained by a compact hyperbolic 3-manifold.

Other articles address wider issues including the arithmetic and analysis of automorphic forms, the geometry of moduli spaces, deformation theory of discrete groups in hyperbolic space and applications in mathematical physics. Several authors provide a review of their general area of study. All reflect something of the field in which Murray Macbeath has worked predominantly during much of his mathematical career and this conference provided clear testimony to the continuing vitality of both.

The editors are very grateful to all who helped in the production of this volume, especially to the referees who performed their traditional act of altruism with skill, dedication and (above all) willingness. For the running of the Birmingham conference we are greatly indebted to the London Mathematical Society, which provided a generous grant, to the University of Birmingham Mathematics department for the welcome and hospitality they extended to us, and especially to Dr. A. H. M. Hoare who organised all the local arrangements. Finally, of course, we thank Murray for making it possible both by furnishing the mathematical focus over the years and by providing us with the perfect excuse for such an enjoyable occasion. We all wish him a long, happy and productive retirement.

Symmetries of modular surfaces

M. Akbas and D. Singerman

To Murray Macbeath on the occasion of his retirement

1. Introduction

Let $\Gamma = \mathrm{PSL}(2, \mathbb{Z})$ be the rational modular group and $\Gamma(N)$, $\Gamma_0(N)$ denote the subgroups represented by the matrices

$$\left\{ \begin{pmatrix} a & b \\ c & d \end{pmatrix} \in \mathrm{SL}(2,\mathbb{Z}) \,\middle|\, \begin{pmatrix} a & b \\ c & d \end{pmatrix} \equiv \pm \begin{pmatrix} 1 & 0 \\ 0 & 1 \end{pmatrix} \bmod N \right\},$$

$$\left\{ \begin{pmatrix} a & b \\ c & d \end{pmatrix} \in \mathrm{SL}(2,\mathbb{Z}) \,\middle|\, c \equiv 0 \bmod N \right\}$$

respectively. Let U denote the upper half-plane and $U^* = U \cup \mathbb{Q} \cup \{\infty\}$. Let $X(N) = U^*/\Gamma(N)$, $X_0(N) = U^*/\Gamma_0(N)$. Then $X(N)$, $X_0(N)$ are compact Riemann surfaces.

A Riemann surface X is symmetric if it admits an anticonformal involution (or as we shall call it, *symmetry*) t. If X has genus g then by Harnack's Theorem, the fixed point set of t consists of k simple closed curves where $0 \leq k \leq g + 1$. We shall call each such simple closed curve a *mirror* of t. The mirrors of a symmetry may either (i) divide the surface X into two homeomorphic components or (ii) not divide X. In case (i) we say that t has species $+k$ and in case (ii) we say that t has species $-k$. It follows that the species of t has a $+$ (respectively $-$) sign if and only if $X/\langle t \rangle$ is orientable (respectively non-orientable). See [3] for an account of the general theory.

As the transformation $z \to -\bar{z}$ normalizes $\Gamma(N)$ and $\Gamma_0(N)$ it induces a symmetry on the surfaces $X(N)$ and $X_0(N)$ (which by abuse of language we call "the symmetry $z \to -\bar{z}$ of $X(N)$, or $X_0(N)$"). The surfaces $X_0(N)$ also admit a symmetry induced by the "Fricke symmetry" $z \to 1/N\bar{z}$ and we also discuss this.

Our aim here is to summarize the work done that enables us to find the species of these symmetries. As the surfaces $X(N)$, $X_0(N)$ also have cusps (at the projections of the parabolic fixed points) another question of interest is to determine the number of cusps on each mirror. The work we report on comes mainly from four sources. The first two ([7], [10]) were concerned with real points on modular curves and are probably not well-known to workers on discrete groups and Riemann surfaces. The second two are from the Ph.D. theses of the first author and Stephen Harding ([1], [4]). Also see [11] for a number-theoretic application of similar ideas.

These questions are closely related to that of determining the signatures of the NEC groups

$$\widehat{\Gamma}(N) = \langle \Gamma(N), z \to -\bar{z} \rangle, \qquad \widehat{\Gamma}_0(N) = \langle \Gamma_0(N), z \to -\bar{z} \rangle,$$

$$\Gamma_F(N) = \langle \Gamma_0(N), z \to 1/N\bar{z} \rangle.$$

We recall that the signature of a NEC group Λ has the form

$$(g;\ \pm;\ [m_1, \ldots, m_r];\ \{(n_{11}, \ldots, n_{1s_1}) \ldots (n_{k1}, \ldots, n_{ks_k})\}$$

where g is the genus of U/Λ, $m_1, \ldots, m_r$ are the periods of Λ and n_{ij} are the link periods. (See [2], [3], [9], [12]). In these references the groups Λ have compact quotient space and so the integers m_i, n_{ij} are finite. In this paper the NEC groups are all commensurable with the modular group and so have parabolic elements. As usual these are represented by infinite periods so that the number of infinite periods in a period cycle correspond to the number of cusps around the hole represented by that period cycle.

EXAMPLE. The extended modular group

$$\widehat{\Gamma} = \widehat{\Gamma}(1) = \widehat{\Gamma}_0(1) = \langle \Gamma, z \to -\bar{z} \rangle \cong \mathrm{PGL}(2, \mathbb{Z}).$$

This is generated by 3 reflections: $c_1 : z \to -\bar{z}$, $c_2 : z \to 1/\bar{z}$, $c_3 : z \to -\bar{z}-1$ or in terms of matrices

$$\begin{pmatrix} 1 & 0 \\ 0 & -1 \end{pmatrix}, \quad \begin{pmatrix} 0 & 1 \\ 1 & 0 \end{pmatrix}, \quad \begin{pmatrix} 1 & 1 \\ 0 & -1 \end{pmatrix}$$

the fundamental domain being bounded by the 3 axes of reflection

$$\mathrm{Re}(z) = 0, \qquad |z| = 1, \qquad \mathrm{Re}(z) = \frac{1}{2}.$$

Its boundary contains elliptic fixed points at i and $(1 + i\sqrt{3})/2$ of orders 2 and 3 and there is a parabolic fixed point at ∞. These correspond to the elements c_1c_2 of order 2, c_2c_3 of order 3 and the parabolic c_1c_3. The signature of $\widehat{\Gamma}$ is $\{0; +; [-]; \{(2, 3, \infty)\}$ and the quotient U^*/Γ is a disc whose boundary contains two branch points and one cusp.

REMARK. If Λ is a NEC group with sense-reversing transformations then we let Λ^+ denote the subgroup of index 2 consisting of the sense-preserving transformations of Λ. Then U/Λ^+ (or U^*/Λ^+ if appropriate) is the canonical double cover of U/Λ (or U^*/Λ). (See [2], 0.1.12). The mirrors of U/Λ^+ are in one-to-one correspondence with the boundary components of U/Λ. As $(\widehat{\Gamma}(N))^+ = \Gamma(N)$, $(\widehat{\Gamma}_0(N))^+ = \Gamma_0(N)$, the mirrors of $X(N)$, (respectively $X_0(N)$), correspond to the boundary components of $X(N) = U^*/\widehat{\Gamma}(N)$ (respectively $\widehat{X}_0(N) = U^*/\widehat{\Gamma}_0(N)$).

2. The mirrors of X(N)

To describe the number of mirrors of $X(N)$ we introduce an arithmetic function $\alpha(N)$ defined as follows: $\alpha(N)$ is the least positive integer such that $2^{\alpha(N)} \equiv \pm 1 \bmod N$. If U_N denotes the groups of units mod N then $\alpha(N)$ is the order of the image of 2 in $U_N/\{\pm 1\}$ so that if $N > 2$, $2\alpha(N)|\phi(N)$, where ϕ is Euler's function. The following Theorems are proved in [1], [7].

THEOREM 1. *The number of mirrors of the symmetry $z \to -\bar{z}$ of $X(N)$ is given by*

$$\begin{cases} \phi(N)/2\alpha(N) & \textit{if } N > 1 \textit{ is odd} \\ \phi(N)/2 & \textit{if } N > 2 \textit{ is even} \\ 1 & \textit{if } N = 2\ . \end{cases}$$

THEOREM 2. *The number of cusps on each mirror is given by*

$$\begin{cases} 2\alpha(N) & N \textit{ odd} \\ 6 & N \equiv 2 \textit{ mod } 4,\ N > 2 \\ 4 & N \equiv 0 \textit{ mod } 4 \\ 3 & N = 2 \end{cases}$$

The proofs in [1] and [7] are rather different. In [1] the proof is algebraic and uses Hoare's Theorem on subgroups of NEC groups [5]. This theorem gives a general method for computing the signature of a subgroup Δ_1 of a NEC group Δ given the signature of Δ and the permutation representation of the generators of Δ on the right Δ_1-cosets. In our case $\Delta = \widehat{\Gamma}$, $\Delta_1 = \widehat{\Gamma}(N)$ and the generators of $\widehat{\Gamma}$ are c_1, c_2, c_3 above.

In [7], a more geometric approach is used, and involves calculating the cusps on the mirrors. We denote the $\Gamma(N)$-orbit of a rational number a/b by $\begin{pmatrix} a \\ b \end{pmatrix}$. The reflection $z \to -\bar{z}$ of $X(N)$ fixes $\begin{pmatrix} a \\ b \end{pmatrix}$ if $\begin{pmatrix} a \\ b \end{pmatrix} = \begin{pmatrix} -a \\ b \end{pmatrix}$. Such a fixed cusp is called a *real cusp.* (For an account of the connection between NEC groups and real algebraic geometry see [2].) For example, if N is odd the real cusps on $X(N)$ are, according to lemma 1 of [7], of the form $\left\{ \begin{pmatrix} u \\ N \end{pmatrix}, \begin{pmatrix} N \\ u \end{pmatrix} \;\middle|\; 1 \le u \le \frac{N-1}{2},\ (u,N) = 1 \right\}$. This gives a total of $\phi(N)$ cusps as implied by Theorems 1 and 2. Two real cusps are joined if they have lifts in U^* which are joined by an axis of reflection of $\widehat{\Gamma}(N)$. This axis then projects to a segment of a mirror on $X(N)$. By finding all such segments we can find all the mirrors and the number of cusps on each mirror.

The sign of the species. Theorem 1 gives the number of mirrors of the reflection $z \to -\bar{z}$ of $X(N)$. We now investigate whether the mirrors separate or do not separate the surface, i.e. whether the species has a + sign or − sign. (See §1). This sign is the same as the sign in the signature of $\widehat{\Gamma}(N)$ and this can be determined by Theorem 2 of [6]. We consider the Schreier coset graph $\mathcal{H} = \mathcal{H}(\widehat{\Gamma}, \widehat{\Gamma}(N), \Phi)$ whose vertices are the $\widehat{\Gamma}(N)$-cosets of $\widehat{\Gamma}$ and where $\Phi = \{c_1, c_2, c_3\}$, the generators of $\widehat{\Gamma}$. We then form the graph $\bar{\mathcal{H}}$ by deleting the loops from $\mathcal{H}$. Then $\widehat{\Gamma}(N)$ is orientable if and only if all circuits of $\bar{\mathcal{H}}$ have even length. Harding [4] investigated the orientability in the case $N = p$ a prime. We describe his method. First of all let

$$A = c_1c_2c_3 = \begin{pmatrix} 0 & 1 \\ 1 & 1 \end{pmatrix}. \qquad \text{Then} \qquad A^k = \begin{pmatrix} u_k & u_{k+1} \\ u_{k+1} & u_{k+2} \end{pmatrix}$$

where u_k is the kth Fibonacci number, $(u_1 = 0, u_2 = 1, u_{k+2} = u_{k+1} + u_k)$. We consider the path that corresponds to the word

$$c_1c_2c_3c_1c_2c_3 \ldots c_1c_2c_3 = A^k.$$

To find the corresponding path in $\bar{\mathcal{H}}$ we search for the loops in this path. These occur if either

$$\text{(i) } \widehat{\Gamma}(p)A^k c_1 = \widehat{\Gamma}(p)c_1, \qquad \text{(ii) } \widehat{\Gamma}(p)A^k c_1 c_2 = \widehat{\Gamma}(p)c_1$$
$$\text{or} \qquad \text{(iii) } \widehat{\Gamma}(p)A^{k+1} = \widehat{\Gamma}(p)c_1 c_2.$$

It is easy to calculate that (i) occurs if and only if $u_{k+1} \equiv 0 \bmod p$, (ii) can occur only if $p = 3$ and (iii) never occurs. Thus if $p > 3$, the only loops in $\mathcal{H}$ occur when $u_{k+1} \equiv 0 \bmod p$. Also $u_k u_{k+2} - u_{k+1}^2 = \det A^k = (-1)^k$. If $u_{k+1} \equiv 0 \bmod p$ then $u_{k+2} \equiv u_k \bmod p$ and so $u_k^2 \equiv (-1)^k \bmod p$. Now suppose that $p \equiv 3 \bmod 4$. Then -1 is not a square mod p so that k is even, $u_k \equiv \pm 1 \bmod p$ and so $A^k \in \widehat{\Gamma}(p)$. A loop does occur at the beginning of the path as $c_1 \in \widehat{\Gamma}(p)$. The path closes again when we have reached $(c_1 c_2 c_3)^k$ with k the first integer such that $u_{k+1} \equiv 0 \bmod p$ and there are no other loops before then. Thus the corresponding circuit in $\bar{\mathcal{H}}$ has length $3k - 1$ which is odd as k is even. Thus we have proved (using [6] Theorem 2)

THEOREM 3. *If $p > 3$ and $p \equiv 3 \bmod 4$ is prime then the mirrors of the symmetry $z \to -\bar{z}$ of $X(p)$ do not separate $X(p)$.*

By drawing the Schreier coset graphs for $p = 2, 3, 5$ Harding showed that $\widehat{X}(2), \widehat{X}(3)$ and $\widehat{X}(5)$ are orientable (so the mirrors do separate in these cases). For other primes $p \equiv 1 \bmod 4$ he considered the element

$$B = c_1 c_2 c_3 (c_1 c_3)^{s-1} = \begin{pmatrix} 0 & 1 \\ 1 & s \end{pmatrix} \qquad (s \in \mathbb{N}).$$

Then
$$B^k = \begin{pmatrix} t_k & t_{k+1} \\ t_{k+1} & t_{k+2} \end{pmatrix}$$

with $t_1 = 0$, $t_2 = 1$, $t_{k+2} = st_{k+1} + t_k$, a generalized Fibonacci sequence. By pursuing an analysis similar to the above Harding showed that $X(p)$ is non-orientable for all primes p with $5 < p < 1000$. Since for non-orientability we only need one circuit in $\mathcal{H}$ of odd length, it seems very likely that $X(p)$ is non-orientable for all primes $p > 5$.

To illustrate the results of this section we give the species of the symmetries $z \to -\bar{z}$ of $X(p)$ for all primes $p < 100$.

p	2	3	5	7	11	13	17	19	23	29	31	37
species	+1	+1	+1	−1	−1	−1	−2	−1	−1	−1	−3	−1

p	41	43	47	53	59	61	67	71	73	79	83	89	97
species	−2	−3	−1	−1	−1	−1	−1	−1	−4	−1	−1	−4	−2

3. The mirrors of $X_0(N)$

The number of mirrors of the symmetry $z \to -\bar{z}$ of $X_0(N)$ and the number of cusps on each mirror were calculated by the first author in [1] using Hoare's Theorem and by Ogg [10] using a technique similar to Jaffee's described previously. The results are as follows. (The notation $m\|n$ means that m is an exact divisor of n, i.e. $m|n$ and $(m, n/m) = 1$, and r is the number of distinct prime factors of N).

THEOREM 4. *The number of mirrors of the symmetry $z \to -\bar{z}$ of $X_0(N)$ and the number of cusps on each mirror is given by the following table*

N	odd	2	$2\|N$, $N > 2$	4	$4\|N$, $N > 4$	$8\|N$
mirrors	2^{r-1}	1	2^{r-2}	1	2^{r-2}	2^{r-1}
cusps	2	2	4	3	6	4

The sign of the species. In [4], Harding calculated the sign of the species of the symmetry $z \to -\bar{z}$ of $X_0(p)$ for all primes p. We describe his method. The group $\widehat{\Gamma} \cong \mathrm{PGL}(2, \mathbb{Z})$ acts transitively on the $p+1$ points of the projective line $GF(p) \cup \{\infty\}$ by

$$\begin{pmatrix} a & b \\ c & d \end{pmatrix} : t \to \frac{at+b}{ct+d}.$$

In particular c_1, c_2, c_3 act as follows:

$$c_1 : t \to 1/t, \qquad c_2 : t \to 1/t, \qquad c_3 : t \to -1-t$$

so that c_1 fixes $0, \infty$, c_2 fixes ± 1, c_3 fixes $-\frac{1}{2}, \infty$. The stabilizer of ∞ has index $p+1$ and contains $\Gamma_0(p)$ and c_1 . Hence $\mathrm{Stab}(\infty) = \widehat{\Gamma}_0(p)$. Thus the vertices of the Schreier coset graph $\mathcal{H}_0(p) = \mathcal{H}(\widehat{\Gamma}, \widehat{\Gamma}_0(p), \Phi)$, where $\Phi = \{c_1, c_2, c_3\}$, can be identified with $GF(p) \cup \{\infty\}$. As before we form the graph $\bar{\mathcal{H}}_0(p)$ obtained by deleting the loops of $\mathcal{H}_0(p)$, and if we find circuits of odd length then the sign in the signature of $\Gamma_0(p)$ is $-$. We first consider the word $c_1 c_2 c_3$. If we begin at $x \in GF(p) - \{O\}$ we get the path

$$\underset{x}{\bullet} \xrightarrow{c_1} \underset{-x}{\bullet} \xrightarrow{c_2} \underset{-x^{-1}}{\bullet} \xrightarrow{c_3} \underset{x^{-1}-1}{\bullet}$$

If $x \neq \pm 1$ then c_2 does not fix $-x$. If $x \neq 0$ or 2 then c_3 does not fix $-x^{-1}$. This path closes if $x = x^{-1} - 1$ or $x = (-1 \pm \sqrt{5})/2$. Thus if the Legendre symbol $\left(\frac{5}{p}\right) = +1$ then we can find x so that the above path is a triangle. We cannot have loops unless $p = 2$. For example, $1 = (-1 \pm \sqrt{5})/2$ is not true in $GF(p)$ if $p > 2$. Thus if $\left(\frac{5}{p}\right) = +1$, $(p > 2)$ then $\bar{\mathcal{H}}_0(p)$ has a triangle. We now consider the word $c_1c_2c_3c_1c_3$.

$$x \xrightarrow{c_1} -x \xrightarrow{c_2} -x^{-1} \xrightarrow{c_3} x^{-1}-1 \xrightarrow{c_1} 1-x^{-1} \xrightarrow{c_3} x^{-1}-2$$

This gives a closed path if $x = x^{-1} - 2$ or $x = -1 \pm \sqrt{2}$. Thus if $\left(\frac{2}{p}\right) = +1$ we can find x giving a closed path. It is now possible to get loops. For example $-x^{-1} = x^{-1} - 1$ gives $x = 2$ but if $p = 7$, $2 \equiv 1 + \sqrt{2}$. However if $\left(\frac{2}{p}\right) = +1$ (equivalent to $p \equiv \pm 1 \bmod 8$) and if $p > 7$ then there are no loops and we have a circuit of odd length in $\bar{\mathcal{H}}_0(p)$.

Thus if $\left(\frac{5}{p}\right) = +1$ or $\left(\frac{2}{p}\right) = +1$ and $p > 7$ then we can find a circuit of odd length. Now we consider the word $c_1c_3c_1c_2c_3(c_1c_3)^4$ of length 13. We find a closed circuit at x if $x = -2 \pm \sqrt{10}$ which exists if $\left(\frac{10}{p}\right) = +1$, and in particular if $\left(\frac{5}{p}\right) = \left(\frac{2}{p}\right) = -1$. We also find that this circuit can only have loops if $p = 3$, 13 or 31. However $\left(\frac{5}{31}\right) = +1$ so this case has been covered already. Hence we can find a circuit of odd length in $\bar{\mathcal{H}}_0(p)$ for all primes p except $p = 2, 3, 5, 7, 13$. If we draw the coset graphs $\bar{\mathcal{H}}_0(p)$ in these cases we see that there are no circuits of odd length. For example if $p = 5$ we get

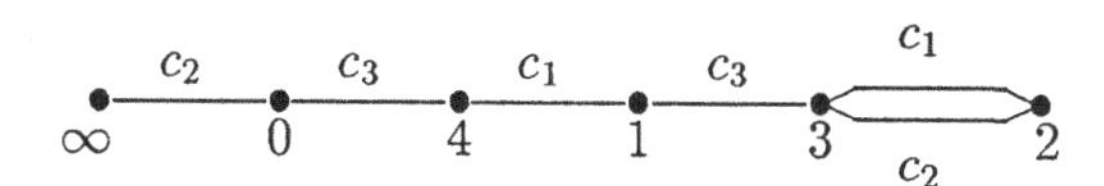

We have therefore proved

THEOREM 5. *If $p = 2, 3, 5, 7, 13$ then the mirrors of the symmetry $z \to -\bar{z}$ of $X_0(p)$ separate the surface. For all other primes the mirrors do not separate.*

Mirrors of the Fricke reflection. The group $\Gamma_0(N)$ is normalized by the Fricke involution $W_N : z \to -1/Nz$ and thus it is also normalized by the reflection $\bar{W}_N : z \to 1/N\bar{z}$ which we shall call the *Fricke reflection*. Therefore $\bar{W}_N$ induces a symmetry $\bar{w}_N$ of $X_0(N)$.

In contrast to the previous cases $\bar{w}_N$ may have mirrors without cusps.

EXAMPLE. $N = 2$. The following diagram shows a fundamental domain for $\Gamma_0(2)$ divided into two by the fixed axis $|z|^2 = \frac{1}{2}$ of $\bar{W}_2$.

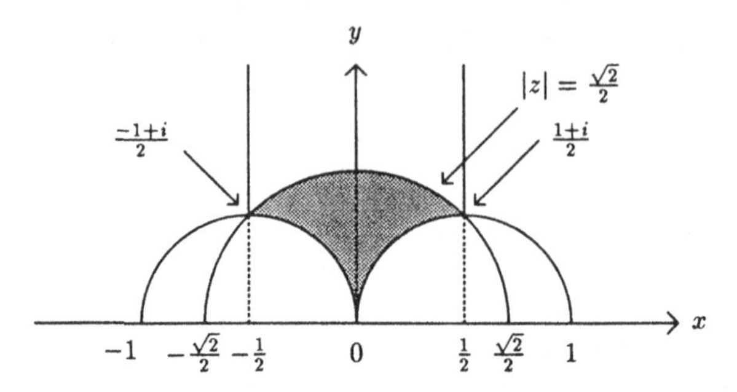

There are two cusps at 0 and ∞ and one orbit of elliptic fixed points of period 2 at $(\pm 1 + i)/2$. Thus the mirror of $\bar{w}_2$ has no cusps (and one branch point corresponding to the elliptic fixed points). More generally we have the following simple result.

THEOREM 6. *$\bar{w}_N$ fixes a cusp on a mirror if and only if N is a perfect square.*

PROOF. If $\bar{w}_n$ fixes a cusp then for some $x \in \mathbb{Q} \cup \{\infty\}$ and $T \in \Gamma_0(N)$, $\bar{W}_N : x \to T(x)$. If

$$T = \begin{pmatrix} a & b \\ cN & d \end{pmatrix} \qquad (ad - bcN = 1)$$

then $S = \bar{W}_N T$ fixes x. Now S reverses orientation. If it were a glide reflection then S^2 would be a hyperbolic element of the modular group fixing a point of $\mathbb{Q} \cup \{\infty\}$ which is impossible so that S is a reflection. As

$$S = \begin{pmatrix} cN & d \\ aN & bN \end{pmatrix}$$

we conclude that $b + c = 0$. As S fixes x, $aNx^2 + 2bNx - d = 0$. The discriminant is $4b^2N^2 + 4adN = 4N$ which must be a perfect square and the result follows. Conversely, if $N = n^2$ then $\bar{W}_N$ fixes $\frac{1}{n}$. ∎

In [1], the first author obtained the following result (c.f. [10]). First we need some notation. Let A be the number of solutions of $x^2 \equiv -1 \bmod n$. If solutions exist then $A = 2^{\sigma+\tau}$ where σ is the number of odd prime divisors of n and $\tau = 0, 1, 2$ according as $4 \nmid n$, $4 \| n$ or $8 | n$. ([8] p.65). We also let k be the order of 4 in U_N, the group of units mod N.

THEOREM 7. *Let $N = n^2$. Then*

(i) If n is odd then $\bar{w}_n$ has A boundary components containing k cusps and $\dfrac{\phi(n)}{2k} + \dfrac{A}{2}$ boundary components with $2k$ cusps.

(ii) If n is even then $\bar{w}_n$ has A boundary components with 1 cusp and $\dfrac{\phi(n)}{2} + \dfrac{A}{2}$ boundary components with 2 cusps.

References

[1] M. AKBAS. PhD Thesis. University of Southampton (1989).

[2] E. BUJALANCE, J. J. ETAYO, J. M. GAMBOA and G. GROMADZKI. Automorphism groups of compact bordered Klein surfaces. *Lecture notes in Math 1439* (Springer-Verlag)

[3] E. BUJALANCE, D. SINGERMAN. The symmetry type of a Riemann surface. *Proc. London Math. Soc. (3)* **51** (1986) 501-519.

[4] S. HARDING. PhD Thesis. University of Southampton (1985).

[5] A. H. M. HOARE. Subgroups of NEC groups and finite permutation groups. *Quart. J. Math. (2)* **41** (1990) 45-59.

[6] A. H. M. HOARE, D. SINGERMAN. The orientability of subgroups of plane groups. *London Math. Soc. Lecture note series* **71** (1982) 221-227.

[7] H. JAFFEE. Degeneration of real elliptic curves. *J. London Math. Soc. (2)* **17** (1978) 19-27.

[8] W. J. LEVEQUE. *Topics in Number Theory, Vol. 1.* (Addison-Wesley 1956).

[9] A. M. MACBEATH. The classification of plane non-euclidean crystallographic groups. *Can. J. Math.* **19** (1967) 1192-1205.

[10] A. OGG. Real points on Shimura curves. *Arithmetic and geometry Vol. 1* Progr. Math. **35** (1983) 277-307.

[11] M. SHEINGORN. Hyperbolic reflections on Pell's equation. *Journal of Number Theory* **33** (1989) 267-285.

[12] H. ZIESCHANG, E. VOGT and H-D. COLDEWEY. *Surfaces and Planar Discontinuous Groups.* Lecture Notes in Math. **835** (Springer, Berlin 1980)

Karadeniz Üniversitesi
Turkey

University of Southampton
Southampton, England

Lifting group actions to covering spaces

M. A. Armstrong

To Murray Macbeath on the occasion of his retirement

Several authors (Bredon [2]; Conner and Raymond [3]; Gottlieb [4]; Rhodes [5]) have considered the following question. Given an action of a topological group G on a space X, together with a covering space $\widetilde{X}_H$ of X, when does this action lift to an action of G on $\widetilde{X}_H$? We propose a systematic approach which unifies and extends previous results. In particular we avoid unnecessary local restrictions on G and X, and we verify the *continuity* of the lifted actions.

Our first task is to fix some notation and terminology. Let X be a path connected, locally path connected space with a chosen base point p, and let $\widetilde{X}_H$ denote the covering space of X which corresponds to the subgroup H of $\pi_1(X,p)$. We shall assume throughout that G is a topological group which acts in a continuous fashion as a group of homeomorphisms of X.

Suppose G also acts on a space Z, and that $f : Z \to X$ is an equivariant map which sends the point q of Z to p. We say that H is $(f,G)-$*invariant* providing the homotopy class

$$\langle f(\gamma).g(\alpha).f(\gamma^{-1})\rangle$$

belongs to H for every group element g in G, loop α based at p in X, and path γ which joins q to $g(q)$ in Z. (When $Z = X$ and f is the identity map, we have a "G-invariant subgroup" in the sense of [1].)

Our main results are as follows.

THEOREM A. *The action of G on X lifts to an action of G on $\widetilde{X}_H$ if and only if there is a path connected, locally path connected space Z, an action of G on Z, and a based equivariant map $f : Z, \{q\} \to X, \{p\}$ such that:*

(i) H is (f, G)-invariant, and

(ii) $f_*(\pi_1(Z, q)) \subseteq H$.

THEOREM B. *Let $f_1 : Z_1, \{q_1\} \to X, \{p\}$, $f_2 : Z_2, \{q_2\} \to X, \{p\}$ be based equivariant maps for which H is (f_i, G) – invariant and contains $f_{i*}(\pi_1(Z, q_i))$, $i = 1, 2$. Then f_1 and f_2 lead to the same lift of G to $\widetilde{X}_H$ if and only if given $g \in G$, and paths γ_i joining q_i to $g(q_i)$ in Z_i, i=1,2, the loop $f_1(\gamma_1).f_2(\gamma_2)^{-1}$ always represents an element of H.*

Before proving the theorems we list some corollaries.

COROLLARY 1. *The action of G lifts to $\widetilde{X}_H$ if some point p of X is fixed by every element of G, and H is invariant under the automorphisms of $\pi_1(X, p)$ induced by elements of G.*

PROOF. Take $Z = \{p\}$, and f to be the inclusion of $\{p\}$ in X. ∎

COROLLARY 2. *The action of G lifts to $\widetilde{X}_H$ if G is path connected and $\omega_*(\pi_1(G, e)) \subseteq H$, where $\omega : G \to X$ is the evaluation map $\omega(g) = g(p)$. In this case the lifted action is unique.*

PROOF. Take $Z = G$, q to be the identity element, the group action as left translation, and f to be the evaluation map. Notice that H is automatically (ω, G)- invariant. If γ is a path which joins e to g in G, and if α is a loop based at p in X, the loops

$$f(\gamma_t).\gamma(t)(\alpha).f(\gamma_t^{-1}), \quad 0 \le t \le 1$$

provide a homotopy from α to $f(\gamma).g(\alpha).f(\gamma^{-1})$. Therefore the latter loop represents an element of H whenever α does so. Here γ_t is the path from e to $\gamma(t)$ defined by $\gamma_t(s) = \gamma(ts)$, $0 \le s \le 1$.

Suppose now we have a lift associated with a map $f : Z \to X$ where H is (f, G)-invariant and contains $f_*(\pi_1(Z, q))$. Given $g \in G$, let γ be a path which joins e to g in G, let σ join q to $g(q)$ in Z, and let β be the path (again from q to $g(q)$) defined by $\beta(s) = \gamma(s)(q)$, $0 \leq s \leq 1$. The loop $f(\beta\sigma^{-1})$ represents an element of H and

$$f\beta(s) = f\gamma(s)(q) = \gamma(s)f(q) = \gamma(s)(p) = \omega\gamma(s), \quad 0 \leq s \leq 1.$$

Therefore $\omega(\gamma)f(\sigma)^{-1}$ represents an element of H. Theorem B now shows us that the lifts determined by $\omega : G \to X$ and $f : Z \to X$ agree. Hence the lifted action is in this case unique. ∎

COROLLARY 3. *The action of G lifts in a unique manner to every covering space of X when G is simply connected.*

PROOF. This is a special case of Corollary 2. ∎

COROLLARY 4. *The action of G lifts in a unique manner to every covering space of X if G is path connected and $\pi_1(X, p)$ has a trivial centre.*

PROOF. This follows from Corollary 2 because the image of ω_* is always contained in the centre of $\pi_1(X, p)$. ∎

COROLLARY 5. *Let G be a discrete group, let S be a set of generators for G and let $\Gamma(G, S)$ denote the corresponding Cayley graph. The action of G lifts to $\widetilde{X}_H$ if there is an equivariant extension $\Omega : \Gamma(G, S) \to X$ of the evaluation map ω such that*

(i) H is (Ω, G)-invariant, and

(ii) $\Omega_*(\pi_1(\Gamma, e)) \subseteq H$.

PROOF. Take $Z = \Gamma(G, S)$, $q = e$ and $f = \Omega$, the action of G on Γ being induced by left translation. ∎

PROOF OF THEOREM A. The necessity of our conditions is clear. When the action does lift, we need only take $Z = \widetilde{X}_H$, with the given action, and f to be the natural projection from $\widetilde{X}_H$ to X.

Now suppose the hypotheses of the theorem are satisfied. Choose a base point $\tilde{p}$ in $\widetilde{X}_H$ which projects to p. Given $\tilde{x}$ in $\widetilde{X}_H$ and g in G, join $\tilde{p}$ to $\tilde{x}$ by a path $\tilde{\alpha}$ in $\widetilde{X}_H$, writing α for its projection into X, and join q to $g(q)$ by a path γ in Z. Form the composite path $f(\gamma)g(\alpha)$ in X, and lift it to a path in $\widetilde{X}_H$ which begins at $\tilde{p}$. The end point of this lifted path is the point $g(\tilde{x})$ of $\widetilde{X}_H$.

If $\tilde{\beta}$ also joins $\tilde{p}$ to $\tilde{x}$ in $\widetilde{X}_H$, and σ joins q to $g(q)$ in Z, then

$$\begin{aligned}\langle f(\gamma)g(\alpha).(f(\sigma)g(\beta))^{-1}\rangle =&\langle f(\gamma).g(\alpha\beta^{-1}).f(\sigma^{-1})\rangle\\ =&\langle f(\gamma).g(\alpha\beta^{-1}).f(\gamma^{-1})\rangle\langle f(\gamma\sigma^{-1})\rangle.\end{aligned}$$

The first part of this product lies in H because $\alpha\beta^{-1}$ represents an element of H, and H is (f, G)-invariant. The second part also belongs to H since $f_*(\pi_1(Z, q)) \subseteq H$. Therefore $f(\gamma)g(\alpha).(f(\sigma)g(\beta))^{-1}$ represents an element of H, and consequently lifts to a *loop* based at $\tilde{p}$ in $\widetilde{X}_H$. So the lifts of $f(\gamma)g(\alpha)$ and $f(\sigma)g(\beta)$ must have the *same end point*, and give the same value for $g(\tilde{x})$. We shall say that paths such as $f(\gamma)g(\alpha)$ and $f(\sigma)g(\beta)$ *represent* the point $g(\tilde{x})$ of $\widetilde{X}_H$).

This procedure does give an action of G on $\widetilde{X}_H$. If e is the identity element of G, take γ to be the constant path at q, when $f(\gamma)$ is the constant path at p and $f(\gamma)e(\alpha)$ is homotopic to α keeping p fixed. Therefore $e(\tilde{x}) = \tilde{x}$. Given $g_1, g_2 \in G$, join q to $g_1(q)$ by γ_1 and to $g_2(q)$ by γ_2, so that $f(\gamma_1)g_1(\alpha)$ represents $g_1(\tilde{x})$ and $f(\gamma_2)g_2(\alpha)$ represents $g_2(\tilde{x})$. Then by construction $f(\gamma_2)g_2(f(\gamma_1)g_1(\alpha))$ represents $g_2(g_1(\tilde{x}))$ and, since $\gamma_2 g_2(\gamma_1)$ joins q to $g_2g_1(q)$ in Z, we see that $f(\gamma_2 g_2(\gamma_1))g_2g_1(\alpha)$ represents $g_2g_1(\tilde{x})$. Now

$$\begin{aligned}f(\gamma_2 g_2(\gamma_1))g_2g_1(\alpha) =&f(\gamma_2)g_2(f(\gamma_1))g_2g_1(\alpha) \quad \text{(because } f \text{ is equivariant)}\\ =&f(\gamma_2)g_2(f(\gamma_1)g_1(\alpha)).\end{aligned}$$

Therefore $g_2g_1(\tilde{x}) = g_2(g_1(\tilde{x}))$, as required.

It only remains to check the *continuity* of this action. We must verify that the function

$$G \times \widetilde{X}_H \to \widetilde{X}_H$$

defined by $(g, \tilde{x}) \to g(\tilde{x})$ is continuous. Begin with a group element $g_0 \in G$, a point $\tilde{x}_0 \in \widetilde{X}_H$ and a neighbourhood $\widetilde{W}$ of $g_0(\tilde{x}_0)$ in $\widetilde{X}_H$. Without loss of generality we may insist that $\widetilde{W}$ project to a *canonical* neighbourhood W of $g_0(x_0)$ in X. That is to say W is an open, path connected neighbourhood of $g_0(x_0)$, and the collection of points "above it" in $\widetilde{X}_H$ consists of pairwise disjoint open sets each of which projects homeomorphically onto W.

In what follows we make repeated use of the continuity of the given action of G on X. As usual we represent the point $g_0(\tilde{x}_0)$ by a path $f(\gamma)g_0(\alpha)$ in X. Cover $g_0(\alpha)$ by a finite number of canonical neighbourhoods which include W and a canonical neighbourhood of $g_0(p)$. Then break up α as a composite path $\alpha = \alpha_1\alpha_2 \dots\dots \alpha_k$ so that each $g_0(\alpha_i)$ is contained inside one of these canonical neighbourhoods, say $g_0(\alpha_i) \subseteq V_i$, $1 \le i \le k$, where V_1 is the chosen canonical neighbourhood of $g_0(p)$ and $V_k = W$. Of course $V_1, \dots\dots, V_k$ need not be distinct. Now choose a path connected neighbourhood N of $g_0(q)$ in $f^{-1}(V_1)$, a canonical neighbourhood V of x_0 in X, and a neighbourhood U of g_0 in G, which are small enough so that:

$$g(q) \in N \quad \text{if } g \in U;$$
$$g(\alpha_i) \subseteq V_i, \ 1 \le i \le k, \quad \text{whenever } g \in U;$$
$$g(x) \in W \quad \text{if } g \in U \text{ and } x \in V.$$

If $\widetilde{V}$ is the neighbourhood of $\tilde{x}_0$ which projects homeomorphically onto V, we shall show that $g(\tilde{x}) \in \widetilde{W}$ provided $g \in U$ and $\tilde{x} \in \widetilde{V}$, thereby completing the argument. Suppose then that g belongs to U, and that $\tilde{x}$ lies in $\widetilde{V}$. Join $g_0(q)$ to $g(q)$ by a path δ in N, and let β be a path which joins x_0 to x in V. Then the path $f(\gamma\delta)g(\alpha\beta)$ represents the point $g(\tilde{x})$. Now $f(\delta)$ is contained in V_1, and $g(\alpha_i)$ always lies in the same canonical neighbourhood V_i as $g_0(\alpha_i)$. Therefore our lift of $f(\gamma\delta)g(\alpha)$ must end in $\widetilde{W}$. Since $g(\beta)$ is contained in W, the lift of $f(\gamma\delta)g(\alpha\beta)$ also ends in $\widetilde{W}$, as required. ∎

PROOF OF THEOREM B. With the usual notation, the lifts of $f_1(\gamma_1)g(\alpha)$ and $f_2(\gamma_2)g(\alpha)$ which begin at $\tilde{p}$ have the same end point precisely when the loop $f_1(\gamma_1)g(\alpha)[f_2(\gamma_2)g(\alpha)]^{-1}$ represents an element of H. ∎

One can of course compare lifts which come from different maps. To this end suppose that $f_1 : Z_1, q_1 \to X, p$ and $f_2 : Z_2, q_2 \to X, p$ satisfy the hypotheses of Theorem B. Given $\tilde{x} \in \widetilde{X}_H$ and $g \in G$ we assume, as usual, that γ_i joins q_i to $g(q_i)$ in Z_i for $i = 1, 2$ and that α is the projection into X of a path which connects $\tilde{p}$ to $\tilde{x}$ in $\widetilde{X}_H$. Lifting $f_1(\gamma_1)g(\alpha)$ and $f_2(\gamma_2)g(\alpha)$ into $\widetilde{X}_H$ and taking the end points of these paths gives two points, say $(g\tilde{x})_1$ and $(g\tilde{x})_2$ respectively. If the given covering is *regular* there will be a deck transformation $d_g \in \pi_1(X, p)/H$ such that $d_g(g\tilde{x})_1 = (g\tilde{x})_2$. The same covering transformation d_g is obtained whichever point $\tilde{x}$ we start from, and the correspondence $g \to d_g$ is a homomorphism from G to $\pi_1(X, p)/H$. We therefore have the following result.

THEOREM C. *With the notation established above, and under the assumption that $\widetilde{X}_H$ is a regular covering space, there is a homomorphism $d : G \to \pi_1(X, p)/H$ such that $(g\tilde{x})_2 = d_g(g\tilde{x})_1$ for every point $\tilde{x} \in \widetilde{X}_H$ and every group element $g \in G$.* ∎

COROLLARY 6. *Let $\widetilde{X}_H$ be a regular covering space of X. If the action of G lifts to $\widetilde{X}_H$, and if the only homomorphism from G to $\pi_1(X,p)/H$ is the trivial homomorphism, then the lifted action is unique.* ∎

References

[1] M. A. ARMSTRONG. Lifting homotopies through fixed points II. *Proc. Roy. Soc. Edinburgh* **96A** (1984) 201-205.

[2] G. E. BREDON. *Introduction to Compact Transformation Groups.* (Academic Press, New York, 1972).

[3] P. E. CONNER AND F. RAYMOND. Actions of compact Lie groups on aspherical manifolds. *Topology of Manifolds.* (Markam Publ., Chicago, 1970) 227-264.

[4] D. H. GOTTLIEB. Lifting actions in fibrations. *Geometric Applications of Homotopy Theory I.* Lecture Notes in Mathematics **657** (Springer-Verlag, Berlin, 1978) 217-254.

[5] F. RHODES. On lifting transformation groups. *Proc. Amer. Math. Soc.* **19** (1968) 905-908.

Department of Mathematical Sciences
University of Durham
Durham DH1 3LE
Email: M.A.Armstrong@uk.ac.durham

A combinatorial approach to the symmetries of M and M–1 Riemann surfaces

E. Bujalance and A. F. Costa

To A. M. Macbeath on the occasion of his retirement

1. Introduction

Let X be a compact Riemann surface of genus g. A *symmetry* of X is an anticonformal involution $T : X \to X$. The topological nature of a symmetry T is determined by properties of its fixed point set $F(T)$. $F(T)$ consists of k disjoint Jordan curves, where $0 \leq k \leq g+1$ (Harnack's theorem). $X - F(T)$ has either one or two components. It consists of one component if $X/\langle T \rangle$ is non-orientable and two components if $X/\langle T \rangle$ is orientable. Let T be a symmetry of X and suppose that in $F(T)$ there are k disjoint Jordan curves, then we shall say (see [4]) that the *species* of T is $+k$ if $X - F(T)$ has two components and $-k$ if $X - F(T)$ has one component.

We shall say that a Riemann surface X is an M (respectively $M - 1$) Riemann surface if it admits a symmetry with $g+1$ fixed curves (respectively g fixed curves). S. M. Natanzon in [9] and [10] announced some properties about the topological nature of the symmetries of M and $M - 1$ Riemann surfaces and, in the hyperelliptic case, the classification of such symmetries up to conjugation in the automorphism group. Later, in [11] and [12] the above results were proved by topological techniques. In this paper we give

Partially supported by DGICYT PB89-0201 and SCIENCE Program CEE 910021 a proof of the results about the topological nature of the symmetries in [9] and [10] by the combinatorial theory of non-euclidean crystallographic groups, introduced by A.M. Macbeath in the $60's$; in this way we obtain some improvements of Natanzon's work in the non-hyperelliptic case. The classification of symmetries up to conjugation in the hyperelliptic case can also be obtained by our techniques. By our methods we can also study the symmetries of Riemann surfaces with less than g fixed curves.

The functorial correspondence between the symmetries of Riemann surfaces and the real forms of algebraic complex curves is well-known (see [1]). In this way all the results in this paper can be translated into the language of real algebraic geometry used by Natanzon in his work.

2. Preliminaries

Let $\mathcal{L}$ denote the group of all conformal and anticonformal automorphisms of the upper-half complex plane D and let $\mathcal{L}^+$ denote the subgroup of index 2 consisting of conformal automorphisms. *A non-euclidean crystallographic group* (or NEC group) is a discrete subgroup Λ of $\mathcal{L}$ with compact quotient space D/Λ (see [8] and [14]). An NEC group contained in $\mathcal{L}^+$ is called a *Fuchsian group*, otherwise it is called a *proper* NEC group. The algebraic and geometric structure of a NEC group is completely determined by its *signature*, which is of the form:

$$(g;\pm;[m_1,\ldots,m_r];\{(n_{11},\ldots,n_{1s_1}),\ldots,(n_{k1},\ldots,n_{ks_k})\})$$

If Λ has the above signature then D/Λ is a compact surface of genus g with k holes; it is orientable if the $+$ sign is used and non-orientable if the $-$ sign is used. The integers $m_1,\ldots,m_r$ are called the *ordinary periods* and represent the branching indices over interior points of D/Λ for the natural projection from D to D/Λ. The k brackets $(n_{i1},\ldots,n_{is_i})$ are the *period cycles* and the integers $n_{i1},\ldots,n_{is_i}$, which are called *link periods*, represent the branching indices around the ith hole.

A group Λ with this signature has a presentation with the following generators:

(i)	x_i,	$i=1,\ldots,r$	
(ii)	c_{ij},	$i=1,\ldots,k$ and $j=0,\ldots,s_i$	
(iii)	e_i,	$i=1,\ldots,k$	
(iv)	a_i, b_i,	$i=1,\ldots,g$	(if the sign is $+$)
	d_i,	$i=1,\ldots,g$	(if the sign is $-$)

subject to the relations,

$$(1) \qquad x_i^{m_i} = 1, \qquad i = 1, \ldots, r$$

$$(2) \qquad c_{is_i} = e_i^{-1} c_{i0} e_i, \qquad i = 1, \ldots, k$$

$$(3) \qquad c_{i,j-1}^2 = c_{ij}^2 = (c_{i,j-1} c_{ij})^{n_{ij-1}}, \qquad i = 1, \ldots, k \text{ and } j = 1, \ldots, s_i$$

$$(4) \quad x_1 \ldots x_r e_1 \ldots e_k a_1 b_1 a_1^{-1} b_1^{-1} \ldots a_g b_g a_g^{-1} b_g^{-1} = 1 \quad \text{if the sign is } +$$

$$x_1 \ldots x_r e_1 \ldots e_k d_1^2 \ldots d_g^2 = 1 \quad \text{if the sign is } -.$$

Every NEC group has a fundamental region whose hyperbolic area depends only on the algebraic structure of the group, and for a group with the above signature the area is given by the following formulae (see [13]):

$$\mu(\Lambda) = 2\pi \left(\alpha g + k - 2 + \sum_{i=1}^{r} (1 - \frac{1}{m_i}) + \frac{1}{2} \sum_{i=1}^{k} \sum_{j=0}^{s_i} (1 - \frac{1}{n_{ij}}) \right)$$

where $\alpha = 1$ if the sign is $-$ and $\alpha = 2$ if the sign is $+$.

Now let X be a compact Riemann surface of genus $g > 1$. Then there is a Fuchsian group Γ with signature $(g; +; [\text{-}]; \{\text{-}\})$ such that $X = D/\Gamma$. If G is a group of automorphisms of X (including orientation-reversing automorphisms) then there exists a NEC group Λ such that Γ is a normal subgroup of Λ and $G \simeq \Lambda/\Gamma$. In the case that G is the group generated by a symmetry of species $+k$ then the NEC group Λ has signature

$$(\tfrac{1}{2}(g + 1 - k); +; [\text{-}]; \{(\text{-}), \overset{k}{\ldots}, (\text{-})\}).$$

If G is generated by a symmetry of species $-k$ then Λ has signature

$$(g + 1 - k; \text{-}; [\text{-}]; \{(\text{-}), \overset{k}{\ldots}, (\text{-})\}).$$

(see [4]). As a consequence of this result we have that if X is a Riemann surface that has a symmetry T of species $+k$ (respectively $-k$) then $X/\langle T \rangle$ is an orientable (respectively non-orientable) Klein surface with k boundary components. Let us remark that if T is a symmetry with $g+1$ fixed curves then $X/\langle T \rangle$ is orientable and then the species of T is $+(g+1)$ and if T has g fixed curves then $X/\langle T \rangle$ is non-orientable and therefore the species is $-g$.

The following result of [2] will be used in the next section:

2.1. If T_1 and T_2 are non-commuting symmetries of X, and k_1 and k_2 are the number of fixed curves of T_1 and T_2 respectively, then $k_1 + k_2 < g + 3$.

Also we will use the following remark proved in [3] and [6]:

2.2. If X is a hyperelliptic orientable Klein surface with non-empty boundary and topological genus $g > 0$, then X has one or two boundary components.

3. Results

We begin by stating an easy group-theoretical lemma that is used in the proof of theorem 3.3:

LEMMA 3.1. *Let D_N be a dihedral group with presentation*

$$D_N = \langle \Phi_1, \Phi_2 : \Phi_1^2 = \Phi_2^2 = (\Phi_1\Phi_2)^N = 1 \rangle.$$

(a) If N is a multiple of 2 not a multiple of 4 there are Φ_1' and Φ_2' in D_N conjugate to Φ_1 and Φ_2 respectively such that $\langle \Phi_1', \Phi_2' \rangle$ is isomorphic to D_2.

(b) If N is a multiple of 4 but not of 8 there are Φ_1' and Φ_2' in D_N conjugate to Φ_1 and Φ_2 respectively such that $\langle \Phi_1', \Phi_2' \rangle$ is isomorphic to D_4, and there are Φ_1'', Φ_1''' conjugates of Φ_1 such that $\langle \Phi_1'', \Phi_1''' \rangle$ is isomorphic to D_2.

(c) If N is a multiple of 8 there are Φ_1', Φ_1'' conjugate to Φ_1 such that $\langle \Phi_1', \Phi_1'' \rangle$ is isomorphic to D_4.

Let X be a Riemann surface of genus $g > 1$ and assume that X admits a symmetry Φ_1 of species $+(g+1)$ or $-g$, i. e. X is an M or $M-1$ Riemann surface. If Φ_2 is another symmetry of X, in the following lemma we shall find the possible species of Φ_2 in the case where the order of $\Phi_1\Phi_2$ is 2 or 4.

LEMMA 3.2. *Let X be a Riemann surface of genus $g > 1$ and Φ_1, Φ_2 two symmetries of X such that the order of $\Phi_1\Phi_2$ is N. Then:*

(1) If the species of Φ_1 is $+(g+1)$ then:

(i) If $N = 2$ the species of Φ_2 is either 0, or $g+1-2t$ where $0 \leq t < \frac{g+1}{2}$ if g is even, or $+2$ if g is odd.

(ii) If $N = 4$ the species of Φ_2 is -1.

(2) If the species of Φ_2 is $-g$ then:

(iii) If $N = 2$ the species of Φ_2 is either $-(g-2t)$ where $0 \leq t < \frac{g}{2}$, or $-(g+1-2t)$ where $0 < t < \frac{g+1}{2}$, or -1.

(iv) If $N = 4$ the species of Φ_2 is -2.

PROOF. Assume $X = D/\Gamma$, then Γ has signature $(g;+;[-];\{-\})$.

CASE 1: Φ_1 has species $+(g+1)$. Then by §2 there is a NEC group such that $[\Gamma_1 : \Gamma] = 2$ and Γ_1 has signature:

$$(3.1.1) \qquad (0;+;[-];\{(-),\overset{g+1}{\ldots},(-)\})$$

If $N = 2$ there is a NEC group Γ^* such that $[\Gamma^* : \Gamma_1] = 2$ and Γ^*/Γ is isomorphic to $\langle\Phi_1, \Phi_2\rangle$. Thus by [5] and the Riemann-Hurwitz formulae Γ^* has one of the following signatures:

(3.1.2) $(0;+;[-];\{(-),\overset{t}{\ldots},(-),(2,\overset{r}{\ldots},2)\})$ with $2t + \frac{r}{2} = g+1$

(3.1.3) $(0;+;[-];\{(-),\overset{(g+3)/2}{\ldots},(-)\})$ with 2 a divisor of $g+1$

(3.1.4) $(1;-;[-];\{(-),\overset{(g+1)/2}{\ldots},(-)\})$ with 2 a divisor of $g+1$

(3.1.5) $(0;+;[2,2];\{(-),\overset{(g+1)/2}{\ldots},(-)\})$ with 2 a divisor of $g+1$

(3.1.6) $(0;+;[2];\{(-),\overset{g/2}{\ldots},(-)\})$ with 2 a divisor of g.

Now if Γ_2 is a NEC group such that Γ_2/Γ is isomorphic to $\langle\Phi_2\rangle$, then we have the following diagram of normal subgroups:

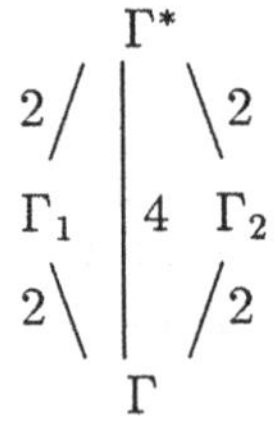

$$\begin{array}{ccccc} & & \Gamma^* & & \\ & {}^2\!\diagup & \big| & \diagdown^2 & \\ \Gamma_1 & & 4 & & \Gamma_2 \\ & {}^2\!\diagdown & \big| & \diagup^2 & \\ & & \Gamma & & \end{array}$$

and so there is an epimorphism $\theta : \Gamma^* \to \langle\Phi_1, \Phi_2\rangle$ such that $\theta^{-1}(\langle\Phi_1\rangle) = \Gamma_1$, $\theta^{-1}(\langle\Phi_2\rangle) = \Gamma_2$ and $\ker\theta = \Gamma$. Since the signatures of Γ, Γ_1 and Γ^* are known then using [5] and [7] we can determine the signature of Γ_2:

—If the signature of Γ_1 is (3.1.2), then Γ_2 has signature
$(g_2;+;[-];\{(-),\overset{w}{\ldots},(-)\})$, where $w = g+1-2t$, $0 \le t < \frac{g+1}{2}$.

—If the signature of Γ_1 is (3.1.3) then Γ_2 has signature $(g_2';+;[-];\{(-),(-)\})$.

—If the signature of Γ_1 is (3.1.4) then Γ_2 has signature $(g_2'';-;[-];\{-\})$.

—If the signature of Γ_1 is (3.1.5) or (3.1.6) then Γ_2 has signature $(g_2''';-;[-];\{-\})$.

It follows that the species of Φ_2 is $g+1-2t$ where $0 \le t < \frac{g+1}{2}$, in the first case, $+2$ in the second case and 0 in the other; note that g must be odd in the cases (3.1.3), (3.1.4) and (3.1.5).

If $N = 4$ then $\langle \Phi_1, \Phi_2 \rangle$ is isomorphic to D_4. The symmetries Φ_1 and $\Phi_2\Phi_1\Phi_2$ have the same species $+(g+1)$ because they are conjugate. Then there is a NEC group Γ^* of signature $(0; +; [-]; \{(2, \overset{2g+2}{\ldots}, 2)\})$ such that Γ^*/Γ is $\langle \Phi_1, \Phi_2\Phi_1\Phi_2 \rangle$. Now there is a Γ' such that Γ'/Γ is $\langle \Phi_1, \Phi_2 \rangle$ isomorphic to D_4, then $[\Gamma' : \Gamma^*] = 2$ and by [5] Γ' has signature $(0; +; [-]; \{(4, 2, \overset{g}{\ldots}, 2, 4)\})$ Hence the only possible epimorphism $\theta' : \Gamma' \to \Gamma'/\Gamma = D_4$ is the following:

$$\theta'(c_1) = \Phi_2, \quad \theta'(c_2) = \Phi_1, \quad \theta'(c_3) = \Phi_2\Phi_1\Phi_2, \quad \theta'(c_4) = \Phi_1,$$
$$\theta'(c_5) = \Phi_2\Phi_1\Phi_2, \quad \ldots, \quad \theta'(c_{g+2}) = \Phi_1 \text{ or } \Phi_2\Phi_1\Phi_2, \quad \theta'(c_{g+3}) = \Phi_2.$$

Then by [5] and [7] the signatures of $(\theta')^{-1}(\langle \Phi_2, \Phi_1\Phi_2\Phi_1 \rangle)$ and $(\theta')^{-1}(\langle \Phi_2 \rangle)$ are $(0; +; [2, \overset{g}{\ldots}, 2]; \{(2,2)\})$ and $(g_2; -; [-], \{(-)\})$ respectively, so the species of Φ_2 is -1.

CASE 2: Φ_1 is a symmetry of species $-g$. In this case there is a NEC group Γ_1 such that $[\Gamma_1 : \Gamma] = 2$ and Γ_1 has signature $(1; -; [-]; \{(-), \overset{g}{\ldots}, (-)\})$.

As in the first case, if $N = 2$ then Γ^* must have one of the following signatures:

$$(0; +; [2]; \{(-), \overset{t}{\ldots}, (-), (2, \overset{r}{\ldots}, 2)\}) \qquad 2t + \frac{r}{2} = g$$
$$(0; +; [-]; \{(-), \overset{t}{\ldots}, (-), (2, \overset{r}{\ldots}, 2)\}) \qquad 2t + \frac{r}{2} = g+1$$
$$(0; +; [-]; \{(-), \overset{(g+3)/2}{\ldots}, (-)\}).$$

Then Γ_2 has signature $(g_2; -; [-]; \{(-), \overset{r/2}{\ldots}, (-)\})$ or $(g_2'; -; [-]; \{(-)\})$. Hence the possible species of Φ_2 are: $-(g-2t)$ with $0 \le t < \frac{g}{2}$, or $-(g-2t+1)$ with $0 < t < \frac{g+1}{2}$, or -1.

If $N = 4$ then Γ' has the signature $(0; +; [-]; \{(2, 4, 2, \overset{g-1}{\ldots}, 2, 4)\})$ and the only possible epimorphism $\theta' : \Gamma' \to \Gamma'/\Gamma = D_4$ is given by

$$\theta'(c_1) = \Phi_2, \quad \theta'(c_2) = \Phi_1\Phi_2\Phi_1, \quad \theta'(c_3) = \Phi_1, \quad \theta'(c_4) = \Phi_2\Phi_1\Phi_2, \ \ldots,$$
$$\theta'(c_{g+1}) = \Phi_2\Phi_1\Phi_2 \text{ or } \Phi_1, \quad \theta'(c_{g+2}) = \Phi_1 \text{ or } \Phi_2\Phi_1\Phi_2, \quad \theta'(c_{g+3}) = \Phi_2.$$

Then the signature of $\theta'^{-1}(\langle \Phi_2, \Phi_1\Phi_2\Phi_1 \rangle)$ is $(0; +; [2, \overset{g-1}{\ldots}, 2]; \{(2,2,2,2)\})$ and the signature of $\theta'^{-1}(\langle \Phi_2 \rangle)$ is $(g'; -; [-]; \{(-)(-)\})$ so that the species of Φ_2 is -2. ∎

First we establish our main result in the case of M Riemann surfaces:

THEOREM 3.3. *Let X be a Riemann surface of genus $g > 1$ having a symmetry Φ_1 that fixes $g+1$ curves.*

(a) If X is hyperelliptic, then:

(i) X has exactly two symmetries of species $+(g+1)$ and the product of such symmetries is the hyperelliptic involution.

(ii) If Φ_2 is another symmetry of X then
—if g is even, Φ_2 has species -1 or 0 or $+1$
—if g is odd, Φ_2 has species -1 or 0 or $+2$.

(b) If X is not hyperelliptic, then:

(i) X has only a symmetry with species $+(g+1)$ which is central in the automorphism group of X.

(ii) If Φ_2 is another symmetry of X then
—if g is even, Φ_2 has species 0 or $+(g+1-2q)$, $0 \leq q < \frac{g+1}{2}$
—if g is odd, Φ_2 has species 0 or $+2$ or $+(g+1-2q)$, $0 \leq q < \frac{g+1}{2}$.

All the species of Φ_2 in the two cases are possible.

PROOF. Assume that X is D/Γ.
(a) (i) If h is the hyperelliptic involution of X then $\Phi' = \Phi_1 h$ is another symmetry of X. The product $\Phi_1\Phi' = h$ has order two and if Γ_h is such that $\Gamma_h/\Gamma = \langle h \rangle$ then the signature of Γ_h is $(0; +; [2, \overset{2g+2}{\ldots}, 2]; \{\text{-}\})$. If Γ^* is such that $\Gamma^*/\Gamma = \langle \Phi_1, \Phi' \rangle$ by the proof of lemma 3.2 the signature of Γ^* is $(0; +; [\text{-}]; \{(2, \overset{2g+2}{\ldots}, 2)\})$. If Γ' is such that $\Gamma'/\Gamma = \langle \Phi' \rangle$ then the signature of Γ' by [12] and [13] must be $(0; +; [\text{-}]; \{(\text{-}), \overset{g+1}{\ldots}, (\text{-})\})$ and Φ' is a symmetry with species $+(g+1)$.

If $\Phi'' \neq \Phi_1$ is a symmetry of species $+(g+1)$ then by 2.1 Φ_1 and Φ'' commute and $\Phi_1\Phi''$ has order two. If Γ^* is such that $\Gamma^*/\Gamma = \langle \Phi_1, \Phi'' \rangle$ then Γ^* has signature $(0; +; [\text{-}]; \{(2, \overset{2g+2}{\ldots}, 2)\})$ and if $\Gamma_{\Phi_1\Phi''}$ is such that $\Gamma_{\Phi_1\Phi''}/\Gamma = \langle \Phi_1\Phi'' \rangle$ then $\Gamma_{\Phi_1\Phi''}$ has signature $(0; +; [2, \overset{2g+2}{\ldots}, 2])$ implying that $\Phi_1\Phi''$ is the hyperelliptic involution. Since the hyperelliptic involution is unique, $\Phi_1\Phi'' = \Phi_1\Phi'$ and $\Phi'' = \Phi'$.

(a) (ii) The order of $\Phi_1\Phi_2$ cannot be a multiple of 8 by lemma 3.1(c) and (a)(i). Since Φ_2 does not have the same species as Φ_1 then the order of $\Phi_1\Phi_2$ is 2 or a multiple of 4. Applying lemma 3.1 we can suppose that this

order is 4 or 2. If we assume that $\Phi_1\Phi_2$ has order 4 then by lemma 3.2 the species of Φ_2 is -1. If the order of $\Phi_1\Phi_2$ is 2 then by 2.2 and 3.2 the possible species for Φ_2 are 0, -1 and $+1$ if g is even and $+2$ if g is odd.

The existence of Riemann surfaces with involutions having the above species is shown by constructing smooth epimorphisms

$$\theta : \Gamma' \to Z_2 + Z_2 + Z_2 \simeq \langle \Phi_1, \Phi_2, \Phi' \rangle \quad \text{or} \quad \theta : \Gamma' \to D_4 \simeq \langle \Phi_1, \Phi_2 \rangle,$$

such that $\theta^{-1}(\Phi_2)$ gives the involution that we want, $\theta^{-1}(\Phi_1)$ gives the maximal symmetry and $\theta^{-1}(\Phi_1\Phi')$ is the hyperelliptic involution.

For example, in order to have a hyperelliptic Riemann surface with a symmetry of species -1 we take a NEC group with signature $(0; +; [-]; \{(4, 2, \overset{g}{\ldots}, 2, 4)\})$ and let θ be the epimorphism in the proof of lemma 3.2 (1)(ii), then $D/\ker\theta$ is a hyperelliptic surface with $\theta^{-1}(\Phi_2)$ a symmetry of species -1.

(b) (i) If Φ' is a maximal symmetry different from Φ_1 then by 2.1, Φ_1 and Φ' commute and then $\Phi_1\Phi'$ has order two. Now by the proof of lemma 3.2, $\Phi_1\Phi'$ must be the hyperelliptic involution, which is a contradiction.

(b) (ii) By lemma 3.1 and (b)(i), $\Phi_1\Phi_2$ must have order two. Now part (ii) is a consequence of lemma 3.2.

To obtain Riemann surfaces of the above type we proceed as in case (a).∎

THEOREM 3.4. *Let X be a Riemann surface of genus $g > 2$ with a symmetry Φ_1 of species $-g$. Then X has no symmetry with species $+s$ or 0., and*

(a) If X is hyperelliptic, then:

(i) X has exactly two $-g$ symmetries and their product is the hyperelliptic involution.

(ii) If Φ_2 is a symmetry with species different from $-g$ then Φ_2 has species either -2 or -1 and both cases occur.

(b) If X is not hyperelliptic, then:

(i) X has only a $-g$ symmetry which is central in the automorphism group of X.

(ii) If Φ_2 is the other symmetry of X the species of Φ_2 must be $-s$ where $0 < s < g$, and every such species occurs.

PROOF. The fact that X cannot have symmetries with species 0 or $+s$ is a consequence of lemma 3.1 and lemma 3.2.

(a) (i) This can be proved in the same way as (a)(i) of theorem 3.3.

(a) (ii) By lemma 3.1 we need consider only the cases where $\Phi_1\Phi_2$ has order 2 or 4. The order 4 case is completely studied in lemma 3.2.

Assume that $\Phi_1\Phi_2$ has order 2. By (a)(i) there exist groups Γ_1 with signature $(1; -; [\text{-}]; \{(\text{-}), \overset{g}{\ldots}, (\text{-})\})$, Γ' with signature $(1; -; [\text{-}]; \{(\text{-}), \overset{g}{\ldots}, (\text{-})\})$, Γ_2 with signature $(g_1; -; [\text{-}]; \{(\text{-}), \overset{\alpha}{\ldots}, (\text{-})\})$ and Γ^* with signature $(0; +; [2]; \{(2, \overset{2g}{\ldots}, 2)\})$ (by the proof of lemma 3.2), and $\Gamma^*/\Gamma = \langle\Phi_1, \Phi'\rangle$, $\Gamma_1/\Gamma = \langle\Phi_1\rangle$, $\Gamma'/\Gamma = \langle\Phi\rangle$, $\Gamma_2/\Gamma = \langle\Phi_2\rangle$.

Since $\Phi_1\Phi_2$ has order two, $\langle\Phi_1, \Phi', \Phi_2\rangle$ is isomorphic to $\mathbb{Z}_2 + \mathbb{Z}_2 + \mathbb{Z}_2$. If Γ^{**} is such that $\Gamma^{**}/\Gamma = \langle\Phi_1, \Phi', \Phi_2\rangle$, by [5] and [7], the signature of Γ^{**} must be $(0; +; [\text{-}]; \{(2, \overset{g+3}{\ldots}, 2)\})$. If $\Gamma_2^* < \Gamma^{**}$ is such that $\Gamma_2^*/\Gamma = \langle\Phi_1, \Phi_2\rangle$ the signature of Γ_2^* is $(0; +; [\text{-}]; \{(\text{-}), \overset{g/2}{\ldots}, (\text{-}), (2,2)\})$ if g is even and $(0; +; [2]; \{(\text{-}), \overset{(g-1)/2}{\ldots}, (\text{-}), (2,2)\})$ if g is odd, and in any case if the signature of Γ_2 is $(g_2; -; [\text{-}]; \{(\text{-})\})$ then the species of Φ_2 is -1.

The following epimorphisms show that the above symmetry occurs for every genus. $\theta : \Gamma^{**} \to \Gamma^{**}/\Gamma \simeq \mathbb{Z}_2 + \mathbb{Z}_2 + \mathbb{Z}_2$ is given by:

$$\theta(c_0) = \Phi_2,\ \theta(c_1) = \Phi_1,\ \theta(c_2) = \Phi',\ \theta(c_3) = \Phi_1,\ \ldots, \theta(c_g) = \Phi' \text{ or } \Phi_1,$$
$$\theta(c_{g+1}) = \Phi_1 \text{ or } \Phi', \quad \theta(c_{g+2}) = \Phi_1\Phi'\Phi_2, \quad \theta(c_{g+3}) = \Phi_2.$$

The proof of part (b) is similar to the proof of theorem 3.3. ∎

REMARK 3.5. By lemma 3.2(iv) we have an example for which the results of theorem 3.4 are not true without the assumption $g > 2$. Theorem 3.4 (b)(ii) gives us, for $g = 2$, a Riemann surface with automorphisms group isomorphic to D_4 having four symmetries of species -2.

References

[1] N. L. ALLING, N. GREENLEAF. *Foundations of the theory of Klein surfaces.* Lect. Notes in Math., **219** (Springer-Verlag, Berlin, etc. 1971).

[2] E. BUJALANCE, A. COSTA, D. SINGERMAN. An application of Hoare's theorem to symmetries of Riemann surfaces. *Preprint*

[3] E. BUJALANCE, J. J. ETAYO, J. M. GAMBOA. Hyperelliptic Klein surfaces. *Quart. J. Math. Oxford (2)* **36** (1985) 141-157.

[4] E. BUJALANCE, D. SINGERMAN. The symmetry type of a Riemann surface. *Proc. London Math. Soc. (3)* **51** (1985) 501-519.

[5] J. A. BUJALANCE Normal subgroups of even index in an NEC group. *Arch. Math.* **49** (1987) 470-478.

[6] B. H. GROSS, J. HARRIS. Real algebraic curves. *Ann. Scient. Ec. Norm. Sup.* **14** (1981) 157-182.

[7] A. H. M. HOARE, D. SINGERMAN. Subgroups of plane groups. *London Math. Soc. Lect. Note Series* **71** (1982) 221-227.

[8] A. M. MACBEATH. The classification of non-euclidean plane crystallographic groups. *Can. J. Math.* **19** (1967) 1192-1205.

[9] S. M. NATANZON. Automorphisms of the Riemann surface of an M-curve. *Functional Anal. Appl.* **12** (1979) 228-229.

[10] S. M. NATANZON. Automorphisms and real forms of a class of complex algebraic curve. *Functional Anal. Appl.* **13** (1979) 148-150.

[11] S. M. NATANZON. Lobachevskian geometry and Automorphisms of Complex M-curves. *Selecta Math. Sovietica* **1** (1981) 81-99.

[12] S. M. NATANZON. Finite groups of homeomorphisms of surfaces and real forms of complex algebraic curves. *Trans Moscow Math. Soc.* (1989) 1-51.

[13] D. SINGERMAN. On the structure of non-euclidean crystallographic groups. *Proc. Camb. Phil. Soc.* **76** (1974) 233-240.

[14] H. C. WILKIE. On non-euclidean crystallographic groups. *Math. Z.* **91** (1966) 87-102.

Departamento de Matemáticas Fundamentales
Facultad de Ciencias
U.N.E.D.
Madrid, Spain

Inequalities for Pell equations and Fuchsian groups

J. H. H. Chalk

Dedicated to Murray Macbeath on the occasion of his retirement

Introduction

Let S denote the solution-set of a diophantine equation of the form

$$D_1(t^2 + Ps^2) - D_2(u^2 + Pv^2) = n, \tag{1}$$

where D_1, D_2, P are positive integers and $n(\neq 0)$ is an integer assumed to be representable by the quaternary quadratic form on the left of (1) with $(t, s, u, v) \in \mathbb{Z}^4$. Some insight into the structure of S will appear from knowledge of the group structure of the solution-set of a special case of (1):

$$(t^2 + Ps^2) - D(u^2 + Pv^2) = +1, \tag{2}$$

if we take $D = D_1D_2$. More generally, we know from the classical theory of indefinite quadratic forms in ≥ 4 variables that

(i) an integer n is representable over $\mathbb{Z}$ if, and only if, it is representable over $\mathbb{Z}_p$ for all primes p

(ii) if n is so representable, then n is representable infinitely often (in particular, card $S = \infty$).

If we introduce the variables x, y in $\mathbb{Z}[i\sqrt{P}]$, by $x = t + i\sqrt{P}s$, $y = u + i\sqrt{P}v$ then (1) and (2) become

$$D_1 x\bar{x} - D_2 y\bar{y} = n, \tag{3}$$

$$x\bar{x} - Dy\bar{y} = 1 \tag{4}$$

respectively; the forms on the left being binary Hermitian.

We shall consider three objectives for S:

I. An effective bound to the 'size' of a nontrivial solution of (4), (i.e. with $y \neq 0$).

II. An effective bound to the size of a solution of (3) using
(a) a group of $\mathbb{Z}[i\sqrt{P}]$-automorphs of $D_1 x\bar{x} - D_2 y\bar{y}$,
(b) number-theoretic techniques.

III. As the group in II(a) is Fuchsian and the bound obtained depends upon the "diameter" of its fundamental region, we conclude with some remarks on the spectral theory of Selberg with reference to Peter Buser's lower bound for the smallest eigenvalue, also dependent on a "diameter" for the closed geodesics on the corresponding Riemann surface.

Section I

For the classical Pell equation $t^2 - Du^2 = 1$, where $D > 0$ is a non-square integer, I. Schur [12] used the Kronecker form of the class-number formula

$$h(D) \log \varepsilon(D) = \sqrt{D} \sum_{1 \leq n < \infty} \left(\frac{D}{n}\right) \frac{1}{n},$$

where $\varepsilon(D) = t + u\sqrt{D}$ and $h(D)$ denotes the number of classes of properly primitive indefinite binary quadratic forms $ax^2 + bxy + cy^2$ of determinant $D = b^2 - 4ac$ to show that

$$\log \log \varepsilon(D) < \tfrac{1}{2} \log D + \log \log D.$$

This estimate is close to being best-possible for m large and D of the form 2^{2m+1} (this idea was extended by E. Landau [11] to units of algebraic number-fields).

For the Pellian equation (4), observe that

$$\Gamma_1(P) = \left\{ \begin{pmatrix} \lambda & d\nu \\ \bar{\nu} & \bar{\lambda} \end{pmatrix} : \quad \lambda, \nu \in \mathbb{Z}[i\sqrt{P}], \quad \lambda\bar{\lambda} - D\nu\bar{\nu} = 1 \right\}$$

is a group which preserves the form f_1, say, in (4) under the mappings

$$\begin{pmatrix} x \\ y \end{pmatrix} = \gamma \begin{pmatrix} x' \\ y' \end{pmatrix}, \qquad \text{for all } \gamma \in \Gamma(P).$$

The corresponding group $\widetilde{\Gamma}_1(P)$ of actions in the complex z-plane

$$z \mapsto \frac{\lambda z + D\nu}{\bar{\nu} z + \bar{\lambda}} \tag{6}$$

is, in fact, Fuchsian of the first kind with principal circle $|z|^2 = D$. Note that the Dirichlet method of defining a fundamental region $\mathcal{F}_1$ in $|z| \leq D$, by the use of the isometric circles $|\bar{\nu}z + \bar{\lambda}| = 1$ of $\widetilde{\Gamma}_1(P)$, introduces the "small" solutions of $\lambda\bar{\lambda} - D\nu\bar{\nu} = 1$ (and, in fact, a set of generators), since the isometric circles have radius $|\nu|^{-1}$ and are orthogonal to the principal circle. Then, by simple geometric considerations of $\mathcal{F}_1$ and a class-number formula (due to Humbert [9]) analogous to that of Schur above, we have the following estimate (cf. [3]).

THEOREM 1. *There exist x, y in $\mathbb{Z}[i\sqrt{P}]$ such that $x\bar{x} - Dy\bar{y} = 1$ with*

$$0 \neq \sqrt{x\bar{x}} < 1 + DP \prod_p \left[1 + \left(\frac{-P}{p}\right)\frac{1}{p}\right] \prod_q \left[1 + \left(\frac{D}{q}\right)\frac{1}{q}\right], \tag{7}$$

where p, q are odd primes with $p|D, q|P$, provided that

$$(D, P) = 1 \text{ or } 2, \quad P \not\equiv 0 \bmod 4, \quad D \not\equiv 0 \bmod 4. \tag{8}$$

REMARK 1. The restrictions in (8) arise from the limitations in the class-number formula (10) below. They can be removed at the expense of a more complicated formula than (7), but for this it is more efficient to regard $x\bar{x} - Dy\bar{y} = n(\tau) = \tau\hat{\tau}$ as the norm of a proper unit of an order of a rational indefinite quaternion algebra and extend the work of M. Eichler [6] on local methods (for details, see [4]).

We outline a proof of Theorem 1 for the case $P = 1$. The formula for the non-Euclidean area $\sigma(\mathcal{F}_1)$ of $\mathcal{F}_1$ has the shape

$$\sigma(\mathcal{F}_1) = 4D \int f_1^{-2}(z, 1)\, dx dy \tag{9}$$

and the analogous class-number formula (G. Humbert [9]) for the h classes of properly primitive binary Hermitian forms is

$$\sum_{1 \leq j \leq h} \int_{\mathcal{F}_j} f_j^{-2}(z, 1)\, dx dy = \frac{\pi}{4} \prod_{p|D, p\neq 2} \left[1 + \left(\frac{-1}{p}\right)\frac{1}{p}\right], \tag{10}$$

where the forms f_j $(1 \leq j \leq h)$ represent the h classes and $\mathcal{F}_j$ the corresponding fundamental region for the $\mathbb{Z}[i]$-automorphs of f_j. Since each term in the sum on the left is ≥ 0, we have

$$\sigma(\mathcal{F}_1) \leq \pi D \prod_{p|D, p\neq 2} \left[1 + \left(\frac{-1}{p}\right)\frac{1}{p}\right], \tag{11}$$

by (9). Now, using Euclidean geometry, the largest circular disc $\mathcal{D}$ with centre $z = 0$ and contained in $\mathcal{F}_1$ has radius $= (|x| - 1)|y|^{-1}$, where we have $x\bar{x} - Dy\bar{y} = 1$ and $|y|$ is minimal. Since $\sigma(\mathcal{F}_1) \geq \sigma(\mathcal{D})$, we obtain

$$0 < |x| - 1 \leq \pi^{-1}\sigma(\mathcal{F}_1) \leq D \prod_{p|D, p\neq 2} \left[1 + \left(\frac{-1}{p}\right)\frac{1}{p}\right], \tag{12}$$

as required.

REMARK 2. For $P = 1$, we note that *if* $D = a^2 + b^2$ $(a, b \in \mathbb{Z})$ then, by inspection of (2), we have a non-trivial solution with

$$(t, s, u, v) = (1, D, a, b), \tag{13}$$

which is roughly comparable in size to that in (12), for D large.

Section II(a)

The properties of the group $\Gamma_1(P)$ with $D = D_1 D_2$ lead to an estimate for (3) based upon the properties of our fundamental region for $\Gamma_1(P)$. More precisely, with $x, y, x_0, y_0, X, Y \in \mathbb{Z}[i\sqrt{P}]$ and conjugates denoted by $'$, note that

$$\begin{pmatrix} D_1 x & D_1 D_2 y \\ y' & D_1 x' \end{pmatrix} \begin{pmatrix} x_0 & D_1 D_2 y_0 \\ y_0' & x_0' \end{pmatrix} = \begin{pmatrix} D_1 X & D_1 D_2 Y \\ Y' & D_1 X' \end{pmatrix},$$

where

$$X = x x_0 + D_2 y y_0', \qquad Y = x_0' y + D_1 x y_0. \tag{14}$$

Thus (a) $(D_1 x x' - D_2 y y')(x_0 x_0' - D y_0 y_0') = D_1 X X' - D_2 Y Y'$

(b)$(D_1 x, y) \mapsto (D_1 X, Y)$, under the action of $\gamma \in \Gamma(P)$, where

$$\gamma = \begin{pmatrix} x_0 & D y_0 \\ y_0' & x_0' \end{pmatrix}.$$

(c) In particular, if $w = D_1 \frac{x}{y}$, $\quad W = D_1 \frac{X}{Y}$, then

$$W = \frac{x_0 w + D y_0}{y_0' w + x_0'} . \tag{15}$$

Note that, by the interchange $(x, y) \to (y, x)$, $(D_1, D_2) \to (D_2, D_1)$, there is no loss of generality if we take $n < 0$. By (c) we see that $\Gamma_1(P)$ acts on the disk $|z| \leq \sqrt{D}$. Suppose now that $\Gamma_1(P)$ is hyperbolic so that we may choose a fundamental region $\mathcal{F}$ for $\Gamma_1(P)$, all of whose points belong to a smaller disk $|z| \leq \rho < \sqrt{D}$, in which case we can ensure that the point $w = D_1 x/y$ is transformed, by a suitable element $\gamma \in \Gamma_1(P)$ into a point $W = D_1 X/Y \in \mathcal{F}$, and so

$$|W| = \left| \frac{D_1 X}{Y} \right| < \rho .$$

Since $D_1 |X|^2 - D_2 |Y|^2 = -|n|$,

$$\max[D_1 |X|^2, D_2 |Y|^2] < \frac{D|n|}{D - \rho^2} \tag{16}$$

in terms of ρ, which we can view as a crude bound for the diameter of the region $\mathcal{F}$.

Section II(b)

Applications of the "circle method" (often referred to as the 'Hardy-Littlewood' method for their pioneering work on diagonal quadratic forms) have, in recent years (cf. [7], [13], [1]) produced effective estimates; the latest of which implies that, when solvable, there is a solution of (1) with

$$\max[|t|, |s|, |u|, |v|] \ll_\varepsilon (D_1 D_2 P)^{5+\varepsilon} |n|^{\frac{1}{2}+\varepsilon} . \tag{17}$$

This is, in fact, a corollary to an asymptotic formula for the number of solutions, when the variables are confined to a large "box" centred at the origin. For $P = 1$, an elementary method due to Hooley (cf. [5]) produced the estimate

$$t^2 + s^2 \ll_\varepsilon (D_1 D_2)^{3+\varepsilon} |n|^\varepsilon , \tag{18}$$

valid for all representable n not too large compared with D_1 and D_2, e.g.

$$n \ll_\varepsilon \min(D_1, D_2)(D_1 D_2)^{3-\varepsilon} \tag{19}$$

would suffice. As before this is a corollary to an asymptotic formula for the numbers of solutions. For $P > 1$, Hooley's method works if the class-number $h(-P)$ of properly primitive binary quadratic forms of discriminant $-P$ satisfies $h(-P) = 1$. Then, we obtain

$$t^2 + Ps^2 \ll (D_1 D_2 P)^{3+\varepsilon} |n|^{\varepsilon}, \tag{20}$$

provided that n is not too large compared with $D_1 D_2 P$. For $P > 1$, $h(-P) > 1$, a weaker asymptotic formula is under investigation.

Section III

Suppose now that Γ is a general hyperbolic Fuchsian group which we may suppose (after a normalization) to act on the unit disk $\mathcal{U} = \{z \in \mathbb{C} : |z| < 1\}$. Let $\mathcal{F}$ denote a Dirichlet fundamental region for Γ; it is, in fact a finite-sided non-Euclidean polygon, the boundary consisting of pairs of equivalent oppositely oriented sides ℓ and $\ell_{\gamma^{-1}}$ which correspond to one another under $\gamma \in \Gamma$. The invariant measures on $\mathcal{U}$ are

$$d\mu(z) = \frac{2}{1-|z|^2}|dz| \text{ and } dA(z) = \left(\frac{2}{1-|z|^2}\right)^2 dx dy \quad (z = x + iy). \tag{21}$$

Now, for any function $f \in \mathcal{C}^2(\mathcal{F})$, the invariant form of Green's identity has the form

$$\int_{\mathcal{F}} \mathcal{D}f \, dA(z) + \int_{\mathcal{F}} f D(f) \, dA(z) = \int_{\partial\mathcal{F}} f \nabla f \, dz, \tag{22}$$

where $$\mathcal{D}f = \left(\frac{1-|z|^2}{2}\right)^2 |\nabla f|^2 \text{ and } Df = \left(\frac{1-|z|^2}{2}\right)^2 \Delta f,$$

$$\left(\Delta = \frac{\partial^2}{\partial x^2} + \frac{\partial^2}{\partial y^2}, \ \nabla f = f_x - i f_y\right)$$

are the Laplace-Beltrami operators. Note that $\nabla f \, dz = \frac{\partial f}{\partial n} \, ds$ is invariant under automorphisms of $\mathcal{U}$. In particular, if f is automorphic under the action of Γ, then

$$\int_{\partial\mathcal{F}} f \nabla f \, dz = 0, \tag{23}$$

by the pairings of the sides of $\mathcal{F}$. More generally, there is a 'nice' spectral theory in the Hilbert space $L_2(\Gamma\backslash\mathcal{U})$ furnished with the inner product (cf. [6] p. 1-12)

$$(f_1, f_2) = \int_{\mathcal{F}} f_1(z)\overline{f_2(z)} \, dA(z). \tag{24}$$

The invariant Laplacian D has a discrete spectrum

$$(i) \qquad 0 = \lambda_0 \leq \lambda_1 \leq \lambda_2 \leq \ldots \leq \lambda_n \leq \ldots \qquad (\lambda_n \to +\infty \text{ as } n \to \infty)$$

and an orthonormal basis $\{\phi_n\}$ of real-valued eigenfunctions which satisfy

$$(ii) \qquad D\phi_n + \lambda_n \phi_n = 0, \qquad \forall n \in \mathbb{N}, \quad \phi_n \in \mathcal{C}^\infty .$$

$$(iii) \qquad L_2(\Gamma \backslash \mathcal{U}) = \bigoplus_{0 \leq n < \infty} \mathbb{C}[\phi_n] .$$

Suppose now that f is a real-valued eigenfunction belonging to λ. Then, by (ii),

$$(-Df, f) = \int_{\mathcal{F}} (-Df) f \, dA(z) = \lambda \int_{\mathcal{F}} f^2 \, dA(z)$$

and by Green's identity, using (23),

$$(-Df, f) = \int_{\mathcal{F}} \mathcal{D}f \, dA(z) = \int_{\mathcal{F}} |\text{grad } f|^2 \, dx dy .$$

Hence

$$(25) \qquad \lambda = \int_{\mathcal{F}} (fx^2 + fy^2) \, dx dy \Big/ \int_{\mathcal{F}} f^2 \left(\frac{2}{1 - |z|^2} \right)^2 dx dy.$$

This is a non-Euclidean form of the classical Rayleigh Quotient and, so far as I know, has not been fully exploited to achieve lower bounds for the eigenvalues λ (however, see [10],12.19 and [2], pp.478, 487). Nevertheless, we do have a lower bound for λ, due to P. Buser [2], in terms of the diameter d of the Riemann surface M for Γ, namely

$$(26) \qquad \lambda_1 > \left(4\pi \sinh \frac{d}{2} \right)^{-2} ,$$

where $d = \sup\{\text{dist}(p, q) : p, q \in M\}$.

Such lower bounds are important in connection with the location of the so-called exceptional zeros of the Selberg zeta function $Z_\Gamma(s)$ for Γ, which has the form (cf. [8], p.66 for details)

$$(27) \qquad Z_\Gamma(s) = \prod_{\{\gamma_0\}} \prod_{k=0}^{\infty} [1 - N(\gamma_0)^{-s-k}] \qquad \text{for } \text{Re}(s) > 1 ;$$

the first product being taken over all the conjugacy classes of Γ, where

$$\gamma \sim \sigma^{-1}\gamma\sigma = \begin{pmatrix} \sqrt{N(\gamma)} & 0 \\ 0 & \dfrac{1}{\sqrt{N(\gamma)}} \end{pmatrix}, \tag{28}$$

γ_0 being a generator of the class. In brief, the remarkable fact is that the eigenvalues λ are directly related to the zeros s of $Z_\Gamma(s)$ by the simple relation:

$$\lambda = s(1-s). \tag{29}$$

It is known that $Z_\Gamma(s)$ is an entire function of s which satisfies the Riemann Hypothesis in the sense that all its non-real zeros reside on the line $\mathrm{Re}(s) = \frac{1}{2}$. Apart from the trivial zeros $s = -n$ ($n \in \mathbb{N}$), there can be at most finitely many exceptional zeros, $s = s_1 < \ldots < s_M$ say with $0 < s_n < 1$, which correspond, by (29), to exceptional eigenvalues with

$$0 < \lambda_j < \frac{1}{4}. \tag{30}$$

Thus a lower bound for λ_1, confirms a zero-free interval for $Z_\Gamma(s)$. If we define a *prime number function* by

$$\pi_\Gamma(x) = \#[\{\gamma\} : N(\gamma) \leq x] \tag{31}$$

then (cf. [8], pp.41 and 113) the asymptotic formula

$$\pi_\Gamma(x) = \ell i(x) + \sum_{1\leq k\leq M} \ell i(x^{s_k}) + \mathbf{O}\left(x^{\frac{3}{4}}(\ell n x)^{-\frac{1}{2}}\right) \quad \text{as } x \to \infty \tag{32}$$

reflects a loss of precision in the absence of a suitable inequality for the s_k, or equivalently the corresponding λ_j in (29).

It is of some interest, in view of Buser's inequality, to recall the example in II(a) and present it in a slightly more general setting. First, we normalize the group $\widetilde{\Gamma}(P)$ to act on the unit disk $|z| \leq 1$ by, for example, writing

$$A = \frac{D_1}{\sqrt{D}}, \quad B = \frac{D_2}{\sqrt{D}}, \quad C = \frac{n}{\sqrt{D}}, \tag{33}$$

where $D_1 D_2 = D$, so that $AB = 1$ and

$$Ax\bar{x} - By\bar{y} = C, \qquad (A > 0, B > 0). \tag{34}$$

Now take any group of elements of the form $\begin{pmatrix} x_0 & y_0 \\ \bar{y}_0 & \bar{x}_0 \end{pmatrix}$, $(x_0, y_0) \in \mathbb{C}^2$ with $|x_0|^2 - |y_0|^2 = 1$ and suppose that $\widetilde{\Gamma}$ is hyperbolic Fuchsian. Then by the same argument as in II(a), there is a solution (X, Y) of (34), in the Γ-orbit of (x, y), with

$$(35) \qquad \max[A|X|^2, B|Y|^2] < \frac{|C|}{1-\rho^2}, \quad (AX\bar{X} - BY\bar{Y} = C),$$

or equivalently,

$$(36) \qquad \max[A|X|^2, B|Y|^2]|C|^{-1} < \sinh^2\left(\frac{\delta}{2}\right) + 1,$$

where

$$\delta = \int_0^\rho \frac{2}{1-r^2}\, dr$$

is the non-Euclidean measure of a "diameter" of the fundamental region $\mathcal{F}$ (as specified in II(a)).

References

[1] R. Ashton. A bound for the least solution of the diophantine equation $ax_1^2 + bx_2^2 + cx_3^2 + dx_4^2 = n$, University of London Ph.D. thesis (1992).

[2] P. Buser. Über den ersten Eigenwert des Laplace-Operators auf kompakten Flächen, *Comment. Math. Helvetici* **54** (1979) 477-493.

[3] J. H. H. Chalk. An estimate for the fundamental solutions of a generalized Pell equation. *Math. Annalen* **132** (1956) 263-276 (cf. *Comptes Rendus*, Paris **244** (1957) 985-988).

[4] J. H. H. Chalk. Units of Indefinite Quaternian Algebras. *Proc. Royal Soc. Edinburgh* **87A** (1980) 111-126.

[5] J. H. H. Chalk. An application of Hooley's method for counting solutions of a diophantine equation. *C. R. Math. Rep. Acad. Sci. Canada* **III** No.2 (1981) 99-103.

[6] M. Eichler. *Lectures on Modular Correspondences*, Tata Institute (1955-6), (1965).

[7] T. Estermann. A new application of the Hardy-Littlewood-Kloostermann method. *Proc. London Math. Soc. (3)* **12** (1962) 425-444.

[8] D. Hejhal. *The Selberg Trace Formula for* PSL(2, $\mathbb{R}$) Vol.1, Springer Lecture Notes, **548**.

[9] G. HUMBERT. Comptes Rendus **166** (1918) 753, **171** (1920) 287,377.
[10] H. IWANIEC. *Spectral Theory of Automorphic Functions Part I.* Lecture Notes, Rutgers University, (Spring 1987) Ch.12.
[11] E. LANDAU. *J. Nachr. Ges. d. Wiss. Göttingen* (1918) 86-87.
[12] I. SCHUR. *J. Nachr. Ges. d. Wiss. Göttingen* (1918) 30-36.
[13] K. S. WILLIAMS. An application of the Hardy-Littlewood method. University of Toronto Ph.D. thesis (1965).

Department of Mathematics
Imperial College
London, SW7 2BZ

The Euler characteristic of graph products and of Coxeter groups

I. M. Chiswell

To Murray Macbeath on the occasion of his retirement

Let Γ be a (finite, simplicial) graph, with vertices $v_1, \ldots, v_n$. Assign to each vertex v_i a group A_i. We define the graph product $G\Gamma$ to be the quotient of the free product $*_{i=1}^{n} A_i$ by the normal subgroup generated by all $[A_i, A_j]$ for which $\{v_i, v_j\}$ is an edge. These groups have been studied by E. R. Green, with particular reference to their residual properties, in [14]. The groups studied by the author in [7] are a special case, in which all the A_i are cyclic. Assuming the Euler characteristics of all the A_i are defined, we give a formula for the Euler characteristic of $G\Gamma$ in terms of those of the A_i. We also make some observations on the virtual cohomological dimension of these groups in special cases, and give an explicit formula for the Euler characteristic of a finitely generated Coxeter group.

1. Graph Products

We shall concentrate on the Euler characteristic defined by H. Bass and by the author ([1], [6]), based on work of Serre and Stallings, so we shall assume that all A_i are in the class $FP(R)$, where R is a commutative ring. Recall that a group G is in $FP(R)$ if the trivial RG-module R has an RG-projective resolution

$$0 \longrightarrow P_d \longrightarrow \cdots \longrightarrow P_0 \longrightarrow R \longrightarrow 0$$

for some d, in which each P_i is finitely generated. We shall denote the Euler characteristic of G by $\chi(G)$, as in §10 of [1] (in [6] this was denoted by $\mu(G)$). We recall some of the properties of χ.

(i) if $1 \longrightarrow N \longrightarrow G \longrightarrow Q \longrightarrow 1$ is a short exact sequence of groups, and N, Q are in $FP(R)$, then $G \in FP(R)$ and $\chi(G) = \chi(N)\chi(Q)$.

(ii) if $G = A *_C B$ is a free product with amalgamation, and A, B, C are all in $FP(R)$, then $G \in FP(R)$ and $\chi(G) = \chi(A) + \chi(B) - \chi(C)$

(iii) if G is a finite group whose order $|G|$ is invertible in R, then $G \in FP(R)$ and $\chi(G) = 1/|G|$.

For the proof of (i) see Theorem 10.9(e) in [1], for (ii) see (1) after Theorem 3 in [6] and (iii) is obtained by a simple modification of Lemma 11 in [6].

If Δ is a subgraph of Γ, with vertices $v_{i_1}, \ldots, v_{i_k}$, put

$$\chi_\Delta = (\chi(A_{i_1}) - 1)(\chi(A_{i_2}) - 1)\ldots(\chi(A_{i_k}) - 1)$$

and define
$$\Sigma_\Gamma = \sum_\Delta \chi_\Delta$$

where the sum is over all complete subgraphs Δ of Γ. It is important to note that the empty subgraph is allowed as a complete subgraph, and that for $\Delta = \emptyset$, $\chi_\Delta = 1$, the empty product.

PROPOSITION 1. *If all A_i are in $FP(R)$, then so is $G\Gamma$, and $\chi(G\Gamma) = \Sigma_\Gamma$.*

PROOF. We use induction on n, the number of vertices of Γ. Assume it is true for graphs with fewer than n vertices. By Lemma 3.20 in [14], we have a decomposition
$$G\Gamma = (A_1 \times GE) *_{GE} GZ$$

where Z is the graph obtained by removing the vertex v_1 and all edges incident with it from Γ, and E is the subgraph of Γ generated by all vertices of Γ adjacent to v_1. Inductively $GE, GZ \in FP(R)$, so $(A_1 \times GE) \in FP(R)$ by Property (i) above, hence $G\Gamma \in FP(R)$ by Property (ii). Moreover,

$$\begin{aligned}\chi(G\Gamma) &= \chi(A_1 \times GE) + \chi(GZ) - \chi(GE)\\ &= \chi(A_1)\chi(GE) + \chi(GZ) - \chi(GE)\\ &= \chi(GE)(\chi(A_1) - 1) + \chi(GZ)\\ &= \Sigma_E(\chi(A_1) - 1) + \Sigma_Z\end{aligned}$$

using the induction hypothesis. Now $\Sigma_E(\chi(A_1) - 1)$ is the sum of the χ_Δ over all complete subgraphs of Γ which contain v_1, while Σ_Z is the sum of χ_Δ over all complete subgraphs of Γ which do not contain v_1. Thus

$$\Sigma_E(\chi(A_1) - 1) + \Sigma_Z = \Sigma_\Gamma$$

as required.

We state two special cases of this formula as corollaries. The first corollary was noted by Droms [13], as a consequence of a result of W. Dicks [11]. One could alternatively use Theorem 10 in [17].

COROLLARY 1. *If $G\Gamma$ is a graph group (i.e. all the A_i are infinite cyclic), then $\chi(G\Gamma) = \sum_{j=0}^{n}(-1)^j n_j$, where n_j is the number of complete subgraphs of Γ with j vertices.*

This follows from Proposition 1 on noting that the infinite cyclic group is in $FP(R)$ for any commutative ring R and has Euler characteristic 0 (see the remarks after Theorem 3 in [7] and §1.4 in [21]). Graph groups have been extensively studied by Droms and others (see [12], [13]).

COROLLARY 2. *If 2 is invertible in R and $G\Gamma$ is a right-angled Coxeter group (i.e. all the A_i are cyclic of order 2), then $\chi(G\Gamma) = \sum_{j=0}^{n}(-\frac{1}{2})^j n_j$, where n_j is as in Corollary 1.*

This follows from Property (iii) above. Of course we have a similar formula whenever all the A_i are isomorphic to a single group. We mention Corollary 2 because it can be obtained by geometric means, a point we shall return to shortly.

There is a result similar to Proposition 1 for the Euler characteristic defined by Brown [5], using Prop.7.3, Ch.IX in [5]. If all A_i are in FH, then so is $G\Gamma$ and the formula $\chi(G\Gamma) = \Sigma_\Gamma$ holds, where now χ denotes the Brown characteristic. In order to make the argument of Prop. 1 work the following result, which generalises Proposition 2 in [7], is needed.

PROPOSITION 2. *In any graph product, if all the A_i are virtually torsion-free, then so is $G\Gamma$.*

PROOF. Let B_i be a torsion-free normal subgroup of finite index in A_i. There is a homomorphism $G\Gamma \longrightarrow \prod_{i=1}^{n} A_i/B_i$ induced by the quotient maps $A_i \longrightarrow A_i/B_i$. Let K_Γ be its kernel. We show by induction on the number n of vertices of Γ that K_Γ is torsion-free. If Γ is a complete graph, then $K_\Gamma = \prod_{i=1}^{n} B_i$ is torsion-free. Otherwise after renumbering we can assume there is some vertex not adjacent to v_1. We then have the decomposition used in Prop. 1:

$$G\Gamma = (A_1 \times GE) *_{GE} GZ.$$

Now $A_1 \times GE$ is the graph product GH, where H is E together with v_1 and all edges of Γ joining v_1 to a vertex of E, and both H and Z have fewer vertices than Γ. Corresponding to this decomposition of $G\Gamma$ we have a decomposition of K_Γ as the fundamental group of a graph of groups using the Bass-Serre Theorem. On noting that $K_\Gamma \cap GH = K_H, K_\Gamma \cap GZ = K_Z$ and $K_\Gamma \cap GE = K_E$, the result follows by using the induction hypothesis on H, Z and E just as in Prop.2 of [7].

2. Coxeter Groups

We now give an explicit formula for the Euler characteristic of a finitely generated Coxeter group. A recursive formula to calculate this was given by Serre [21: Prop.16(c)]. Such a group can be described by a finite graph Γ with a label $\phi(e)$ attached to each edge e, which is an integer ≥ 2. Let $x_1, \ldots, x_n$ be the vertices of Γ.

The corresponding Coxeter group C_Γ is defined to be the group with generators $x_1, \ldots, x_n$ and relators x_i^2 for $1 \leq i \leq n$, together with $(x_i x_j)^{\phi(\{x_i, x_j\})}$, for each edge $\{x_i, x_j\}$ of Γ. This method of specifying a Coxeter group is used in [10] and [19], and is the most useful here. It differs from the usual one [9; 9.2], in that edges with label 2 are included, and we do not use edges $\{x_i, x_j\}$ with label ∞ to indicate the absence of a relator which is a power of $x_i x_j$. We recall that, if Δ is a subgraph of Γ with the induced labelling, then C_Δ is a subgroup of C_Γ in the obvious way (see [4; Ch.4, 1.8]). If $G\Gamma$ is a right-angled Coxeter group as in Corollary 2, it coincides with the group C_Γ (using the same graph Γ), where each edge is given label 2.

Before stating our formula we need a combinatorial lemma which is presumably well-known, but we are unable to give a reference, so a proof is included. We define $\lambda_{k,r}$ to be the number of chains of the form $\emptyset = E_0 \underset{\neq}{\subset} E_1 \underset{\neq}{\subset} \ldots \underset{\neq}{\subset} E_r = E$ in a set E with k elements. It is not difficult to see that $\lambda_{k,r} = r!\, S(k,r)$, where $S(k,r)$ is the Stirling number of the second kind, that is, the number of equivalence relations on a k-set with r equivalence classes, and that $r!\, S(k,r)$ is the number of surjections from a k-set to an r-set.

LEMMA. *For any $k \geq 0$, we have $\sum_{r=0}^{k}(-1)^r \lambda_{k,r} = (-1)^k$.*

PROOF. The proof is by induction on k, and it is obviously true when $k = 0$. Suppose $k \geq 1$, so that $\lambda_{k,0} = 0$. If $r \geq 1$, then by considering the possibilities for $l = |E_{r-1}|$ in a chain of subsets as above, we obtain $\lambda_{k,r} = \sum_{l=r-1}^{k-1} \binom{k}{l} \lambda_{l,r-1}$. Thus

$$\begin{aligned}
\sum_{r=0}^{k}(-1)^r \lambda_{k,r} &= \sum_{r=1}^{k}(-1)^r \lambda_{k,r} \\
&= \sum_{r=1}^{k}(-1)^r \sum_{l=r-1}^{k-1} \binom{k}{l} \lambda_{l,r-1} \\
&= \sum_{l=0}^{k-1} \sum_{r=1}^{l+1}(-1)^r \binom{k}{l} \lambda_{l,r-1} \\
&= -\sum_{l=0}^{k-1} \binom{k}{l} \sum_{r=0}^{l}(-1)^r \lambda_{l,r} \\
&= -\sum_{l=0}^{k-1} \binom{k}{l} (-1)^l \quad \text{(induction hypothesis)} \\
&= (-1)^k \quad \text{(binomial theorem)}
\end{aligned}$$

as required.

PROPOSITION 3. *Suppose that, for all subgraphs Δ of Γ such that C_Δ is finite, $|C_\Delta|$ is invertible in R. Then $C_\Gamma \in FP(R)$, and*

$$\chi(C_\Gamma) = \sum_{\Delta} \frac{(-1)^{|\Delta|}}{|C_\Delta|}$$

where the sum is over all subgraphs Δ of Γ such that C_Δ is finite. (Here $|\Delta|$ means the number of vertices of Δ, the empty subgraph is allowed as a subgraph and $C_\emptyset$ is the trivial group).

PROOF. According to §14 in [10], there is a contractible C_Γ-complex Y on which C_Γ acts properly. The quotient Y/C_Γ is a simplicial complex, in which an r-simplex is of the form $\emptyset \subseteq \Delta_0 \underset{\neq}{\subset} \Delta_1 \underset{\neq}{\subset} \ldots \underset{\neq}{\subset} \Delta_r = \Delta$, where Δ is a subgraph of Γ such that C_Δ is finite, and the Δ_i are subgraphs of Δ.

Moreover, up to conjugacy, the stabilizer of a lift of this simplex to Y is the subgroup C_{Δ_0} of C_Γ. If $s = |\Delta_0|$ and $i = |\Delta|$, then the number of

r-simplices (starting with Δ_0 and ending with Δ) is $\lambda_{i-s,r}$. Therefore, by Theorem 2 in [6], $C_\Gamma \in FP(R)$ and

$$\chi(C_\Gamma) = \sum_{i=0}^{n} \sum_{\substack{|\Delta|=i \\ C_\Delta \text{ is finite}}} \left(\sum_{s=0}^{i} \sum_{\substack{\Delta_0 \subseteq \Delta \\ |\Delta_0|=s}} \frac{1}{|C_{\Delta_0}|} \sum_{r=0}^{i-s} \lambda_{i-s,r}(-1)^r \right)$$

$$= \sum_{i=0}^{n} \sum_{\substack{|\Delta|=i \\ C_\Delta \text{ is finite}}} \left(\sum_{s=0}^{i} \sum_{\substack{\Delta_0 \subseteq \Delta \\ |\Delta_0|=s}} \frac{1}{|C_{\Delta_0}|} (-1)^{i-s} \right)$$

(using the lemma)

$$= \sum_{\Delta} (-1)^{\Delta} \left(\sum_{\Delta_0 \subseteq \Delta} \frac{(-1)^{|\Delta_0|}}{|C_{\Delta_0}|} \right)$$

But substituting $t = 1$ in the formula in Exercise 26(e) on p.45 of [4] gives

$$\sum_{\Delta_0 \subseteq \Delta} \frac{(-1)^{|\Delta_0|}}{|C_{\Delta_0}|} = \frac{1}{|C_\Delta|}$$

and the proposition follows.

The author is grateful to the referee for the final step in the proof of Prop. 3, which results in a simplification of the original formula, making it considerably easier to use.

If $|C_\Delta|$ is finite, then Δ must be a complete subgraph of Γ (otherwise one can find an infinite dihedral group in C_Δ). In order for Proposition 3 to be useful one needs a list of the finite Coxeter groups. This is provided by §9.3 and Table 10 in [9].

In the case of a right-angled Coxeter group, the subgraphs Δ of Γ for which $|C_\Delta|$ is finite are precisely the complete subgraphs of Γ, because if Δ is complete, then C_Δ is the direct product of $|\Delta|$ cyclic groups of order 2, so has order $2^{|\Delta|}$. Thus the formula of Cor. 2 follows from Prop. 3.

A finitely generated Coxeter group C_Γ is called *aspherical* by Pride and Stöhr [19] if Γ contains no triangles Δ such that C_Δ is finite. In this case the formula of Proposition 3 simplifies to

$$\chi(C_\Gamma) = 1 - \frac{n}{2} + \sum_{e} \frac{1}{2\phi(e)}$$

where the sum is over all edges e (recall that edges are unoriented and are viewed as unordered pairs of vertices, n is the number of vertices of Γ and ϕ is the labelling function). This formula is valid without the assumption in [19] that Γ has no isolated vertices. It can also be obtained from their results. They obtain a presentation closely related to that of C_Γ which is concise and CA (combinatorially aspherical) in the sense of [8], and the Euler characteristic of such a group is given by the following proposition; it is then easy to calculate $\chi(C_\Gamma)$ using the properties of χ listed at the beginning.

PROPOSITION 4. *Let $G = < x_1, \ldots, x_m; r_1^{n_1}, \ldots, r_s^{n_s} >$ be a concise, finite CA presentation of a group G, where r_i is not a proper power in the free group on $x_1, \ldots, x_m$ for $1 \le i \le s$. Assume the commutative ring R is such that all n_i are invertible in R. Then $G \in FP(R)$, and*

$$\chi(G) = 1 - m + \sum_{i=1}^{s} \frac{1}{n_i}.$$

PROOF. This is a straightforward generalisation of the argument used to prove Theorem 4 in [6]. From the argument of [8:Prop. 1.2], we obtain an exact sequence

$$0 \longrightarrow \oplus_{i=1}^{s} \mathbb{Z}[G/C_i] \longrightarrow \oplus_{i=1}^{m} \mathbb{Z}G \longrightarrow \mathbb{Z}G \longrightarrow \mathbb{Z} \longrightarrow 0$$

where C_i is the subgroup of G generated by r_i. We then apply the functor $R \otimes_{\mathbb{Z}} -$ to obtain an exact sequence of RG-modules and proceed as in [6]. One needs to know that r_i has order n_i in G, and this is proved in [15].

3. Virtual Cohomological Dimension

In the situation of Prop. 2 it is natural to ask what is the virtual cohomological dimension (vcd) of $G\Gamma$ in terms of $vcd(A_i)$. We can say nothing in general except to note that part of Prop.4 in [7] generalises.

PROPOSITION 5. *In any graph product, if all A_i have finite vcd, then so does $G\Gamma$.*

PROOF. If Γ is a complete graph then $G\Gamma = \prod_{i=1}^{n} A_i$, and the result follows easily (take a torsion-free subgroup B_i of finite index in A_i, so $\prod_{i=1}^{n} B_i$ is of finite index in $\prod_{i=1}^{n} A_i$, and has finite cd using Prop.6 in [21] and induction). Otherwise we have the decomposition of $G\Gamma$ as a free product with amalgamation used in Props 1 and 2, and the result follows by induction on n, just as in Prop. 4 of [7].

In [7] there is an ill-considered conjecture about $vcd(G\Gamma)$ in the case that all the A_i are cyclic. The author thanks J-P. Serre and T. Januszkiewicz for showing him counterexamples. The simplest example is the right-angled Coxeter group G with presentation

$$< x_1, \ldots, x_n; x_i^2 = 1 \ (1 \leq i \leq n), \ [x_i, x_{i+1}] = 1 \ (1 \leq i \leq n) >$$

where $n \geq 5$ and in the commutator relations indices are taken modulo n. It follows from Corollary 3 in [19] that G has vcd equal to 2, while the number $w(G)$ in [7] is equal to 1. In this case it is particularly easy to see G is an NEC group. A faithful action as a properly discontinuous group of isometries of the real hyperbolic plane, with compact fundamental domain, is obtained by letting the x_i act as reflections in the sides of a regular hyperbolic n-gon whose interior angles are right angles (this result is attributed to Dyck in §5.3 of [9]). For simple ways to construct such an n-gon see [2], [3; §7.16] or [20; §3].

We finish with an example of a sequence of right-angled Coxeter groups for which the conjecture gives the correct answer, in the hope that it might eventually provide insight into a correct description of the vcd. Let G_n be the group with presentation

$$< x_1, \ldots, x_n; x_i^2 = 1 \ (1 \leq i \leq n), \ [x_i, x_j] = 1 \ (\text{for } |i - j| \geq 2) > .$$

Let X_n denote the graph X_{G_n} defined in [7]. Its vertices are $v_1, \ldots, v_{n-1}$, where $v_i = \{x_i, x_{i+1}\}$, and there are edges joining v_i to v_j for $|i - j| \geq 3$. The number $w(G_n)$ in [7] is the maximum possible number of vertices in a complete subgraph of X_n.

Now X_n has a complete subgraph with vertices $v_1, v_4, v_7, \ldots, v_{3k+1}$, where

$$k = \left[\frac{n-2}{3}\right]$$

($[x]$ denotes the integer part of x). This has $k + 1 = [\frac{n+1}{3}]$ vertices, so $w(G_n) \geq [\frac{n+1}{3}]$. (It is not difficult to see directly that $w(G_n) = [\frac{n+1}{3}]$, but this is not needed).

Let $\phi_n : G_n \longrightarrow C_2$ be the homomorphism taking each x_i to the generator of C_2, the cyclic group of order two, and let K_n be the kernel of ϕ_n. We apply the Reidemeister-Schreier process to K_n, using the transversal $\{1, x_1\}$. We have (using $\overline{x}$ to denote the representative of x in this transversal):

Generators $\quad x_j(\overline{x}_j)^{-1}$, *i.e.* $z_j = x_j x_1^{-1}, \quad j = 1, \ldots, n$

and $y_j = x_1 x_j, \quad j = 1, \ldots, n.$

$$\text{Relators} \qquad x_i^2 = (1.x_i)(\overline{1.x_i})^{-1}(\overline{1.x_i})x_i$$
$$= (x_i x_1^{-1})(x_1 x_i) = z_i y_i \qquad (1 \le i \le n)$$
$$x_1 x_i^2 x_1^{-1} = (x_1 x_i^{-1})(x_i x_1^{-1}) = y_i z_i \qquad (1 \le i \le n)$$

$$[x_i, x_j] = x_i x_j x_i^{-1} x_j^{-1}$$
$$= (x_i x_1^{-1})(x_1 x_j)(x_1 x_i)^{-1}(x_1 x_j^{-1})$$
$$= z_i y_j y_i^{-1} z_j^{-1}, \quad \text{for } |i-j| \ge 2$$

$$x_1[x_i, x_j]x_1^{-1} = (x_1 x_i)(x_j x_1^{-1})(x_1 x_i^{-1})(x_j^{-1} x_1^{-1})$$
$$= y_i z_j z_i^{-1} y_j^{-1}, \quad \text{for } |i-j| \ge 2$$

Also, $z_1 = 1$. Hence K_n has a presentation with

Generators $\quad y_2, \ldots, y_n$

Relations $\quad y_3^2 = \ldots = y_n^2 = 1$

$[y_i, y_j] = 1 \quad$ for $3 \le i, j \le n$ and $|i-j| \ge 2$

$(y_2^{-1} y_j)^2 = 1$, for $j > 3$.

(This can be generalised to arbitrary Coxeter groups. See [19], where their presentation (2.4) can be obtained without assuming the Coxeter group is aspherical).

The last relations can be rewritten as $y_j y_2 y_j^{-1} = y_2^{-1}$. Hence, if $n \ge 4$, we have a decomposition

$$K_n = G_{n-2} *_{G_{n-3}} (C_\infty \rtimes G_{n-3})$$

where G_{n-2} is generated by $y_3, \ldots, y_n$, G_{n-3} is generated by $y_4, \ldots, y_n$, C_∞ is an infinite cyclic group generated by y_2, and G_{n-3} acts on C_∞ via the map $\phi_{n-3} : G_{n-3} \longrightarrow C_2$.

By Prop.2 (or indeed by Prop.2 in [7]), G_{n-3} has a torsion-free subgroup H of finite index. Then by Prop. 6 in [21], $cd(C_\infty \rtimes H) \le 1 + cd(H)$ (since $cd(C_\infty) = 1$), hence $vcd((C_\infty \rtimes G_{n-3}) \le 1 + vcd(G_{n-3})$. Therefore, by [21] Prop. 15,

$$vcd(G_n) \le \max\{vcd(G_{n-2}), 1 + vcd(G_{n-3})\}.$$

PROPOSITION 6. *For all $n \geq 1$, $vcd(G_n) = [\frac{n+1}{3}] = w(G_n)$.*

PROOF. The proof is by induction on n. If $n = 1$, then G_n is cyclic of order 2, so $vcd(G_1) = 0$. For $n = 2$, G_n is the infinite dihedral group, which has an infinite cyclic subgroup of finite index, so $vcd(G_2) = 1$. Also, G_3 is isomorphic to $((C_2 \times C_2) * C_2)$, which is free by finite, so $vcd(G_3) = 1$. (The kernel of the obvious homomorphism onto $C_2 \times C_2 \times C_2$ is free — see, for example, [18]. For the general characterisation of free by finite groups, see the references in Example 2 after VIII.11.2 in [5]). Also, the graph X_1 is empty, X_2 consists of a single vertex and X_3 of two isolated vertices. It is now easily checked that the Proposition holds for $n \leq 3$.

If $n \geq 4$, then by induction and the inequality preceding the Proposition,

$$\begin{aligned} vcd(G_n) &\leq \max\{[\tfrac{n-1}{3}], 1 + [\tfrac{n-2}{3}]\} \\ &= 1 + [\tfrac{n-2}{3}] = [\tfrac{n+1}{3}]. \end{aligned}$$

But by Prop. 4 in [7], $vcd(G_n) \geq w(G_n)$, and as noted above $w(G_n) \geq [\frac{n+1}{3}]$, and the result follows.

We note that the automorphism group of G_n has been studied by James [16]. This paper also contains a discussion of the connection between transitive representations of G_n and cell decompositions of n-manifolds. The automorphism group of a general Coxeter group is studied by Tits in the immediately following paper [22].

References

[1] H. Bass, "Euler characteristics and characters of discrete groups," *Invent. Math.* **35** (1976), 155–196.

[2] A. F. Beardon, "Hyperbolic polygons and Fuchsian groups", *J. London Math. Soc. (2)* **20** 1979, 247–254.

[3] A. F. Beardon, *The geometry of discrete groups.* Graduate texts in Mathematics **91**. New York, Heidelberg, Berlin: Springer-Verlag 1983.

[4] N. Bourbaki, *Groupes et algèbres de Lie*, Chapitres 4,5 et 6. Paris: Hermann 1968.

[5] K. S. Brown, *Cohomology of groups.* Springer-Verlag, New York 1982.

[6] I. M. Chiswell, "Euler characteristics of groups", *Math. Z.* **147** (1976), 1–11.

[7] I. M. Chiswell, "Right-angled Coxeter groups". In: Low-dimensional topology and Kleinian groups, (ed. D.B.A.Epstein), London Math. Soc. Lecture Notes **111** 297-304. Cambridge: University Press 1986.

[8] I. M. Chiswell, D. J. Collins and J. Huebschmann, "Aspherical group presentations", *Math. Z.* **178** (1981), 1–36.
[9] H. S. M. Coxeter and W. O. J. Moser, *Generators and relations for discrete groups.* Berlin, Heidelberg: Springer-Verlag 1980.
[10] M. W. Davis, "Groups generated by reflections and aspherical manifolds not covered by Euclidean space", *Annals of Math.* **117** (1983), 293–324.
[11] W. Dicks, "An exact sequence for rings of polynomials in partly commuting indeterminates", *J. Pure Appl. Algebra* **22** (1981), 215–228.
[12] C. Droms, "Graph groups, coherence and three-manifolds", *J. Algebra* **106** (1987), 484–489.
[13] C. Droms, "Subgroups of graph groups", *J. Algebra* **110** (1987), 519–522.
[14] E. R. Green, *Graph products of groups.* Ph.D.Thesis, University of Leeds 1990.
[15] J. Huebschmann, "Cohomology theory of aspherical groups and of small cancellation groups", *J. Pure Appl. Algebra* **14** (1979), 137–143.
[16] L. D. James, "Complexes and Coxeter groups-operations and outer automorphisms", *J. Algebra* **113** (1988), 339–345.
[17] K. Kim and F. Roush, "Homology of certain algebras defined by graphs", *J. Pure Appl. Algebra* **17** (1980), 179–186.
[18] R. C. Lyndon, "Two notes on Rankin's book on the Modular Group", *J. Austral. Math. Soc.* **16** (1973), 454–457.
[19] S. J. Pride and R. Stöhr, "The (co)homology of aspherical Coxeter groups", *J. London Math. Soc. (2)* **42** (1990), 49–63.
[20] G. P. Scott, "Subgroups of surface groups are almost geometric", *J. London Math. Soc. (2)* **17** (1978), 555–565.
[21] J-P. Serre, "Cohomologie des groupes discrets". In: Prospects in mathematics, Annals of Mathematics Studies **70**, pp 77-169. Princeton: University Press 1971.
[22] J. Tits, "Sur le groupe des automorphismes de certains groupes de Coxeter", *J. Algebra* **113** (1988), 346–357.

School of Mathematical Sciences
Queen Mary and Westfield College
University of London
Mile End Road
London E1 4NS.

Infinite families of automorphism groups of Riemann surfaces

Marston D. E. Conder and Ravi S. Kulkarni

Dedicated to A. M. Macbeath with much respect

1. Introduction

Let a, b be two rational numbers, with $a > 0$. Consider the sequence $N_{a,b} : g \mapsto ag + b$, for $g = 2, 3, \ldots$. We say that this sequence is *admissible* if for infinitely many values of g the number $ag + b$ is the order of an automorphism group G of a compact Riemann surface X_g of genus g. When this occurs, the pair (X_g, G), or simply X_g or G, is said to *belong to* $N_{a,b}$.

A cocompact Fuchsian group gives rise to an admissible sequence of the form $a(g-1)$, where a is a positive rational number which depends only on the group. For example the $\{2,3,7\}$–triangle group gives rise to the sequence $84(g-1)$; see Macbeath [Mc] for an early thorough discussion of these aspects. Conversely, it was shown in [K] that every admissible sequence of the form $a(g-1)$ arises from a fixed finite number of cocompact Fuchsian groups.

There are a few known admissible sequences $N_{a,b}$ where $a + b \neq 0$. Wiman showed that $4g + 2$ is the largest order of a cyclic automorphism group of a compact Riemann surface of genus g, and he also exhibited such surfaces for every g; see [W], [H]. Accola and independently Maclachlan showed that for infinitely many values of g the largest order of an automorphism group of a compact Riemann surface of genus g is $8g+8$. They also exhibited for every $g \geq 2$, a Riemann surface admitting an automorphism group of order $8g+8$; see [A], [Ml]. It is easy to see that this group contains a cyclic subgroup of order $2g+2$. A partial explanation of such families was given in [K], where it was proved that *if $ag+b$ is an admissible sequence with $a+b \neq 0$, and G is a group belonging to it, then G contains a cyclic subgroup whose index is bounded above by a number depending only on a and b, and moreover, $a+b$ must be positive and $\frac{2b}{a}$ must be an integer.* In this paper we further analyse such families where $N_{a,b}$ is *large* in the technical sense that for all but finitely many values of g we have $ag+b > 4g-4$. Our main result is the following:

THEOREM. *Let $N_{a,b}$ be a large admissible sequence with $a+b \neq 0$. Then $\frac{2b}{a}$ is a non-negative integer and either*

(1) $a = 4, 5, 6$ or 8, or

(2) $a = \dfrac{4c}{c-1}$ for some integer c where $c = 4$ or $c \geq 6$, in which case a is not an integer.

We also exhibit several families of automorphism groups with the values of a restricted as above. In particular we show that the sequence $N_{8,b}$ is admissible whenever b is an odd positive multiple of 8, extending the examples described in [A], [Ml]. A natural question of what values can be taken by b (for given a) leads to various construction problems in the theory of groups.

2. Signatures of large automorphism groups

Let G be an automorphism group of a compact Riemann surface X_g of genus g. We say G is *large* if its order is strictly greater than $4(g-1)$. Suppose X_g/G has genus h with b branch points, and let the b branching indices be $n_1, n_2, \ldots, n_b$, in increasing order.

2.1 PROPOSITION. *If G is large then $h = 0$, and $3 \leq b \leq 4$. If $b = 4$, then $n_1 = n_2 = 2$, while if $b = 3$ then $n_1 \leq 5$. More precisely, only the following possibilities for the branching indices can occur:*

(A) Four branch points:

(a) One infinite family: $(2,2,2,n)$ for $n \geq 3$,

(b) Other cases: $(2,2,3,n)$ for $3 \leq n \leq 5$;

(B) Three branch points:

(a) One doubly-infinite family: $(2, m, n)$ *for* $3 \le m \le n$, *with* $n \ge 7$ *if* $m = 3$, *and* $n \ge 5$ *if* $m = 4$,

(b) Five singly-infinite families: $(3,3,n)$ *for* $n \ge 4$, $(3,4,n)$ *for* $n \ge 4$, $(3,5,n)$ *for* $n \ge 5$, $(3,6,n)$ *for* $n \ge 6$, *and* $(4,4,n)$ *for* $n \ge 4$,

(c) Other cases: $(3,7,n)$ *for* $7 \le n \le 41$, $(3,8,n)$ *for* $8 \le n \le 23$, $(3,9,n)$ *for* $9 \le n \le 17$, $(3,10,n)$ *for* $10 \le n \le 14$, $(3,11,n)$ *for* $11 \le n \le 13$, $(4,5,n)$ *for* $5 \le n \le 19$, $(4,6,n)$ *for* $6 \le n \le 11$, $(4,7,n)$ *for* $7 \le n \le 9$, $(5,5,n)$ *for* $5 \le n \le 9$, *and* $(5,6,n)$ *for* $6 \le n \le 7$.

PROOF. These are easy consequences of the Riemann-Hurwitz formula

$$2g - 2 = |G|\Big(2h - 2 + \Sigma_i(1 - \frac{1}{n_i})\Big). \tag{2.1.1}$$

We seek the values of h and n_i so that the order of G is greater than $4(g-1)$, which means the multiple of $|G|$ on the right hand side of (2.1.1) must lie strictly between 0 and $\frac{1}{2}$. In particular, if $h \ge 2$ then obviously this cannot happen, while if $h = 1$ then (2.1.1) implies that at least one n_i must occur, but then since each n_i is at least 2, again the restriction is violated. Hence h must be 0. Similarly (with $h = 0$) the formula forces b to be at least 3, and as each n_i is at least 2, also b is at most 4.

Now suppose the number of branch points is 4. Then (2.1.1) reduces to

$$2g - 2 = |G|\Big(2 - \frac{1}{n_1} - \frac{1}{n_2} - \frac{1}{n_3} - \frac{1}{n_4}\Big), \tag{2.1.2}$$

and we seek the values of n_i so that $\frac{3}{2} < \frac{1}{n_1} + \frac{1}{n_2} + \frac{1}{n_3} + \frac{1}{n_4} < 2$. Since the n_i are in increasing order, this implies $n_1 = n_2 = 2$, and $\frac{1}{2} < \frac{1}{n_3} + \frac{1}{n_4} < 1$. If $n_3 = 2$ then the value of n_4 is unrestricted (except for the fact that it cannot be 2), while on the other hand, if $n_3 = 3$ then the value of n_4 must lie between 3 and 5 for the inequality to hold. This deals with case (A).

Now suppose the number of branch points is 3. Then (2.1.1) reduces to

$$2g - 2 = |G|\Big(1 - \frac{1}{n_1} - \frac{1}{n_2} - \frac{1}{n_3}\Big), \tag{2.1.3}$$

and we seek the values of n_i so that $\frac{1}{2} < \frac{1}{n_1} + \frac{1}{n_2} + \frac{1}{n_3} < 1$. If $n_1 \ge 6$ clearly the inequality cannot hold, so $2 \le n_1 \le 5$. When $n_1 = 2$ the values of n_2

and n_3 are unrestricted (except for those which make the right hand side of (2.1.3) non-positive), giving the doubly-infinite family of case (B), subcase (a). The remaining subcases are easily obtained in a similar fashion. ∎

3. A lemma on groups with a cyclic subgroup of bounded index

The following lemma is useful in eliminating possibilities for the branch data of an admissible sequence $N_{a,b}$ with $a+b \neq 0$. We would like to thank Peter Neumann for the elegant use of Schur's theorem on the transfer in its proof.

3.1 LEMMA. *Suppose p, q and d are positive integers, such that p and q are relatively prime. Then there are only finitely many finite groups which can be generated by two elements x and y of orders p and q respectively such that their product xy generates a subgroup of index at most d.*

PROOF. Let G be any finite group which can be generated by two elements x and y of orders p and q respectively, with the product $w = xy$ generating a (cyclic) subgroup H of index at most d.

We claim that G contains a central cyclic subgroup of index at most $d!$. Indeed consider the natural permutation representation of G on right cosets of H (by right multiplication): the kernel of this representation is a normal subgroup K of index at most $d!$ in G, and because it is a subgroup of H, it is also cyclic, generated say by z (where z is a power of w). Further, as z is centralized by xy we have $x^{-1}zx = yzy^{-1} = z^s$ for some s, giving $z = x^{-p}zx^p = z^{s^p}$ and $z = y^q zy^{-q} = z^{s^q}$, so that $s^p \equiv s^q \equiv 1 \bmod |z|$ (where $|z|$ denotes the order of z). But p and q are coprime, so $s \equiv 1 \bmod |z|$, and therefore z is centralized by both x and y. Thus K is central in G.

Next, let m be the index of the center $Z(G)$ in G. This is at most $d!$ (as $Z(G)$ contains K), and by a theorem of Schur, the transfer of G into $Z(G)$ is an endomorphism of $Z(G)$ taking each element to its mth power, so that the order of every element of the commutator subgroup G' is a divisor of m (see [R; Chapter 10]).

On the other hand, the Abelian factor group G/G' is generated by two elements of orders dividing p and q, so the order of G/G' is a divisor of pq, and in particular, G' contains w^{pq}. It follows that $w^{pqm} = 1$, so the order of H is at most pqm, and thus $|G| = |G : H||H| \leq dpqm \leq dpqd!$. As there are only finitely many finite groups of any given order, this completes the proof. ∎

3.2 COROLLARY. *If p and q are relatively prime integers, then in any admissible sequence $N_{a,b}$ with $a+b \neq 0$ there can be only finitely many groups having branch data of the type $(0; p, q, n)$ for some n.*

PROOF. Assume the contrary. Any group G in $N_{a,b}$ with branch data $(0; p, q, n)$ is generated by two elements x and y of orders p and q respectively, with product xy of order n, and the Riemann-Hurwitz formula (2.1.3) may be rewritten as

$$\frac{1}{n} = 1 - \frac{1}{p} - \frac{1}{q} - \frac{2g-2}{ag+b}, \tag{3.2.1}$$

since $|G| = ag + b$. Now as n tends to ∞, the order of G (and hence also the genus g) must increase, and taking the limit of both sides of (3.2.1) gives $0 = 1 - \frac{1}{p} - \frac{1}{q} - \frac{2}{a}$. Note that this determines the value of a, but more importantly, it implies $\frac{1}{n} = \frac{2}{a} - \frac{2g-2}{ag+b}$, which in turn rearranges and simplifies to give

$$\frac{|G|}{n} = \frac{ag+b}{n} = \frac{2(a+b)}{a}, \tag{3.2.2}$$

fixing the index in G of the cyclic subgroup generated by xy. The lemma now provides the required contradiction. ∎

(Note: a similar argument is used in [K] to bound the index of one or more cyclic subgroups of any group G belonging to an admissible sequence $N_{a,b}$ with $a + b \neq 0$, and this will be used again below.)

4. Proof of the Theorem

Suppose $N_{a,b}$ is a large admissible sequence with $a + b \neq 0$. As may be seen from the Riemann-Hurwitz formula (and explained in detail in [K]), at least one of the branching indices in the signatures of the Fuchsian groups uniformizing the groups belonging to $N_{a,b}$ must tend to ∞ as the genus g tends to ∞. Hence for large genera we need only consider the signatures listed below:

(i) $(0; 2, 2, 2, n)$ for $n \geq 3$,

(ii) $(0; 2, m, n)$ for $3 \leq m \leq n$, with $n \geq 7$ if $m = 3$, and $n \geq 5$ if $m = 4$,

(iii) $(0; 3, m, n)$ for $3 \leq m \leq n$ and $m \leq 6$, with $n \geq 4$ if $m = 3$,

(iv) $(0; 4, 4, n)$ for $n \geq 4$.

In cases (i) and (iv) the Riemann-Hurwitz formula gives $a = 4$, and similarly in the subcases $(3,3,n)$ and $(3,6,n)$ of case (iii) the values of a are 6 and 4 respectively. On the other hand Corollary 3.2 precludes the possibilities $(3,4,n)$ and $(3,5,n)$ in case (iii), along with the subcases of (ii) in which m is odd.

Thus we are left with those signatures of case (ii) in which m is even, say $m = 2c$ where $c \geq 2$. Here the Riemann-Hurwitz formula gives $a = \frac{4c}{c-1}$, and in particular $a = 8, 6$ or 5 when $c = 2, 3$ or 5 respectively, while in all other cases a is a non-integer.

Next let G be any group belonging to $N_{a,b}$ with branch data corresponding to one of the signatures listed above. The same argument as used in the proof of Corollary 3.2 (and given in more detail in [K]) shows that $\frac{2(a+b)}{a}$ is the sum of the indices in G of one or two of its cyclic subgroups, and so $\frac{2b}{a}$ is an integer. Also unless G itself is cyclic, this sum is at least 2, and therefore $b \geq 0$.

So finally, assume $b < 0$, with G cyclic (of order equal to one of the branching indices). Then $4g - 4 < ag + b = |G| < 4g + 2$ by Wiman's theorem, giving $a = 4$ and $b = -2$, and $|G| = 4g - 2$. In particular, we find G must have branch data of one of the types $(2,2,2,4g-2)$, $(3,6,4g-2)$ or $(4,4,4g-2)$, noting that the type $(2,m,4g-2)$ fails to satisfy the Riemann-Hurwitz formula. On the other hand, since G is cyclic its order is bounded above by the least common multiple of all but one of its branching indices, so that $|G| = 2, 6$ or 4 respectively in these cases, contradicting the assumption that $N_{a,b}$ is admissible. Thus $b \geq 0$ always, and the proof is complete. ∎

5. Examples and constructions

The largest value of a provided by our theorem is 8, corresponding to groups with signature $(0; 2, 4, n)$ for $n \geq 5$, and a necessary condition for the admissibility of a sequence $N_{8,b}$ is that b is a non-negative multiple of 4. The examples in [A] and [Ml] show the sequences $N_{8,8}$ and $N_{8,24}$ are admissible, along with a few other cases where b is an odd multiple of 8. However these and many others are consequences of the following results.

5.1 General construction. *Suppose H is any finite group that can be generated by two elements x and y of orders 2 and 4 respectively, such that y is not an element of the (index 2) subgroup generated by xy and y^2. Also let K be a cyclic group of arbitrary finite order, generated say by z.*

Now form the semi-direct product (or split extension) KH of K by H, with conjugation of K by x and y defined by $xzx^{-1} = yzy^{-1} = z^{-1}$, and let G be the subgroup generated by the elements $X = zx$ and $Y = y$.

It is easy to see that X and Y have orders 2 and 4 respectively. Also if s is the order of xy then $(XY)^s = z^s(xy)^s = z^s$ because z is centralized by xy, and the order of XY is equal to sm, where m is the order of z^s. The Riemann-Hurwitz formula gives

$$2g - 2 = |G|\left(1 - \frac{1}{2} - \frac{1}{4} - \frac{1}{sm}\right)$$

and thus $|G| = 8g - 8 + 4k$, where $k = \frac{|G|}{sm}$, the index in G of the cyclic subgroup generated by XY.

In particular, this index $k = |G : \langle XY \rangle|$ is bounded, whereas the order of G is not (since the order of z^s can be arbitrarily large), and therefore the sequence $N_{8,4k-8}$ is admissible. For example, if the order of K itself is m (chosen to be coprime to s), then z^s generates K and in that case $G = KH$, so that $k = \frac{|G|}{sm} = \frac{|H|}{s}$, which is the index in H of the cyclic subgroup generated by xy. Similarly if the order of z is chosen to be sm (an arbitrary multiple of s), and z^s generates the largest subgroup of K contained in G, again k will be the index of $\langle xy \rangle$ in H. In other cases the value of k depends on the choice of the group H, as will be evident in the examples given below.

(5.1.1) If H is the dihedral group $D_4 = \langle x, y \mid x^2 = y^4 = (xy)^2 = 1 \rangle$ of order 8, then the order of z may be chosen freely, giving $k = 4$ in each case, and we obtain the groups of order $8g + 8$ (and arbitrary genus g) in the Accola-Maclachlan sequence $N_{8,8}$.

(5.1.2) If H is the octahedral group $S_4 = \langle x, y \mid x^2 = y^4 = (xy)^3 = 1 \rangle$ of order 24, and K is cyclic of order $3m$ for any positive integer m, we have $k = 8$ and obtain groups of order $8g + 24$ ($= 24m$) in the sequence $N_{8,24}$, for all g divisible by 3; *cf.* [A], [Ml].

(5.1.3) If instead H is taken as the group $\langle x, y \mid x^2 = y^4 = (xy)^4 = (x^{-1}y^{-1}xy)^2 = 1 \rangle$, which is an extension of the Klein 4-group $Z_2 \times Z_2$ by the group $Z_2 \times Z_4$, while K is cyclic of order $4m$ for any positive integer m, again we have $k = 8$ and obtain groups of order $8g + 24$ in the sequence $N_{8,24}$, but this time for all $g \equiv 1 \bmod 4$ (as $8g + 24 = 32m$).

(5.1.4) For each positive integer d in turn, let H be the group $\langle x, y \mid x^2 = y^4 = (xy)^4 = (xyx^{-1}y^{-1})^d = 1 \rangle$, which is an extension of $Z_d \times Z_d$ by $Z_2 \times Z_4$, of order $8d^2$. (Note: the elements $xyx^{-1}y^{-1}$ and $x^{-1}y^{-1}xy$ generate the normal Abelian subgroup of order d^2.) With K cyclic of order $4m$ as above, the construction gives $k = 2d^2$ and a family of groups of orders $8g + 8d^2 - 8$ for the appropriate g in each case. In particular, the case $d = 1$ exhibits the admissibility of the sequence $N_{8,0}$.

(5.1.5) For any integer prime congruent to 1 modulo 4, let λ be a primitive 4th root of 1 modulo p, and form the semi-direct product of a cyclic group $\langle w \rangle$ of order p by the group $Z_2 \times Z_4 = \langle u, v \mid u^2 = v^4 = u^{-1}v^{-1}uv = 1 \rangle$, with conjugation defined by $uwu^{-1} = w^{-1}$ and $vwv^{-1} = w^{\lambda}$. Now take H to be the group generated by the elements $x = wu$ (of order 2) and $y = v$ (of order 4), and apply the same construction as used above. This time the product xy has order 4 since $1 + \lambda + \lambda^2 + \lambda^3 = 0$, and the group H has order $8p$ since $xyx^{-1}y^{-1} = w^{1-\lambda}$ (which has order p). In particular, the index of $\langle xy \rangle$ in H is $2p$, and thus we obtain the admissibility of the sequence $N_{8,8p-8}$ for all such p.

Note that in all these examples the index k is even, and b is a multiple of 8. In fact this is always the case:

5.2 PROPOSITION. *If the sequence $N_{8,b}$ is admissible then b is divisible by 8.*

PROOF. Suppose b is an odd multiple of 4. Let G be a group in $N_{8,b}$ which has generators x and y of orders 2 and 4 respectively, such that xy generates a (cyclic) subgroup H of order n and index d in G. By the observations made earlier, $|G| = 8g - 8 + 4d$ (where g is the genus of the associated surface), and so $b = 4d - 8$, implying that d is odd.

As in the proof of Lemma 3.1, the core of H is a normal subgroup K of index at most $d!$ in G, generated by some power of xy. Now consider $C_G(K) = \{g \in G \mid gw = wg, \forall w \in K\}$, the centralizer of K in G. This too is a normal subgroup of G, and because it contains H, it also has odd index in G. In particular, the factor group $G/C_G(K)$ has odd order (and therefore no elements of order 2), so both x and y lie in $C_G(K)$, and thus K is central in G. By Schur's theorem it now follows that if m is the index of the centre $Z(G)$ in G, then $(xy)^{4m} = 1$ (since $(xy)^4$ lies in the commutator subgroup G'), so $|H| \leq 4m \leq 4d!$ and therefore $|G| = |G : H||H| \leq 4dd!$.

Again this bounds the order of a group in the sequence $N_{8,b}$, contradicting admissibility. ∎

More importantly, we can also prove the following:

5.3 PROPOSITION. *For every odd positive integer c, the sequence $N_{8,8c}$ is admissible.*

PROOF. Let N be the direct product of two cyclic groups generated by (commuting) elements α and β, each of order $2d$, where $c = 2d - 1$, and form the semi-direct product of N by the dihedral group $D_4 = \langle u, v \mid u^2 = v^4 = (uv)^2 = 1\rangle$ of order 8, with conjugation defined by $u\alpha^i\beta^j u^{-1} = \alpha^j\beta^i$ and $v\alpha^i\beta^j v^{-1} = \alpha^{-j}\beta^i$ for $0 \leq i, j < 2d$. Take $x = u$ and $y = \beta v$, which are easily seen to have orders 2 and 4 respectively, and let H be the group they generate. As $xy = u\beta v = \alpha uv$, we find $(xy)^2 = \alpha uv\alpha uv = \alpha u\beta vuv = \alpha^2(uv)^2 = \alpha^2$, and it follows that H has order $8d^2$ (with a normal subgroup of order d^2 generated by α^2 and β^2). In particular, as xy has order $2d$, the index of xy in H is $4d$, and so our construction provides groups of order $8g + 16d - 8$, that is, $8g + 8c$, for infinitely many values of the genus g. ∎

All of the examples to which we applied our construction above are soluble, but there are certainly other cases where the resulting groups are insoluble. For instance H may be taken as the symmetric group S_5, which is generated by the odd permutations $x = (1,2)$ and $y = (2,3,4,5)$, and in this case we obtain another illustration of the admissibility of the sequence $N_{8,88}$.

The profusion of examples is explained by the fact that the free product $Z_2 * Z_4 = \langle x, y \mid x^2 = y^4 = 1\rangle$ is SQ-universal, which means that every countable group is isomorphic to a subgroup of some quotient of $Z_2 * Z_4$. Indeed for every $n \geq 5$, even the $(2,4,n)$ triangle group $\langle x, y \mid x^2 = y^4 = (xy)^n = 1\rangle$ is SQ-universal (see [N]); also it can easily be shown that the symmetric group S_n is itself a quotient of $Z_2 * Z_4$ for all but finitely many positive integers n. Thus there are plenty of examples for H to choose from.

Finally we note that similar constructions may be applied in the cases of smaller values of the parameter a, and again there are infinitely many possibilities to choose from because of the SQ-universality of the associated triangle groups in each case. On the other hand, the admissibility of a number of sequences $N_{a,b}$ for small values of b can be ruled out by elementary group-theoretic means, but we shall not pursue that matter here.

Acknowledgments. The first author would like to thank the University of Auckland Research Committee and the N.Z. Lottery Grants Board for their support. The second author gratefully acknowledges a partial support from a National Science Foundation (USA) grant and a PSC–CUNY award.

References

[A] R. D. M. Accola, On the number of automorphisms of a closed Riemann surface, *Trans. Amer. Math. Soc.* **131** (1968), 398–408.

[H] W. J. Harvey, Cyclic groups of automorphisms of a compact Riemann surface, *Quart. J. Math. (Oxford)* **17** (1966), 86–97.

[K] R. S. Kulkarni, Infinite Families of Surface Symmetries, *Israel J. Math.*, to appear.

[Mc] A. M. Macbeath, On a theorem of Hurwitz, *Proc. Glasgow Math. Assoc.* **5** (1961), 90–96.

[Ml] C. Maclachlan, A bound for the number of automorphisms of a compact Riemann surface, *J. London Math. Soc.* **44** (1969), 265–272.

[N] P. M. Neumann, The SQ-universality of some finitely presented groups, *J. Australian Math. Soc.* **16** (1973), 1–6.

[R] D. J. S. Robinson, *A Course in the Theory of Groups.* Springer-Verlag (New York), 1982.

[W] A. Wiman, Über die hyperelliptischen Curven und diejenigen vom Geschlechte $p = 3$ welche eindeutige Transformationen in sich zulassen, *Bihang Till Kongl. Svenska Vetenskaps-Akademiens Handlingar (Stockholm)* vol. 21 (1895-6), pp. 1–23.

Department of Mathematics & Statistics
University of Auckland
Private Bag 92019
Auckland, New Zealand

Department of Mathematics
Graduate Center, CUNY
33 West 42nd Street
New York NY 10036, USA

Planar hyperelliptic Klein surfaces and fundamental regions of N. E. C. groups

A. F. Costa* and E. Martínez*

A compact Klein surface X is a compact surface with a dianalytic structure. We shall say that X is *hyperelliptic* if there exists an involution ϕ of X such that $X/\langle\phi\rangle$ has algebraic genus 0. A planar Klein surface has topological genus 0 and k boundary components (we shall assume $k \geq 3$). If we denote the hyperbolic plane by $\mathcal{D}$, a compact Klein surface X can be expressed as $X = \mathcal{D}/\Gamma$ where Γ is a non-Euclidean crystallographic group (*NEC* group for short, see [6]). Hyperelliptic surfaces have been characterized by means of *NEC* groups in the paper [2]. It is not easy to decide using the main result of [2] if a given planar Klein surface is hyperelliptic. In this note we characterize the hyperellipticity of planar Klein surfaces by means of geometric conditions. In this way we obtain information about the points in the Teichmüller space that correspond to hyperelliptic Klein surfaces.

1. Preliminaries.

A *NEC* group Γ is a discrete group of isometries of the hyperbolic plane $\mathcal{D}$ with quotient space $\mathcal{D}/\Gamma$ compact. *NEC* groups are classified according to their signature, which has the form

$$(g; \pm; [m_1, \ldots, m_r]; \{(n_{11}, \ldots, n_{1s_1}), \ldots, (n_{k1}, \ldots, n_{ks_k})\})$$

where the numbers m_i and n_{ij} are integers greater than or equal to 2, and g,

* Partially supported by CICYT, PB89-0201.

r and k are non-negative integers. This signature determines a presentation of the group Γ, given by the following data:

Generators:		Relations:	
x_i	$(i = 1, \ldots, r)$	$x_i^{m_i} = 1$	
e_i	$(i = 1, ..., k)$	$e_i^{-1} c_{i0} e_i c_{is_i} = 1$	
c_{ij}	$(i = 1, ..., k)$ $(j = 1, ..., s_i)$	$(c_{ij-1})^2 = (c_{ij})^2 = (c_{ij-1}c_{ij})^{n_{ij}} = 1$	
$a_i,\ b_i$	$(i = 1, ..., g)$	$\prod_1^r x_i \prod_1^k e_i \prod_1^g [a_i, b_i] = 1$	$(*)$
d_i	$(i = 1, ..., g)$	$\prod_1^r x_i \prod_1^k e_i \prod_1^g d_i^2 = 1$	$(**)$

where $(*)$ or $(**)$ occurs according to whether the sign of the signature is '+' or '$-$'. The x_i are elliptic, the e_i are generally hyperbolic, the a_i, b_i are hyperbolic, the c_{ij} are reflections and the d_i are glide reflections.

Let X be a compact Klein surface with genus g and k boundary components: Then X can be expressed as $\mathcal{D}/\Gamma$ where Γ has signature

$$(g; \pm; [\text{-}]; \{(\text{-}), \overset{k}{\ldots}, (\text{-})\}).$$

Such a group is said to be a *surface group*. We are concerned with planar surfaces (spheres with holes), so $g = 0$ and we shall assume $k \geq 3$.

Let Γ be a planar surface group, then Γ admits a *Wilkie fundamental region* W, see [8], that is a hyperbolic polygon with sides labelled

$$\epsilon_1,\ \gamma_1,\ \epsilon_1',\ \epsilon_2,\ \gamma_2,\ \epsilon_2',\ \ldots,\ \epsilon_k,\ \gamma_k,\ \epsilon_k'$$

and angles

$$\langle \epsilon_i, \gamma_i \rangle + \langle \gamma_i, \epsilon_1' \rangle = \pi\ ; \qquad \langle \epsilon_1', \epsilon_2 \rangle + \langle \epsilon_2', \epsilon_3 \rangle + \ldots + \langle \epsilon_k', \epsilon_1 \rangle = 2\pi.$$

The edges of W are identified by $e_i(\epsilon_i') = \epsilon_i$ $(i = 1, \ldots, k)$ and the reflection c_i of Γ fixes γ_i $(i = 1, \ldots, k)$. Each region W is obtained by choosing a point P in the interior of $\mathcal{D}/\Gamma$ and cutting by geodesics joining P with each boundary component in $\mathcal{D}/\Gamma$. Given a point P in the interior of $\mathcal{D}/\Gamma$, let

M_P be a Wilkie fundamental polygon obtained by cutting $\mathcal{D}/\Gamma$ by geodesics having minimal length from P to each boundary component (thus $\langle \epsilon_i, \gamma_i \rangle = \langle \gamma_i, \epsilon_1' \rangle = \pi/2$). Then M_P is a Wilkie region with minimal perimeter among the Wilkie polygons obtained cutting from P. We shall call M_P a *minimal* Wilkie polygon. Without loss of generality we may assume that M_P is a *convex* polygon (see [7]).

2. Rectangular Polygonal Fundamental Regions.

Let Γ be a NEC group with signature $(0; +; [-]; \{(-), \overset{k}{\ldots}, (-)\})$. In this section we shall describe the construction of a fundamental region for Γ that is a polygon with right angles from a minimal convex Wilkie polygon of Γ.

First we shall describe the fundamental step in our construction. Assume that we have a fundamental region for Γ that is a polygon Q having in its boundary sides labelled with the following letters:

$$\ldots \delta_1, \ \ldots, \ \varphi_2, \ \delta_2, \ \varphi_2', \ \ldots$$

and that there exists a hyperbolic element e in Γ, such that $e(\varphi_2') = \varphi_2$ and reflections c_1 and c_2 such that $c_1(\delta_1) = \delta_1$ and $c_2(\delta_2) = \delta_2$. Let μ be the intersection of Q with the common orthogonal line to δ_1 and δ_2. Suppose that μ cuts Q in an arc joining δ_1 with δ_2. Let us call δ_1' and δ_1'', δ_2' and δ_2'' the segments on δ_1 and δ_2 such that if we cut Q by μ we obtain two polygons Q_1 and Q_2 with sides labelled:

$$\ldots \delta_1', \ \mu, \ \delta_2', \ \varphi_2', \ \ldots$$

for Q_1 and

$$\delta_1'', \ \ldots, \ \varphi_2, \ \delta_2'', \ \mu$$

for Q_2. Note that with this labelling Q has the labelled sides

$$\ldots \delta_1' \cup \delta_1'', \ \ldots, \ \varphi_2, \ \delta_2'' \cup \delta_2', \ \varphi_2', \ \ldots$$

Let Q' be the polygon $Q_1 \cup e^{-1}(Q_2)$ that has labelled sides:

$$\ldots \delta_1', \ \mu, \ \delta_2' \cup e^{-1}(\delta_2''), \ e^{-1}(\mu), \ e^{-1}(\delta_1''), \ \ldots$$

We shall call $C_{\delta_2\delta_1}(Q)$ the polygon Q' relabelled as follows:

- replace δ_1' and $e^{-1}(\delta_1'')$ by δ_1 and δ_1'
- replace μ and $e^{-1}(\mu)$ by μ and μ'
- replace $\delta_2' \cup e^{-1}(\delta_2'')$ by δ_2 and δ_2'

Let s be a side in Q' different from the above ones:

- if s is in Q_1 give it the same label as in Q_1
- if $s = e^{-1}(s')$ with s' in Q_2 give s the same label that s' has in Q_2

We remark that $C_{\delta_2\delta_1}(Q)$ is a fundamental region for Γ. Now we shall apply this procedure to a minimal convex Wilkie polygon for Γ, M_P.

We define:

$$R(M_P) = C_{\gamma_k\gamma_{k-1}} C_{\gamma_{k-1}\gamma_{k-2}} \cdots C_{\gamma_2\gamma_1}(M_P).$$

Note that every common orthogonal line to two sides γ_i and γ_{i-1} cuts M_P in an arc μ_{i-1} joining γ_i with γ_{i-1} because of the convexity of M_P. Moreover by straightforward geometrical arguments $\{\mu_1, \ldots, \mu_{k-1}\}$ is a set of disjoint arcs. This property of M_P allows the operation $C_{\gamma_k\gamma_{k-1}} C_{\gamma_{k-1}\gamma_{k-2}} \cdots C_{\gamma_2\gamma_1}$. Then $R(M_P)$ is a fundamental region for Γ and it is a right angled polygon with $4(k-1)$ sides. See Figures 1 to 5, for the case $k = 3$.

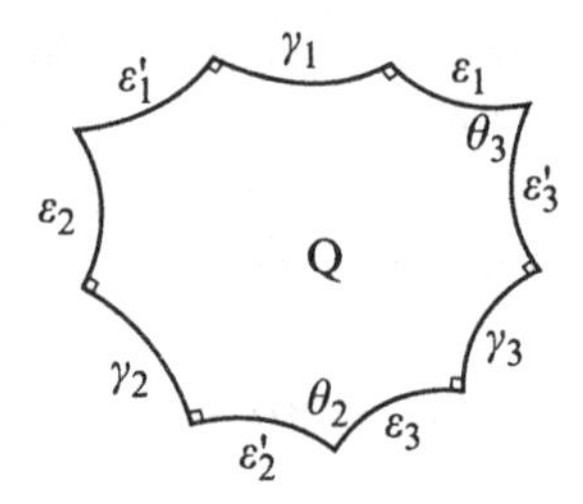

Figure 1 - The polygon $Q = M_P$

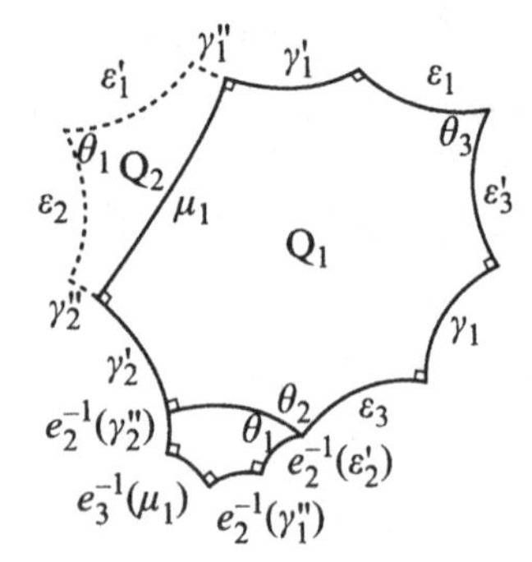

Figure 2 - The polygon $Q' = Q_1 \cup e_2^{-1}(Q_2)$

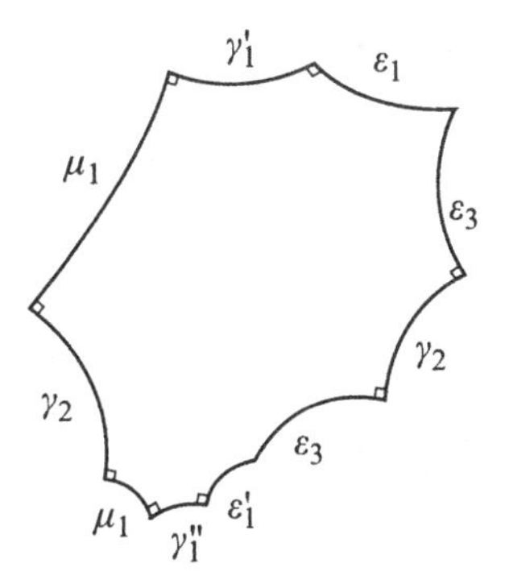

Figure 3 - The relabelled polygon $C_{\gamma_2\gamma_1}(Q)$

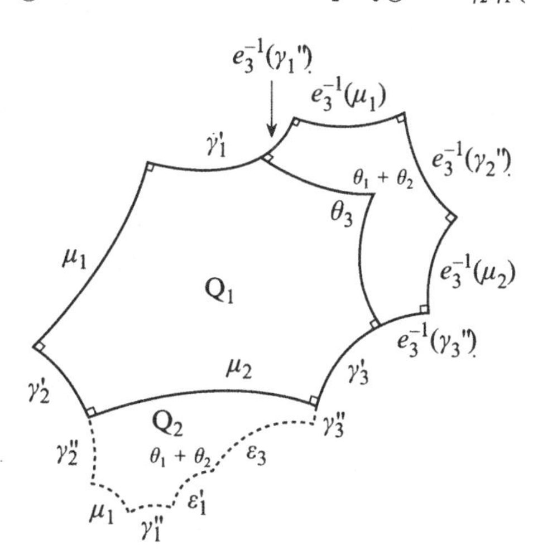

Figure 4 - The polygon $Q_1 \cup e_3^{-1}(Q_2)$

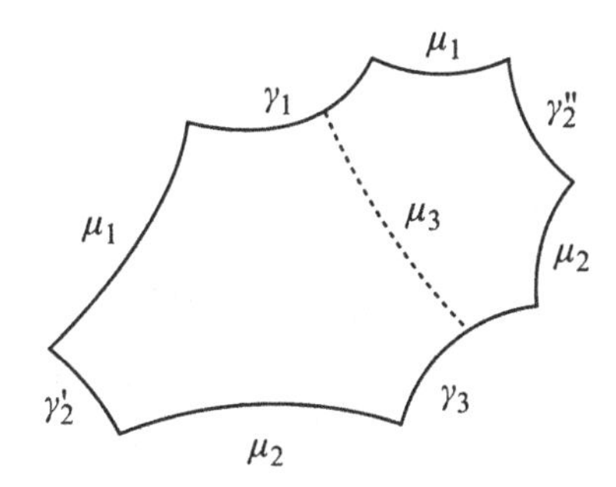

Figure 5 - The relabelled polygon $C_{\gamma_3\gamma_2}C_{\gamma_2\gamma_1}(Q) = R(M_P)$

3. Hyperelliptic Planar Klein Surfaces and $R(M_P)$.

THEOREM 3.1. *Let Γ be a NEC group with signature $(0;+;[\text{-}];\{(\text{-}),\overset{k}{\ldots},(\text{-})\})$ and let M_P be a minimal convex Wilkie polygon. Then $\mathcal{D}/\Gamma$ is hyperelliptic if and only if $R(M_P)$ is symmetric with respect to the common perpendicular of γ_1 and γ_k.* PROOF. Let $\mathcal{D}/\Gamma$ be hyperelliptic and M_P be a minimal Wilkie polygon of Γ. By [2] there exists a *NEC* group Γ_1 with $[\Gamma_1 : \Gamma] = 2$ and signature $(0;+;[\text{-}];\{(2,\overset{2k}{\ldots},2)\})$. Γ_1 is generated by $2k$ reflections. We may suppose that these reflections are labelled $\bar{c}_1, \tilde{c}_1, \ldots, \bar{c}_k, \tilde{c}_k$, and satisfy the following relations

$$(\bar{c}_i\tilde{c}_i)^2 = (\tilde{c}_i\bar{c}_{i+1})^2 = (\tilde{c}_k\bar{c}_1)^2 = 1, \qquad (i = 1, \ldots, k-1)$$

Let R_1 be the right-angled fundamental region of Γ_1 with $2k$ sides labelled $\gamma_1,\ \mu_1,\ \ldots,\ \gamma_k,\ \mu_k$, where γ_i and μ_i lie on the axes of $\bar{c}_i$ and $\tilde{c}_i$, respectively.

Let O be a point in R_1 congruent by Γ to the vertex of M_P corresponding to an interior point in $\mathcal{D}/\Gamma$. For $i = 1, \ldots, k$, let ν_i be the orthogonal line to γ_i from O. The side γ_i is divided by ν_i into two segments denoted by $\check{\gamma}_i$ and $\hat{\gamma}_i$. Then, R_1 becomes divided into k pentagons (one of them can be degenerated to a point if O is in the boundary of R_1). We reflect the pentagon containing μ_i in μ_i, $i = 1, \ldots, k$, to obtain a new region M with sides labelled

$$\ldots,\ \tilde{c}_{i-1}(\nu_i),\ \tilde{c}_i\check{\gamma}_i) \cup \gamma_i \cup \tilde{c}_i(\hat{\gamma}_i),\ \tilde{c}(\nu_i),\ \tilde{c}(\nu_{i+1}),\ \ldots$$

Let us observe that

$$(\tilde{c}_{i-1}\tilde{c}_i)(\tilde{c}_i(\nu_i)) = \tilde{c}_{i-1}(\nu_i) \qquad (i = 2, \ldots, k)$$

Let $e_1 = \tilde{c}_k\tilde{c}_1$, $e_i = \tilde{c}_{i-1}\tilde{c}_i$, $c_i = \bar{c}_i$; we then have the following relations

$$e_1c_1 = \tilde{c}_k\tilde{c}_1\bar{c}_1 = \tilde{c}_k\bar{c}_1\tilde{c}_1 = \bar{c}_1\tilde{c}_k\tilde{c}_1 = c_1e_1$$

$$e_ic_i = \tilde{c}_{i-1}\tilde{c}_i\bar{c}_i = \tilde{c}_{i-1}\bar{c}_i\tilde{c}_i = \bar{c}_i\tilde{c}_{i-1}\tilde{c}_i = c_ie_i$$

$$e_1 \ldots e_k = (\tilde{c}_k\tilde{c}_1)(\tilde{c}_1\tilde{c}_2)\ldots(\tilde{c}_{k-1}\tilde{c}_k) = 1$$

Then M is a fundamental region of Γ. To finish we must prove that M is congruent to M_P. Let $p : \mathcal{D} \to \mathcal{D}/\Gamma$ be the canonical projection. It is enough to show that the $p(\nu_i)$ are geodesics of minimal length from $p(O)$ to the boundary components.

Assume that $O \in \overset{\circ}{R}_1$. Suppose that l_i is a geodesic from $p(O)$ to a boundary component of $\mathcal{D}/\Gamma$ and that the length of l_i is smaller than the length of the corresponding $p(\nu_i)$. Let $\tilde{l}_i$ be the lifting of l_i to $\mathcal{D}$ containing O. Then $\tilde{l}_i$ lies in a hyperbolic line containing O and cutting orthogonally the sides of γ_i congruent by Γ_1. Assume $\tilde{l}_i = \lambda_1 \cup \lambda_2 \cup \ldots \cup \lambda_r$ where $\overset{\circ}{R}_t \subset g_t(\overset{\circ}{R}_1)$ and $g_t \in \Gamma_1$ ($r > 1$ because $O \in \overset{\circ}{R}_1$). Then $l = \lambda_1 \cup g_2^{-1}(\lambda_2) \cup \ldots \cup g_r^{-1}(\lambda_r)$ is a polygonal line in R_1 from O orthogonal to γ_i. Hence $p(l)$ is homotopically equivalent to $p(\nu_i)$ and length $(p(\nu_i)) >$ length$(p(l))$ =length(l_i) which is absurd because $p(l)$ is not a geodesic.

If O is in the boundary of R_1 the argument is similar bearing in mind that in this case there exist two minimal geodesics joining $p(O)$ with each boundary component in $\mathcal{D}/\Gamma$. Then M may be different from M_P. In any case $R(M)$ is congruent to $R(M_P)$ because the common orthogonals which are cut in order to obtain $R(M)$ and $R(M_P)$ both project onto $p(\mu_1), \ldots, p(\mu_k)$.

Finally by construction $R(M_P) = R_1 \cup \tilde{c}_k(R_1)$ so that $R(M_P)$ verifies the condition in the Theorem.

Now, let $X = \mathcal{D}/\Gamma$ be a planar Klein surface, and let Γ have fundamental region M_P with labelling

$$\epsilon_1,\ \gamma_1,\ \epsilon_1',\ \epsilon_2,\ \gamma_2,\ ;\epsilon_2',\ \ldots,\ \epsilon_k,\ \gamma_k,\ \epsilon_k'$$

If $R(M_P)$ is symmetric with respect to the common orthogonal μ_k of γ_1 and γ_k, the region R_1

$$\hat{\gamma}_1,\ \mu_1,\ \gamma_2,\ \mu_2,\ \ldots,\ \check{\gamma}_k,\ \mu_k$$

is right-angled. Let $\bar{c}_i$ and $\tilde{c}_i$ be the reflections on these sides. The group Γ_1 generated by $\bar{c}_i$ and $\tilde{c}_i$ is a *NEC* group with signature $(0; +; [-]; \{(2, \overset{2k}{\ldots}, 2)\})$ and it has R_1 as a fundamental region, because $R_1 \cup \tilde{c}_k(R_1) = R(M_P)$ and we have $[\Gamma_1 : \Gamma] = 2$. Hence $\mathcal{D}/\Gamma$ is hyperelliptic. ∎

COROLLARY 3.2. *Assume that $\mathcal{D}/\Gamma$ is a planar Klein surface that is hyperelliptic and that P is a point in $\mathcal{D}/\Gamma$ that is not a fixed point of the hyperelliptic involution. Then M_P is unique up to congruence and the hyperelliptic involution is also unique.*

We remark that the uniqueness of the hyperelliptic involution for Klein surfaces is a well known fact (see [3]).

PROOF. Using the notation of the proof of the Theorem, the $p(\mu_i)$, $i = 1, \ldots, k$, do not cut the $p(\nu_j)$, $j = 1, \ldots, k$, which implies that the homotopy classes of the $p(\mu_i)$ are completely determined by the $p(\nu_j)$. Since the

$p(\nu_j)$ are geodesics and orthogonal to the boundary components they are completely determined. Hence the way of cutting $\mathcal{D}/\Gamma$ in order to obtain M_P is unique. For similar reasons $p(\nu_1), \ldots, p(\nu_k)$ (and hence the hyperelliptic involution) are completely determined by $p(\mu_1), \ldots, p(\mu_k)$. ∎

COROLLARY 3.3. *The set of points corresponding to hyperelliptic surfaces in the Teichmüller space T of a planar Klein surface with k boundary components consists of a submanifold of dimension $2k-3$ with $\frac{(k-1)!}{2}$ connected components.*

PROOF. Let $\mathcal{D}/\Gamma$ be a planar surface and P a point in the interior of $\mathcal{D}/\Gamma$. Let M_P be a minimal Wilkie polygon for Γ. By the same argument as in Corollary 3.2, the geodesic arcs $p(\nu_j)$ are completely determined and these arcs give two cyclic permutations, ϵ and ϵ^{-1}, on the set of boundary components of $\mathcal{D}/\Gamma$: $\epsilon(C_i) = C_j$ if there exists an arc $p(\nu_t)$ from the boundary component C_i to the boundary component C_j. The polygon $R(M_P)$ can be constructed if we know the lengths of $4(k-1)-3$ sides (see [4]). In $R(M_P)$ there are $k-1$ pairs of identified sides so it is enough to know $3(k-1)-3$ lengths in order to construct $R(M_P)$. The Klein surfaces admitting a minimal Wilkie polygon inducing the permutations ϵ and ϵ^{-1} are parameterized by $3k-6$ lengths. Let us call U_ϵ the subspace of T representing such surfaces. The existence of minimal Wilkie polygons implies that the union of all the U_ϵ is T. By Corollary 3.2 there are no points corresponding to hyperelliptic surfaces in the intersection of two U_ϵ. The condition on the symmetry of $R(M_P)$ in Theorem 3.1 represents the equality of $k-3$ pairs of sides, which implies that the points giving hyperelliptic surfaces form a submanifold of T of dimension $2k-3$ contained in U_ϵ. There are $\frac{(k-1)!}{2}$ such U_ϵ, one for each ϵ and ϵ^{-1}. ∎

The above result contrasts with the situation in Teichmüller theory for Riemann surfaces where for genus bigger than 2 the hyperelliptic locus has infinitely many connected components (see [5]).

The connected components of the hyperelliptic locus in Corollary 3.3 give just one connected component on the corresponding Moduli space (see [1]). This fact can easily be proved from Corollary 3.3 using an automorphism α of a group Γ uniformizing a planar surface with k boundary components which permutes the canonical generators of Γ that are reflections (see §1). Such an automorphism α can be defined in the following way in terms of the generators of Γ defined in §1:

$$\alpha(c_1) = e_1 c_2 e_1^{-1},\ \alpha(c_2) = c_1,\ \alpha(c_i) = c_i,\ \ i = 3, \dots, k$$

$$\alpha(e_1) = e_1 e_2 e_1^{-1},\ \alpha(e_2) = e_1,\ \alpha(e_i) = e_i,\ \ i = 3, \dots, k.$$

The authors would like to thank the referee and the editors for their comments and suggestions.

References

[1] E. Bujalance, A. F. Costa, S. M. Natanzon, D. Singerman. Involutions of compact Klein surfaces, to appear in *Math. Z.*

[2] E. Bujalance, J. J. Etayo, J. M. Gamboa. Hyperelliptic Klein surfaces, *Quart. J. Math. Oxford* (2) **36** (1985) 141-157.

[3] E. Bujalance, J. J. Etayo, J. M. Gamboa, G. Gromadzki. Automorphism Groups of Compact Bordered Klein Surfaces, *Lecture Notes in Mathematics 1439* (Springer-Verlag, Berlin, 1990)

[4] J. J. Etayo, E. Martínez. Hyperbolic polygons and NEC groups, *Math. Proc. Camb. Phil. Soc.* **104** (1988) 261-272.

[5] S. Kravetz. On the geometry of Teichmüller spaces and the structure of their modular groups, *Ann. Acad. Sci. Fenn.* Ser. AI **278** (1959) 1-35.

[6] A. M. Macbeath. The classification of non- euclidean crystallographic groups, *Can. J. Math.* **19** (1967) 1192-1205.

[7] E. Martínez. Convex fundamental regions for NEC groups, *Arch. Math.* **47** (1986) 457-464.

[8] H. C. Wilkie. On non-euclidean crystallographic groups, *Math. Z.* **91** (1966) 87-102.

Antonio F. Costa and Ernesto Martínez
Departamento de Matemáticas Fundamentales
Facultad de Ciencias
U.N.E.D.
28040 Madrid, Spain

An example of an infinite group

M. Edjvet

Dedicated to A. M. Macbeath on the occasion of his retirement

Introduction

The group $(2, 3, 7; q)$ is defined by the presentation

$$\langle a, b \mid a^2, b^3, (ab)^7, [a, b]^q \rangle.$$

In [3] it is shown, with the possible exception of $q = 11$, that $(2, 3, 7; q)$ is infinite if and only if $q \geq 9$. In [2], using different methods, the same result is obtained again with this same (but 'unlikely') possible exception. In this note we fill the gap by showing that $(2, 3, 7; 11)$ is indeed infinite.

The group $G^{3,7,k}$ is defined by the presentation

$$\langle x, y, z \mid x^2,\ y^2,\ z^2,\ (xy)^2,\ (yz)^3,\ (zx)^7,\ (xyz)^k \rangle.$$

It has been observed in [3] that for $k = 2q$ even, $(2, 3, 7; q)$ has index 2 in $G^{3,7,k}$. Coxeter has conjectured that $G^{3,7,k}$ is infinite if and only if $k \geq 18$ (it is now known, see [3] and references there, that this group is finite for $k \leq 17$). This note together with the above mentioned results confirm the conjecture for k even. Our approach is to use the curvature argument from [1] together with the methods of [3]. Indeed our work can be viewed as an appendix to [3] and we will assume that the reader is familiar with its contents.

1. Henceforth **P** shall denote the presentation $\langle a, b \mid a^2, b^3, (ab)^7, [a,b]^{11} \rangle$ and G the group defined by **P**. Let A and B be the groups C_2 and C_3 defined by the presentations $\langle a \mid a^2 \rangle$ and $\langle b \mid b^3 \rangle$, and let $\alpha = (ab)^7$ and $\beta = [a,b]^{11}$. We show that **P** is *quasi-aspherical* over $A * B$, and so G is infinite [3].

If **P** is not quasi-aspherical over $A * B$ then there is a non-empty reduced connected picture Π over **P** – we show that this is impossible.

Recall briefly that Π is a tessellation of S^2 each of whose corners is labelled by one of $\{a, b, b^{-1}\}$. Reading the labels round a vertex or a region in the clockwise direction gives the label of that vertex or region. An *α-vertex* (*β-vertex* respectively) is one whose label is a cyclic permutation of $(ab)^7$ ($[a,b]^{11}$ respectively). Each vertex of Π is either an α-vertex or a β-vertex. The corner labels of any particular region in Π must all belong to either A or B and their product must give the identity element. The following facts were noted in [3]: it can be assumed without any loss that each A-region has degree 2; there are always 2 edges joining either an α-vertex to an α-vertex or a β-vertex to a β-vertex; there are always either 2 or 4 edges joining an α-vertex to a β-vertex.

Identify all the edges of Π that share the same initial and terminal vertices. This gives a tessellation Π' of S^2 whose regions each have degree at least 3 and whose labels are equal to 1 in $\langle b \mid b^3 \rangle$. It follows further from the above that

$$4 \leq \deg(\alpha\text{-vertex}) \leq 7$$

and

$$11 \leq \deg(\beta\text{-vertex}) \leq 22.$$

Now form Γ from Π' by triangulating all those regions Δ of Π' that have degree at least 4 and contain at least one β-vertex. The method of triangulation is to pick a β-vertex v of Δ and insert an edge from v to each of the other vertices of Δ not adjacent to v.

It is this tessellation Γ we analyse. Our contradiction is obtained via curvature arguments of the sort used in [1] for example. Give each corner at a vertex of degree d the angle $\frac{2\pi}{d}$. The curvature, $c(\Delta)$, of a region Δ of degree k and whose vertices have degrees d_i $(1 \leq i \leq k)$ is then

$$c(\Delta) = (2-k)\pi + 2\pi \sum_{i=1}^{k} \frac{1}{d_i} = \alpha\pi, \quad \text{say.} \qquad (*)$$

We write $c(\Delta) = [d_1, ..., d_k;\ \alpha\pi]$ to denote this situation.

It follows from Euler's formula that the sum of the curvatures of the regions of Γ is 4π. What we do is compensate for regions of positive curvature, wherever they occur in Γ, with negatively curved neighbouring regions, thus showing that this total of 4π cannot be attained.

In order to conduct our analysis of Γ we introduce some notation and concepts.

In diagrams an α-vertex (β-vertex, respectively) of degree k (l, respectively) is denoted by Ⓚ ($\boxed{l}$, respectively). Occasionally we shall omit k and l if the degree of the vertex is not needed.

A β-vertex of Γ of degree 11 can be adjacent only to α-vertices. Such a vertex, together with its 11 adjacent regions is called a *β-wheel*:

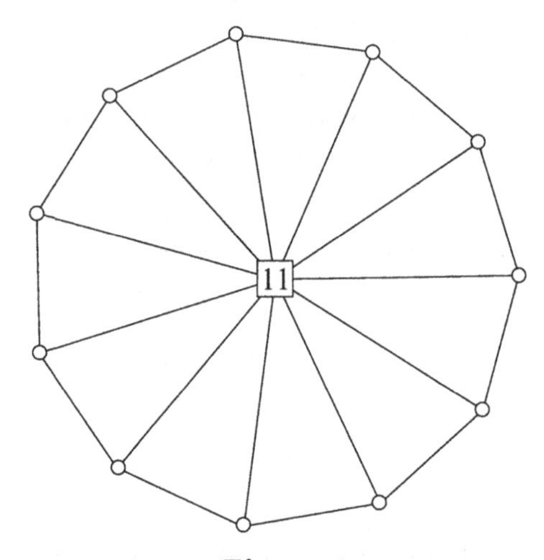

Figure 1

The following observations will be useful in what follows:

(a) β-wheels consist entirely of 3-gons;

(b) if two β-vertices are adjacent (in Γ) then both have degree > 11;

(c) if Δ is a region of Γ of degree > 3 then $c(\Delta) < 0$;

(d) in passing from Π' to Γ it is possible to create a new region of positive curvature as the diagram below illustrates, and we must take this into account when compensating;

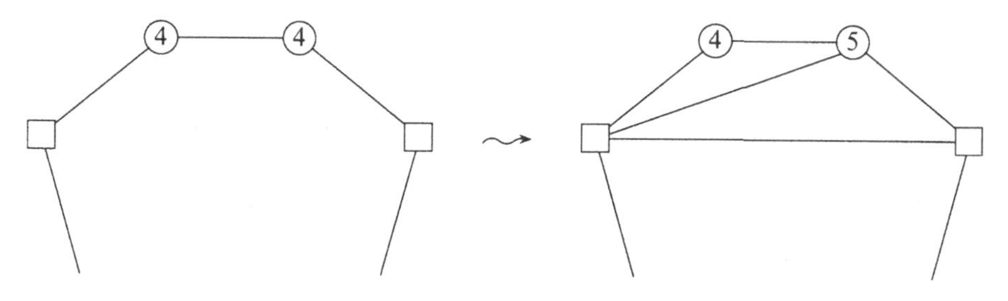

Figure 2

(*e*) any α-vertex of Γ of degree 5 is adjacent to at most 3 other α-vertices.

2. In this final section we list the possible regions of Γ of positive curvature and indicate the compensating regions.

If Δ is a region of Γ of positive curvature then deg $\Delta = 3$ and $c(\Delta)$ is one of the following:

$$\begin{aligned} &\left[4,4,k;\tfrac{2\pi}{k}\right] \qquad (k \geq 12) \\ &\left[4,5,k;\tfrac{2\pi}{k}-\tfrac{\pi}{10}\right] \qquad (12 \leq k < 20) \\ &\left[5,6,7;\tfrac{2\pi}{105}\right] \\ &\left[5,6,6;\tfrac{\pi}{15}\right] \\ &\left[5,5,7;\tfrac{3\pi}{35}\right] \\ &\left[5,5,6;\tfrac{2\pi}{15}\right] \\ \text{and} \quad &\left[5,5,5;\tfrac{\pi}{5}\right]. \end{aligned}$$

There are two types of region of positive curvature; those that share an edge in Γ with a β-wheel W and require W to compensate for its positive curvature – a so-called *satellite* of W, and the rest that are not satellites of any β-wheel. We deal first with the latter type.

In what follows we list the possible ways a region of Γ of positive curvature can occur. Note that a compensating face P of negative curvature can be required for *at most two* regions of positive curvature and accordingly contributes $\frac{1}{2}c(P)$ on each occasion. We will indicate when only $\frac{1}{2}c(P)$ is being used.

(i) $(\mathbf{4,r,k})$ $(\mathbf{4 \leq r \leq 5})$

$c(\Delta) \leq \left[4, 4, 12; \frac{\pi}{6}\right]$

Δ_1 is compensating region

$c(\Delta_1) \leq \left[4, 12, 12; -\frac{\pi}{6}\right]$

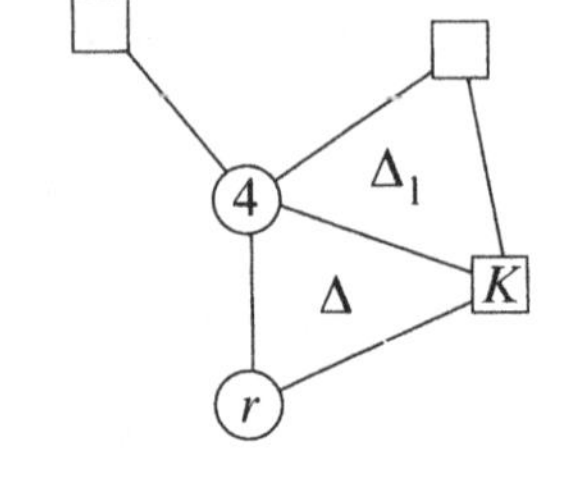

Figure 3 - (i)

(ii) $(\mathbf{5, r_1, r_2})$ $(\mathbf{5 \leq r_1 \leq 6,\ 6 \leq r_2 \leq 7})$

$c(\Delta) \leq \left[5, 5, 6; \frac{2\pi}{15}\right]$

Δ_1 is compensating region

$c(\Delta_1) \leq \left[5, 12, 12; -\frac{4\pi}{15}\right]$

Use only half of $c(\Delta_1)$ to compensate.

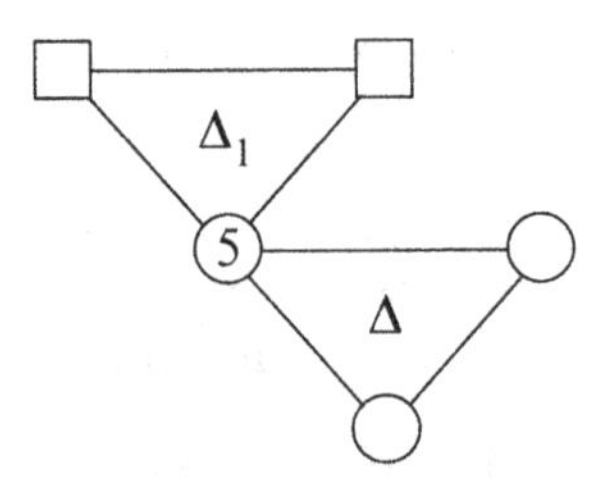

Figure 4 - (ii)

(iii) $(\mathbf{5, r, 7})$ $(\mathbf{5 \leq r \leq 6})$

$c(\Delta) \leq \left[5, 5, 7; \frac{3\pi}{35}\right]$

Δ_1 is compensating region

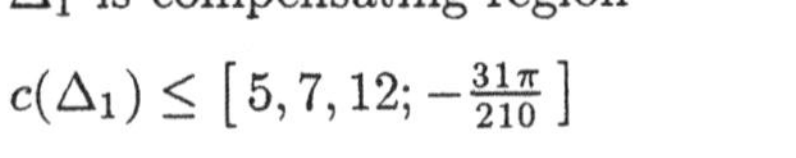

$c(\Delta_1) \leq \left[5, 7, 12; -\frac{31\pi}{210}\right]$

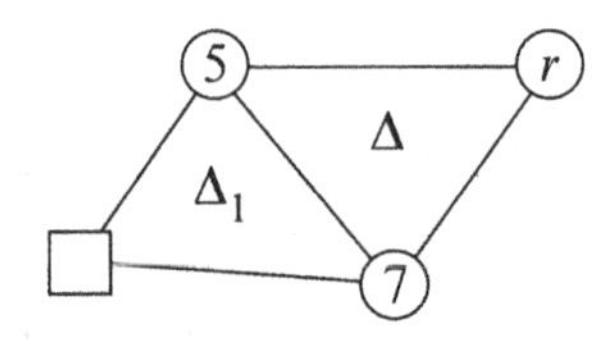

Figure 5 - (iii)

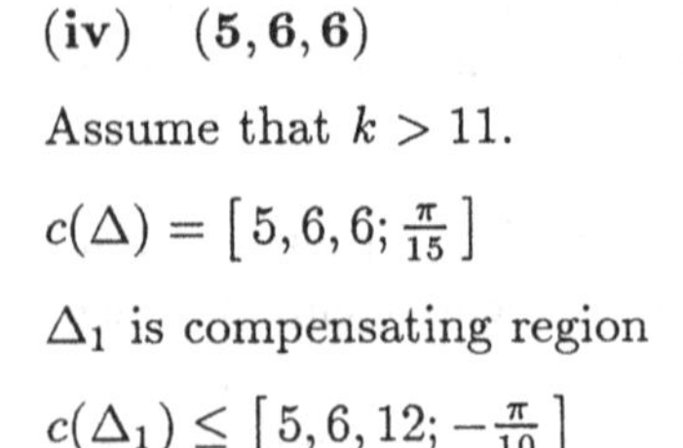

(iv) $(\mathbf{5, 6, 6})$

Assume that $k > 11$.

$c(\Delta) = \left[5, 6, 6; \frac{\pi}{15}\right]$

Δ_1 is compensating region

$c(\Delta_1) \leq \left[5, 6, 12; -\frac{\pi}{10}\right]$

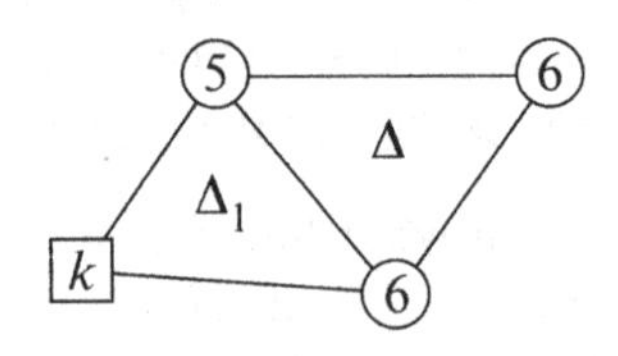

Figure 6 - (iv)

(v) $\mathbf{(5,5,6)}$

We know one of k_2, k_3 must be at least 12; let $k_3 > 11$. Assume that at least one of k_1, k_2 is greater than 11, and use the corresponding region Δ_1 or Δ_2 to compensate together with Δ_3.

$c(\Delta) = [5,5,6; \frac{2\pi}{15}]$

$c(\Delta_1) \le [5,5,12; -\frac{\pi}{30}]$

$c(\Delta_2) \le [5,6,12; -\frac{\pi}{10}]$

$c(\Delta_3) \le [5,6,12; -\frac{\pi}{10}]$

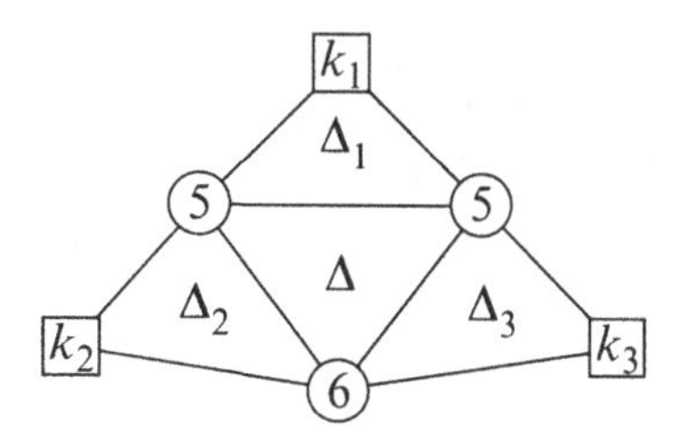

Figure 7 - (v)

(vi) $\mathbf{(5,5,5)}$

$c(\Delta) = \frac{\pi}{5}$

Δ_1 and Δ_2 are compensating regions.

$c(\Delta_1),\ c(\Delta_2) \le [5,12,12; -\frac{4\pi}{15}]$

Use half of each of $c(\Delta_1)$, $c(\Delta_2)$.

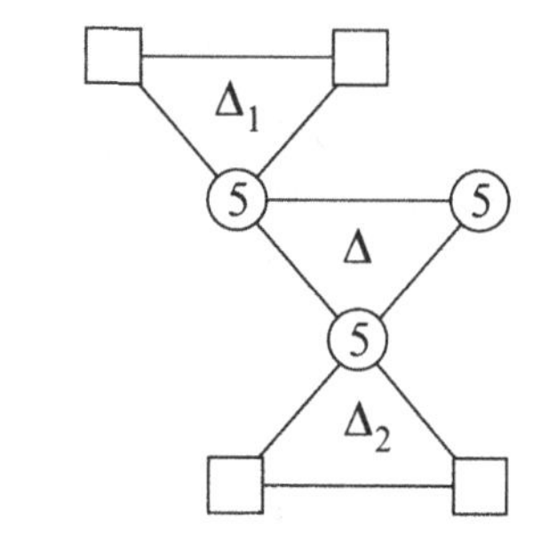

Figure 8 - (vi)

(vii) $\mathbf{(5,5,5)}$

$c(\Delta) = \frac{\pi}{5}$

Δ_i $(1 \le i \le 4)$ are compensating regions.

$c(\Delta_1),\ c(\Delta_2) \le [4,12,12; -\frac{\pi}{6}]$

$c(\Delta_3),\ c(\Delta_4) \le [4,5,12; \frac{\pi}{15}]$

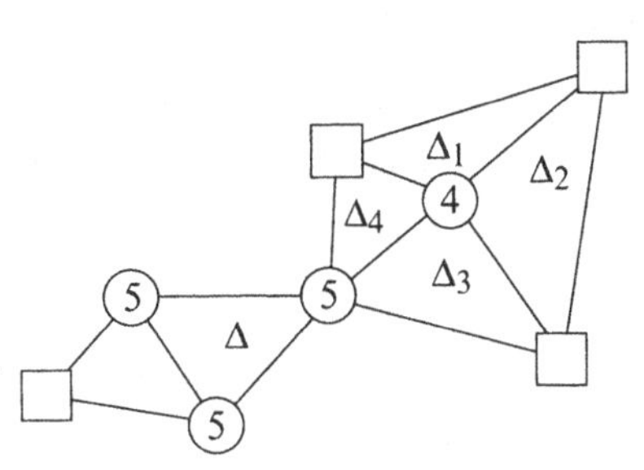

Figure 9 - (vii)

(viii) (5, 5, 5)

Assume each α-vertex has degree ≥ 5.
Assume each β-vertex has degree ≥ 12.

$c(\Delta) = \frac{\pi}{5}$

Δ_i $(1 \leq i \leq 9)$ are compensating regions.

$c(\Delta_i) \leq \left[5, 5, 12; -\frac{\pi}{30}\right]$

Use all of $c(\Delta_j)$ $(1 \leq j \leq 3)$.

Use half of $c(\Delta_l)$ $(4 \leq l \leq 9)$.

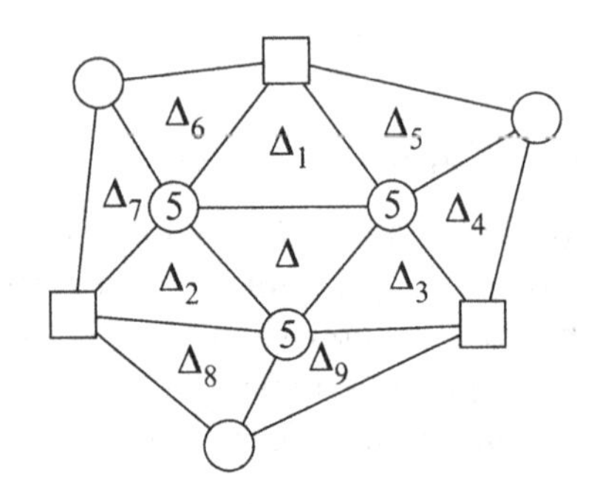

Figure 10 - (viii)

We now list all possible satellites in Γ.

(S1) $c(\Delta) = \frac{\pi}{15}$

Δ is a satellite of 2 β-wheels and contributes $\frac{\pi}{30}$ to each.

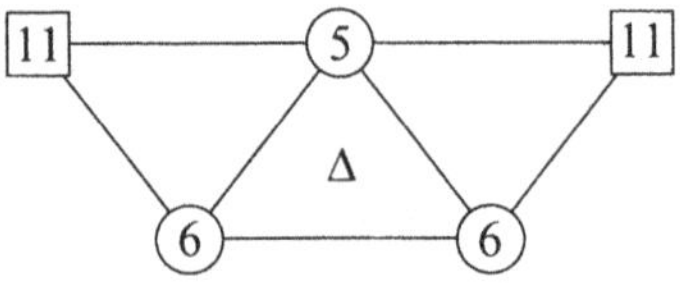

Figure 11 - (S1)

(S2) $c(\Delta) = \frac{2\pi}{15}$

Δ_1 is compensating region.

$c(\Delta_1) \leq \left[5, 6, 12; -\frac{\pi}{10}\right]$

Δ is a satellite of 2 β-wheels and, with Δ_1 compensating, contributes at most $\frac{1}{2}\left(\frac{2\pi}{15} - \frac{\pi}{10}\right) = \frac{\pi}{60}$ to each.

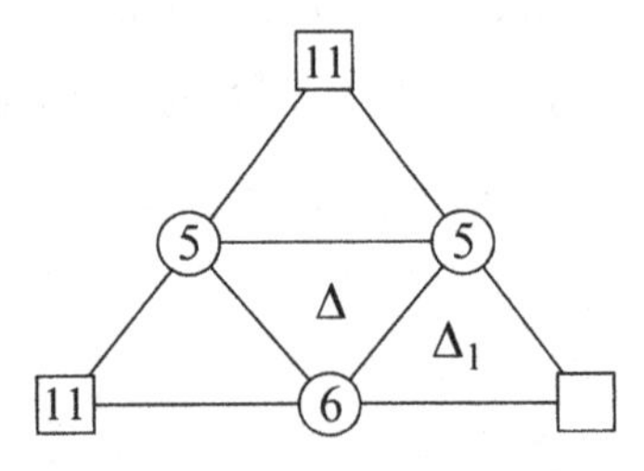

Figure 12 - (S2)

(S3) $c(\Delta) = \frac{\pi}{5}$

Δ is a satellite of 3 β-wheels and contributes $\frac{\pi}{15}$ to each.

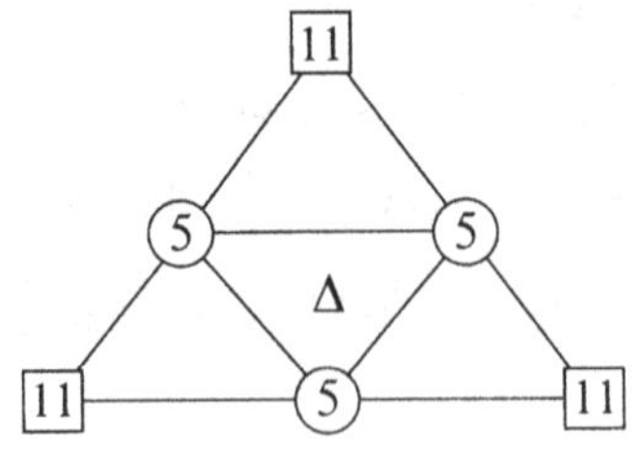

Figure 13 - (S3)

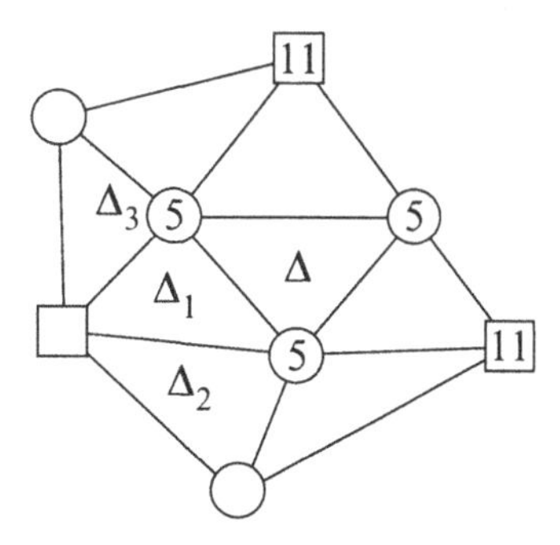

Figure 14 - (S4)

(S4) $c(\Delta) = \frac{\pi}{5}$

$\Delta_i (1 \leq i \leq 3)$ are compensating regions.

$c(\Delta_i) \leq \left[5, 5, 12; \frac{-\pi}{30}\right]$

Use all of $c(\Delta_1)$ and half of each of $c(\Delta_2)$ and $c(\Delta_3)$ to compensate.

Δ is a satellite of 2 β-wheels and contributes a net total of $\frac{1}{2}\left(\frac{\pi}{5} - \frac{\pi}{30}\right) - \frac{\pi}{60} = \frac{\pi}{15}$ to each.

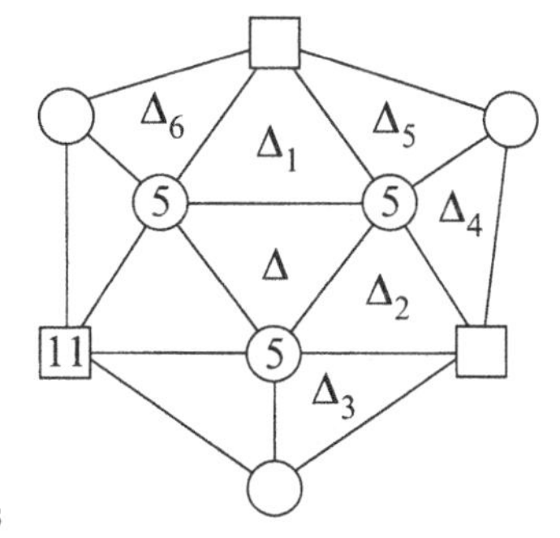

Figure 15 - (S5)

(S5) $c(\Delta) = \frac{\pi}{5}$

$c(\Delta_i) \leq \left[5, 5, 12; \frac{-\pi}{30}\right]$

Use all of $c(\Delta_1)$, $c(\Delta_2)$ and half of each of $c(\Delta_j)$ $(3 \leq j \leq 6)$ to compensate.

Δ is a satellite of one β-wheel and contributes a net total of $\frac{\pi}{5} - \frac{\pi}{15} - \frac{1}{2}\left(\frac{2\pi}{15}\right) = \frac{\pi}{15}$ to it.

Let W be a β-wheel with at least one satellite. The total curvature of W together with its satellites we denote by $c(\widehat{W})$. We finish by showing that $c(\widehat{W}) \leq 0$.

Observe that $c(W) \leq 11\left[5, 5, 11; -\frac{\pi}{55}\right] = -\frac{\pi}{5}$ and that if we change the degree of an α-vertex of W from 5 to 6 then $c(\widehat{W})$ is altered by $2(-\frac{\pi}{15})$, whereas the maximum positive contribution to $c(\widehat{W})$ is $2(\frac{\pi}{30})$ in the case of two satellites of type S1. The net effect on $c(\widehat{W})$ is therefore negative, so it can be assumed without any loss that the satellites of W are of type S3, S4 or S5.

If W contains an α-vertex of degree 6 then $c(W) \leq -\frac{\pi}{5} - \frac{2\pi}{15} = -\frac{\pi}{3}$. Now by assumption W has at most 5 satellites and so $c(\widehat{W}) \leq -\frac{\pi}{3} + 5\left(\frac{\pi}{15}\right) = 0$.

If every α-vertex of W has degree 5 then, given that W has at least one satellite, any attempt at labelling shows that the following situation must occur.

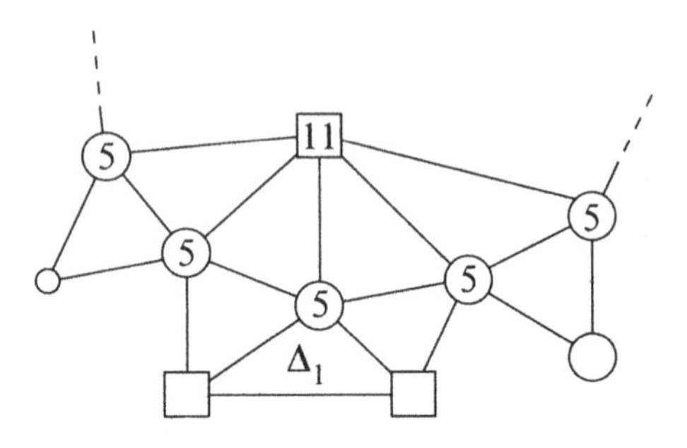

Figure 16

In this case we add $c(\Delta_1)$ to $c(\widehat{W})$ by way of compensation to obtain a curvature of at most

$$11\left[5,5,11;-\tfrac{\pi}{55}\right]+5\left(\tfrac{\pi}{15}\right)+\left[5,12,12;-\tfrac{4\pi}{15}\right]<0.$$

References

[1] M. EDJVET. Equations over groups and a theorem of Higman, Neumann and Neumann. *Proc. Lond. Math. Soc.* (3) **62** (1991) 563 - 589.

[2] D. F. HOLT AND W. PLESKEN. A cohomological criterion for a finitely presented group to be infinite. (Preprint 1990).

[3] J. HOWIE AND R. M. THOMAS. On the asphericity of presentations for the groups $(2, 3, p; q)$ and a conjecture of Coxeter. *Journal of Algebra* to appear.

Department of Mathematics
University of Nottingham
University Park
Nottingham NG7 2RD

Moduli of Riemann surfaces with symmetry

G. González-Díez and W. J. Harvey

To Murray Macbeath on the occasion of his retirement

The moduli space $\mathcal{M}_g$ of Riemann surfaces with genus $g \geq 2$ contains an important subset corresponding to surfaces admitting non-trivial automorphisms. In this paper, we study certain irreducible subvarieties $\mathcal{M}_g(G)$ of this singular set, which are characterised by the specification of a finite group G of mapping-classes whose action on a surface S is fixed geometrically. In the special case when the quotient surface S/G is the sphere, we describe a holomorphic parameter function λ which extends the classical λ-function of elliptic modular theory, and which induces a birational isomorphism between the normalisation of $\mathcal{M}_g(G)$ and a certain naturally defined quotient of a configuration space $\mathbb{C}^n - \Delta$ where Δ is the discriminant set $\{z_i = z_j, \text{ for some } i \neq j\}$. Thus $\mathcal{M}_g(G)$ is always a unirational variety. We also show that in general $\mathcal{M}_g(G)$ is distinct from its normalisation, and construct a (coarse) modular family of G-symmetric surfaces over the latter space.

1. Teichmüller spaces and modular groups

First we introduce some of the necessary formalism. Let H_0 be a subgoup of the group $Aut(S_0)$ of automorphisms of a closed surface S_0 of genus $g \geq 2$; by a famous theorem of Hurwitz (see e.g. [19]), $Aut(S_0)$ is finite of order at most $84(g-1)$. We shall later concentrate on the case where the quotient surface S_0/H_0 is $\mathbb{P}^1$, the Riemann sphere, but results in §1 and §2 apply without this restriction.

DEFINITION. A *Riemann Surface with H_0-symmetry* is a pair (S, H) comprising a Riemann surface S with $H < Aut(S)$ such that (S_0, H_0) and (S, H) are topologically conjugate by some homeomorphism θ: $S_0 \to S$.

Two surfaces (S, H), (S', H') with H_0-symmetry are *H_0-isomorphic* if there is a biholomorphic mapping ϕ: $S \to S'$ such that $H' = \phi H \phi^{-1}$.

NOTATION. An H_0-isomorphism class is denoted by $\{S, H\}$ and the set of all H_0-isomorphism classes of surfaces with H_0-symmetry is denoted by $\widetilde{\mathcal{M}_g}(H_0)$.

We shall also need to consider the weaker equivalence relation of (non-equivariant) isomorphism for surfaces (S, H), (S', H') with H_0-symmetry; here there must be a biholomorphic mapping $\phi : S \to S'$ as before, but it is no longer required to satisfy the condition $H' = \phi H \phi^{-1}$. We shall denote by $\mathcal{M}_g(H_0)$ the set of all isomorphism classes of surfaces with H_0-symmetry.

There is a natural surjection $\widetilde{\mathcal{M}_g}(H_0) \to \mathcal{M}_g(H_0)$ between these two sets. Our primary purpose is to provide complex analytic structures for them which make this mapping a morphism of analytic spaces. Our approach rests on well-known results of Teichmüller theory which we now discuss briefly. Good references for the facts we need are [9], [22]. More details of our methods are given in earlier papers [14, 20].

Let T_g be the *Teichmüller space* of S_0. A point $t \in T_g$ is an equivalence class $[S, \theta]$, where $\theta : S_0 \to S$ is a *marking* homeomorphism, and two marked pairs (S, θ), (S', θ') are equivalent iff there is a biholomorphic $f : S \to S'$ such that θ' is isotopic to $f \circ \theta$.

If $\boldsymbol{b} = \{b_1, \ldots, b_n\}$ is a finite subset of S_0, and $S_0^* = S_0 - \boldsymbol{b}$ denotes the surface punctured at $\boldsymbol{b}$, then the (stronger) equivalence relation obtained by requiring the isotopy between θ' and $f \circ \theta$ to fix the points of $\boldsymbol{b}$ determines the Teichmüller space $T_{g,n}$ of S_0^*, $n \geq 1$.

The group of mapping classes $Mod(S_0)$, viewed as the path components of the group of homeomorphisms of S_0, is denoted Mod_g if S_0 has genus g (or $Mod_{g,n}$ for S_0^*). This group operates on T_g (or on $T_{g,n}$) by the rule

$$[S, \theta] \overset{\mathbf{f}}{\longmapsto} [S, \theta \circ f]$$

By fundamental results of Bers [3], there is a canonical representation of each $T_{g,n}$ as a bounded domain in some $\mathbb{C}^N$, with $N = 3g - 3 + n$. Furthermore, the action of $Mod_{g,n}$ is by holomorphic isomorphisms and properly discontinuous [18], [20].

We shall regard a subgroup $H_0 \subseteq Aut(S_0)$ as tantamount to a subgroup of $Mod(S_0)$, since by a theorem of Hurwitz an automorphism of S_0 that is homotopic to the identity must be trivial. By a result which goes back to W. Fenchel and J. Nielsen, the fixed point set in T_g of any such finite group $G \subset Mod(S_0)$ is a (complex) submanifold denoted by $T_g(G)$.[1] In the present terminology it was reformulated in [14] as follows.

THEOREM A. *$T_g(H_0)$ is the set of Teichmüller points $[S, \theta]$ such that S possesses a group of automorphisms H conjugate to H_0 by means of the homeomorphism θ: $S_0 \to S$.*

Because the action of the modular group on T_g is properly discontinuous, the quotient *moduli space* $\mathcal{M}_g$ carries an induced structure of complex analytic V- manifold, for which the canonical projection map p: $T_g \to \mathcal{M}_g$ is holomorphic. In fact $\mathcal{M}_g$ is a projective variety; it is worth noting that the $\underline{\lambda}$-functions which we describe later fit in naturally with the projective embedding originally constructed by Baily [1] using Jacobi varieties and the Lefschetz embedding theorem.

COROLLARY. *$\mathcal{M}_g(H_0)$ is the image of $T_g(H_0)$ under the projection p.*

The submanifold $T_g(H_0)$ is itself a Teichmüller space. To see this, let the quotient surface $R_0 = S_0/H_0$ have genus γ, let $\boldsymbol{b} = \{b_1, \ldots, b_r\}$ be the point set over which the projection $S_0 \to R_0$ is ramified and denote by $T_{\gamma,r}$ the Teichmüller space of the punctured surface $R_0^* = R_0 - \boldsymbol{b}$.

For each $[S,\theta] \in T_g(H_0)$, write R for the quotient surface S/H and R^* for the corresponding unramified subsurface. Then $\theta : S_0 \to S$ induces a homeomorphism $\theta^* : R_0^* \to R^*$, which defines a rule

$$[S,\theta] \stackrel{\psi}{\longmapsto} [R^*,\theta^*].$$

At the level of Teichmüller spaces, this is a bijection.

[1] The fact that $T_g(G)$ is non-empty for all finite G was proved by S. Kerckhoff [17].

THEOREM B. *The spaces $T_g(H_0)$ and $T_{\gamma,r}$ are biholomorphically equivalent via the mapping ψ.*

For a proof, see [18], [14], [23].

Not every element in Mod_g stabilizes $T_g(H_0)$. The modular group permutes the various finite subgroups H_0 by conjugation and the relevant group for our purposes is the *relative modular group* with respect to H_0, which is defined as the subgroup of those mapping classes that do stabilise $T_g(H_0)$; this is the normaliser of H_0 in Mod_g (see[20]). We denote it by $Mod_g(H_0)$.

For each $[S,\theta] \in T_g(H_0)$ with a marked symmetry group $H = \theta H_0 \theta^{-1}$, the rule $[S,\theta] \to \{S,H\}$ defines a mapping from $T_g(H_0)$ into $\widetilde{\mathcal{M}}_g(H_0)$ which we shall denote by π_1. This map is clearly surjective.

Let f be an H_0-equivariant homeomorphism of S_0 representing an element $\mathbf{f}$ of $\mathcal{M}_g(H_0)$. Then $\mathbf{f}([\mathbf{S},\theta]) = [\mathbf{S},\theta \circ \mathbf{f}]$ has marked symmetry group H_f, obtained *via* $H_f = (\theta \circ f)H_0(\theta \circ f)^{-1} = \theta H_0 \theta^{-1}$. Notice that the underlying Riemann surface S and its automorphism group H are unchanged: the change of marking by $\mathbf{f}$ induces a complementary change of marking for H. This implies that the images under π_1 of $[S,\theta]$ and $\mathbf{f}([\mathbf{S},\theta])$ coincide in $\widetilde{\mathcal{M}}_g(H_0)$.

Suppose now that we have two pairs (S_1,H_1), (S_2,H_2) of surfaces with H_0-symmetry, related by a biholomorphic isomorphism $\phi:\ S_1 \to S_2$ such that $H_2 = \phi H_1 \phi^{-1}$. Choose two markings $\theta_j:\ S_0 \to S_j$, $j = 1,2$, so that each $[S_j,H_j]$ is a Teichmüller point in $T_g(H_0)$ lying over $\{S_j,H_j\}$. Then there is a homeomorphism $f:\ S_0 \to S_0$ making the diagram

$$\begin{array}{ccc} S_0 & \xrightarrow{\theta_1} & S_1 \\ f\downarrow & & \downarrow\phi \\ S_0 & \xrightarrow{\theta_2} & S_2 \end{array}$$

commute and compatible with $H_0(=\theta_j^{-1}H_j\theta_j)$. Therefore f determines an element of $Mod_g(H_0)$ and we have proved the following statement.

PROPOSITION 1. *The mapping $\pi_1:\ T_g(H_0) \to \widetilde{\mathcal{M}}_g(H_0)$ induces a natural bijection between the quotient space $T_g(H_0)/Mod_g(H_0)$ and $\widetilde{\mathcal{M}}_g(H_0)$.*

Since $T_g(H_0) \cong T_{\gamma,r}$ is a bounded domain in $\mathbb{C}^m$ where $m = 3\gamma - 3 + r$, it follows that $\widetilde{\mathcal{M}}_g(H_0)$ is a complex V-manifold of dimension m.

2. The relationship between $\widetilde{\mathcal{M}_g}(H_0)$ and $\mathcal{M}_g(H_0)$.

The first aim of this section is to prove that $\widetilde{\mathcal{M}_g}(H_0)$ is the normalisation of $\mathcal{M}_g(H_0)$. We shall use [11] as a basic reference for analytic spaces.

THEOREM 1. *$\mathcal{M}_g(H_0)$ is an irreducible subvariety of $\mathcal{M}_g$ and $\widetilde{\mathcal{M}_g}(H_0)$ is its normalisation.*

PROOF. Because $Mod_g(H_0)$ acts discontinuously with finite isotropy groups on $T_g(H_0) \cong T_{\gamma,r}$, a domain in $\mathbb{C}^m$, it follows from a theorem of Cartan [4] that $\widetilde{\mathcal{M}_g}(H_0)$ is a normal complex space. Also, by the discontinuity of Mod_g on T_g, the family of submanifolds $\{\mathbf{h}(T_g(H_0)),\ \mathbf{h} \in Mod_g\}$ is *locally finite*, that is each point of $T_g(H_0)$ has a neighbourhood in T_g intersecting only finitely many distinct subvarieties $\mathbf{h}(T_g(H_0))$.

We have already defined the natural mapping π: $\widetilde{\mathcal{M}_g}(H_0) \to \mathcal{M}_g$ whose image is precisely $\mathcal{M}_g(H_0)$. Thus if we check that:

(1) π is closed,

(2) π has finite fibres,

(3) π is injective outside a proper subvariety,

then by the Proper Mapping Theorem and the definition of normalisation the theorem will be proved.

Let us prove (2) and (3). The diagram below summarises the situation; the map $\pi_2 = \pi \circ \pi_1 : T_g(H_0) \to \mathcal{M}_g(H_0)$ is the restriction of p to $T_g(H_0)$.

$$\begin{array}{ccccc} T_g(H_0) & & \hookrightarrow & & T_g \\ \downarrow \pi_1 & & & & \downarrow p \\ \widetilde{\mathcal{M}_g}(H_0) & \xrightarrow{\pi} & \mathcal{M}_g(H_0) & \hookrightarrow & \mathcal{M}_g \end{array}$$

Two points in $T_g(H_0)$ with the same image in $\mathcal{M}_g(H_0)$ are of the form $[S,\ \theta]$ and $[S,\ \theta \circ h]$ with $\mathbf{h} \in Mod_g$. By Theorem A, $H = \theta H_0 \theta^{-1}$ and $H' = (\theta \circ h)H_0(\theta \circ h)^{-1}$ are both subgroups of $Aut(S)$. Now, if $[S,\ \theta]$ and $[S,\ \theta \circ h]$ have different images in $\widetilde{\mathcal{M}_g}(H_0)$, then $\mathbf{h} \notin Mod_g(H_0)$ and so $\mathbf{h}H_0\mathbf{h}^{-1} \neq H_0$. Hence necessarily $H \neq H'$. Since $Aut(S)$ is finite, there are only finitely many possibilities for $[S,\ \theta \circ h]$, which proves (2).

This argument also shows that π fails to be injective only on the π_1-image of intersections $T_g(H_0) \cap \mathbf{h}(T_g(H_0))$ with $\mathbf{h} \in Mod_g - Mod_g(H_0)$. By the local finiteness this is a subvariety of $\widetilde{\mathcal{M}_g}(H_0)$, which proves (3).

For completeness we sketch the elementary property (1). Referring to the diagram above, it is sufficient to prove that if C is a closed subset of $T_g(H_0)$ then $\pi_2(C)$ is closed in $\mathcal{M}_g(H_0)$ or, equivalently, that the union of all $\mathbf{h}(C)$, $\mathbf{h} \in Mod_g$, is closed in T_g. Suppose that $y = \lim \mathbf{h}_n(x_n)$ with $x_n \in C$ and $\mathbf{h}_n \in Mod_g$. Taking N_y a small enough open set in T_g containing y such that N_y intersects only finitely many sets $\mathbf{h}(C)$, there is then a single set $\mathbf{h}_0(C)$ which contains an infinite subsequence of the $\{\mathbf{h}_n(x_n)\}$. Thus we have a sequence of points $x'_n \in C$ with $\mathbf{h}_0(x'_n) \to y$. But $T_g(H_0)$ is closed in T_g and $\mathbf{h}_0$ is an isometry in the Teichmüller metric, so it follows that $x'_n \to x \in C$. This completes the verification of the property (1). ∎

We next address the question whether $\widetilde{\mathcal{M}_g}(H_0)$ is biholomorphic to $\mathcal{M}_g(H_0)$. From the proof of the theorem we can see that these spaces are different if and only if there is a surface S whose automorphism group contains two subgroups H, H' that are conjugate topologically but not holomorphically. This situation occurs, for instance, when there is a surface S, which admits a larger group G of automorphisms containing a pair of (conjugate) subgroups H, H' such that $\langle H, H' \rangle = K$ is a proper subgroup of G with H, H' not conjugate in K. Usually a deformation of S may then be constructed which preserves the K-symmetry but destroys the G-symmetry. Examples are readily produced using the fact that for any finite group G there exist Riemann surfaces S with G as a group of automorphisms and such that the quotient surface S/G has arbitrarily given genus γ; see for instance [12]. An elementary example of this type is given later in this section (example 1).

Provided that the Teichmüller space $T_g(K)$ is not a point (the case $T_{0,3}$) and is *not* in the small list of types for which there is an isomorphism between Teichmüller spaces of surfaces with different signatures, we may conclude that the space $T_g(G)$ is *properly* contained in $T_g(K)$ for any proper overgroup $G > K$. This list is as follows (see [22] p.129 for details):

$$T_{0,6} \cong T_{2,0}\,, \qquad T_{0,5} \cong T_{1,2}\,, \qquad T_{0,4} \cong T_{1,1}\,.$$

The implication is that in general the locus of points $[S, \theta] \in T_g(K)$, with S admitting two automorphism groups $\theta H \theta^{-1}$ and $\theta H' \theta^{-1}$ which are conjugate only in some larger group than $\theta K \theta^{-1}$, forms an analytic subset Z of strictly lower dimension. Therefore the restriction of the mapping $\pi : \widetilde{\mathcal{M}_g}(H) \to \mathcal{M}_g(H)$ to the π_1-image of $T_g(K)$ is not injective since outside $\pi_1(Z)$ one has $\pi([S, H]) = \pi([S, H'])$. Thus π is not biholomorphic, and the variety $\mathcal{M}_g(H)$ is non-normal at all points in the image of $\pi_1(T_g(K) - Z)$.

Furthermore, using elementary facts on analytic spaces, we can conclude that since this subset of the non-normal points of $\mathcal{M}_g(H)$ is Zariski-open in $\mathcal{M}_g(K)$, and therefore dense, the whole subvariety $\mathcal{M}_g(K) = \pi_1(T_g(K)$ is non-normal because the non-normal set is necessarily closed ([11], p.128).

These arguments prove the following result.

THEOREM 2. *The modular subvariety $\mathcal{M}_g(H_0)$ is in general distinct from its normalisation $\widetilde{\mathcal{M}_g}(H_0)$.*

As illustration, we give two examples.

EXAMPLE 1. Let F_{2p} be the compact (Fermat) Riemann surface with affine algebraic equation

$$x^{2p} + y^{2p} = 1 ,$$

let H (respectively H') be the cyclic group generated by the involution $(x, y) \to (-x, y)$ (respectively by $(x, y) \to (x, -y)$) and let $K = \langle H, H' \rangle$. Then H and H' are not conjugate in K but in $G = Aut(F_{2p})$ they are conjugate by the automorphism $\alpha(x, y) = (y, x)$. Now F_{2p}/K has genus > 2; in fact F_{2p}/K is isomorphic to the surface F_p with equation $x^p + y^p = 1$, the isomorphism being given in affine coordinates by $\phi(x, y) = (x^2, y^2)$; F_p has genus $(p-1)(p-2)/2$ which is > 2 for $p \geq 4$.

Thus, by the discussion preceding theorem 2, the modular subvariety $\mathcal{M}_g(H)$, $g = (p-1)(2p-1)$, is not normal; in fact the point representing the Fermat surface F_{2p} is a non-normal point of this modular subvariety.

On the other hand, for certain types of surface with automorphism, the modular subvariety $\mathcal{M}_g(H)$ *is* itself normal.

EXAMPLE 2. Let S be a hyperelliptic surface of genus g, with $J : S \to S$ the hyperelliptic involution. Since J is the unique automorphism of S with order 2 having quotient $S/\langle J \rangle \equiv \mathbb{P}^1$, we obtain $\widetilde{\mathcal{M}_g}(\langle J \rangle) = \mathcal{M}_g(\langle J \rangle)$.

REMARK. Since $\mathcal{M}_g$ is a projective variety, the G.A.G.A. Principle implies that our complex-analytic results remain valid within the framework of complex algebraic geometry. Thus $\mathcal{M}_g(G)$ is also an irreducible algebraic subvariety of $\mathcal{M}_g$ by Chow's Theorem ([11] p.184). Furthermore, since the algebraic normalisation of a projective variety is again projective ([13] p.232) and therefore analytic, it follows from the uniqueness of the normalisation ([11] p.164) that $\widetilde{\mathcal{M}_g}(G)$ is also the algebraic normalisation of $\mathcal{M}_g(G)$.

3. The case of tori: Legendre's modular function.

We review the classical theory of moduli for elliptic curves from the point of view developed in the previous section. In genus 1, the Teichmüller space T_1 is the upper half plane U: any Riemann surface of genus 1 may be expressed as a complex torus $E = E_\tau = \mathbb{C}/\Lambda(\tau)$ with $\Lambda(\tau) = \mathbb{Z} + \mathbb{Z}\tau$ a lattice subgroup of the additive group $\mathbb{C}$ and $\tau \in U$. There is a standard involutory automorphism $J : E \to E$, given by the symmetry $z \to -z$ of $\mathbb{C}$ and so, writing $H = \langle J \rangle$, we have that $T_1(H) = T_1$. The quotient E/H is the projective line $\mathbb{P}^1$, with four ramification points $a_1, \ldots, a_4$ corresponding to the four fixed points of J (which are the points of order 2, the orbits of $0, \frac{\tau}{2}, \frac{1+\tau}{2}$, and $\frac{1}{2}$ under $\Lambda(\tau)$).

Let the orbit of the origin $\underline{0}$ be chosen as a base point of E and write $T_{1,1}$ for $T(E - \underline{0})$: this procedure renders the (flat) homogeneous space E into a hyperbolic surface, thereby placing the theory of moduli for E within the framework of Teichmüller spaces. Theorem B now captures the identification of T-spaces, $T_{1,1} \cong T_{0,4}$, in our earlier list.

This space may be identified with the upper half-plane U by associating to $\tau \in U$ the Teichmüller pair $[E_\tau, f_\tau]$, $f_\tau : E_i \to E_\tau$, where $E_i = \mathbb{C}/\Lambda(i)$ has been chosen as reference surface and f_τ is the projection of the real linear homeomorphism $L_\tau : \mathbb{C} \to \mathbb{C}$ which sends $1, i$ to $1, \tau$ respectively (see e.g. [22] 2.1.8).

Similarly the fact that $T_{1,1}(H)$ is the whole of $T_{1,1}$ implies, by the definition of relative modular group given in §1, that $Mod_1(H)$ is the modular group of genus 1, $SL(2, \mathbb{Z})$, so we have

$$\mathcal{M}_1(J) \equiv \mathcal{M}_1 \equiv U/SL(2, \mathbb{Z}) \, .$$

Here $SL(2, \mathbb{Z})$ acts by $A \cdot \tau = \dfrac{a\tau + b}{c\tau + d}$, where $A = \begin{pmatrix} a & b \\ c & d \end{pmatrix}$. We identify τ with $[E_\tau, f_\tau]$ and A with the homeomorphism $f_A : E_i \to E_i$ characterised by the real-linear map $L_A(1) = ci + d, \quad L_A(i) = ai + b$. Then the Teichmüller modular group acts by the rule

$$f_A \cdot (E_\tau, f_\tau) = (E_\tau, f_\tau \circ f_A) = (E_{A\cdot\tau}, h_A \circ f_\tau \circ f_A)$$

where $h_A : E_\tau \to E_{A\cdot\tau}$ is the isomorphism induced by $h_A(z) = (c\tau + d)^{-1} z$. To see that this is a genuine group action, one checks directly that $h_A \circ f_\tau \circ f_A$ is just $f_{A\cdot\tau}$, so we have $f_A \cdot (E_\tau, f_\tau) = (E_{A\cdot\tau}, f_{A\cdot\tau})$ as it should be.

Next we focus attention on the level-2 congruence subgroup $\Gamma(2)$, comprising matrices $A \equiv Id \pmod 2$ in $SL_2(\mathbb{Z})$. The involution J corresponds to the central element $\begin{pmatrix} -1 & 0 \\ 0 & -1 \end{pmatrix}$ of $\Gamma(2)$, which fixes every point of U. The quotient group $P\Gamma(2)$ acts faithfully on U and (see for instance [5], [16]) this action is free and discontinuous. In classical vein, we define a complex function λ on U by the rule

$$\lambda(\tau) = \{\wp(a_1), \wp(a_2); \wp(a_3), \wp(a_4)\},$$

where $\wp(z) = \wp_\tau(z)$ is the *Weierstrass* $\wp$*- function* of the lattice $\Lambda(\tau)$ and $\{-,-;-,-\}$ denotes *cross-ratio*; λ is the *Legendre modular function*, which is automorphic with respect to $P\Gamma(2)$ and induces an isomorphism

$$\lambda : \; U/P\Gamma(2) \to \mathbb{C} - \{0,1\}.$$

The famous modular invariant $j(\tau)$ may then be written as an invariant (degree 6) rational function of λ.

We shall need the following description of this classical theory in terms of the *universal family* $\mathbf{E}$ of tori over U; a brief account appears in [26]. This is a fibre space $\mathbf{E} = (U \times \mathbb{C})/\mathbb{Z}^2$ over U where $\mathbb{Z}^2$ acts on $U \times \mathbb{C}$ by $(n,m) \cdot (\tau; z) = (\tau; z + n + m\tau)$, so that the fibre over τ is precisely E_τ.

Corresponding to the four fixed points of J, we have the following four holomorphic sections of the family $\mathbf{E} \to U$,

$$s_1(\tau) = 0, \qquad s_2(\tau) = \frac{\tau}{2}, \qquad s_3(\tau) = \frac{1+\tau}{2}, \qquad s_4(\tau) = \frac{1}{2}.$$

By normalising the $\wp$-function we obtain a meromorphic function $x(\tau, z)$ on $\mathbf{E}$ which when restricted to each fibre gives rise to a function $x_\tau : E_\tau \to \mathbb{P}^1$, having these four points as branch points and with corresponding branch values

$$x_\tau(s_1(\tau)) = \infty, \quad x_\tau(s_2(\tau)) = 1, \quad x_\tau(s_3(\tau)) = 0, \quad x_\tau(s_4(\tau)) = \lambda(\tau).$$

Finally, the congruence group $\Gamma(2)$ can be characterised as the group of matrices A such that the corresponding mapping classes f_A introduced above preserve each of these four points; and $SL(2,\mathbb{Z})/\Gamma(2)$ is isomorphic to the subgroup stabilising s_1 of the symmetric group Σ_4 which permutes the $\{s_j\}$. This description will become relevant later on.

4. Moduli of Galois coverings of $\mathbb{P}^1$: higher genus λ-functions.

In this section we generalise the preceding one to the situation where instead of the elliptic involution one has a surface S_0 of genus $g \geq 2$, possessing a group of automorphisms H_0 such that the quotient surface is $\mathbb{P}^1$.

Let S be a Riemann surface occurring in $T_g(H_0)$ with automorphism group H as in §1; thus the projection $S \to S/H = \mathbb{P}^1$ is ramified over points $b_1, \ldots, b_r \in \mathbb{P}^1$. We assume that $r \geq 4$, so that $T_g(H_0)$ has dim> 0. Let β be the unique Möbius transformation which sends b_1, b_2, b_r to $0, 1, \infty$ respectively. The λ-function we shall associate to S is defined by

$$\lambda_1 = \beta \circ \phi(b_{r-1}), \ \ldots, \ \lambda_{r-3} = \beta \circ \phi(b_3).$$

To make this definition unambiguous, we shall proceed more formally by considering the *Teichmüller curve* $\pi : V_g \to T_g$. This is a holomorphic fibre space over T_g whose fibre over a point $t = (S_t, \theta_t) \in T_g$ is a Riemann surface S_t homeomorphic to S_0. The group H_0 also acts as a group of biholomorphic self-mappings of V_g, and the fibre S_t over each point of $T_g(H_0)$ is preserved by H_0: the action of H_0 in the fibre S_t is obtained by conjugating H_0 with the homeomorphism $\theta_t : S_0 \to S_t$ to obtain the group $H_t = \theta_t H_0 {\theta_t}^{-1}$.

We write $V_g(H_0)$ for the restriction of the Teichmüller curve lying over $T_g(H_0)$. Our aim now is to construct a holomorphic morphism Φ from the family $V_g(H_0)$ onto $\overline{V}_{0,r}$, the Teichmüller family of r-fold pointed spheres, which restricts on each fibre of π to the projection $\phi_t : S_t \to \mathbb{P}^1_t = S_t/H_t$. This requires an enumeration of the ramification points of the ϕ_t that depends analytically on $t \in T_g(H_0)$. To this end we reformulate Theorems A and B as in [14] in terms of uniformisation and Fuchsian group theory. We use as reference point the surface $S_0 = U/\Gamma_0$ with automorphism group H_0, so that there is an exact sequence of groups

$$1 \longrightarrow \Gamma_0 \xrightarrow{\iota} \Gamma \longrightarrow H_0 \longrightarrow 1 \tag{4.1}$$

with Γ Fuchsian, cocompact and given by the following presentation:

$$\Gamma = \langle X_1, \ldots, X_r : \ X_1 X_2 \ldots X_r = Id, \ {X_j}^{n_j} = Id \rangle;$$

here the X_j are elliptic Möbius transformations of U of order n_j, each fixing a point whose Γ-orbit is a ramification point b_j of the projection $\phi_0 : \ S_0 \to \mathbb{P}^1 = U/\Gamma$. The inclusion $\iota_* : T(\Gamma) \hookrightarrow T(\Gamma_0)$ identifies the image set $T_g(H_0)$ with the space $T(\Gamma) \cong T_{0,r}$ as in Theorem B of §1, and there is a corresponding identification between the Bers fibre space $F_g(H_0)$, which is the universal covering of $V_g(H_0)$, and the Bers space $F(\Gamma)$ for $V(\Gamma)$;[2] note that $V(\Gamma) \cong \overline{V}_{0,r}$. Following the methods of Bers, Earle and Kra, we observe that the operation of conjugation by any generator X_j defines an automorphism α_j of the space $F(\Gamma)$. Each α_j lies in the *extended modular group*, $mod\,\Gamma$, defined by lifting to $F(\Gamma)$ the action of $Mod\,\Gamma$ on $V(\Gamma)$; an exact sequence

$$1 \longrightarrow \Gamma \longrightarrow mod\,\Gamma \longrightarrow Mod\,\Gamma \longrightarrow 1 \tag{4.2}$$

describes the relationship between these groups (see [7], [22] for more details). Because this action is holomorphic the fixed point set of $\langle \alpha_j \rangle$ is a complex submanifold A_j of $F(\Gamma)$ identified with $F_g(H_0)$ as above. Furthermore A_j intersects each fibre $U_t = \widetilde{S}_t$ of the Bers fibration $\tilde{\pi} : F_g \to T_g$ lying over $t \in T_g(H_0)$ transversely in precisely one point – the fixed point of the deformed group element $X_j(t) \in \Gamma_t$ acting on U_t. This permits us to define a marked set of r disjoint sections $s_1(t), \ldots, s_r(t)$ of the family $V(\Gamma) \to T(\Gamma)$, which induces a rule for labelling the ramification set

$$b_t = \{s_1(t), \ldots, s_r(t)\} \subseteq \mathbb{P}^1_t \qquad \text{for } t \in T_g(H_0).$$

and a mapping $\Phi : V_g(H_0) \to \overline{V}_{0,r}$ as desired. It follows from our construction that Φ is H_0-invariant.

PROPOSITION 2. *The mapping Φ induces a biholomorphic isomorphism of Teichmüller families between $V_g(H_0)/H_0$ over $T_g(H_0)$ and $\overline{V}_{0,r}$ over $T_{0,r}$.*

PROOF. By the theory in [7] of holomorphic sections of $V(\Gamma) \to T(\Gamma)$, it follows that Φ is holomorphic and locally injective. Surjectivity is shown by constructing for each point of $V_{0,r}$ via the ramification data of $\Phi : S_0 \to R_0$ an essentially unique representation of a marked Riemann surface over $\mathbb{P}^1$, which lies in the space of quasi-conformal Fuchsian deformations of the sequence (4.1). Passage from local isomorphism to global, valid because $T_g(H_0)$ is a cell, completes the argument. ∎

Next we employ the Möbius action of the group $PSL_2(\mathbb{C})$ on $\mathbb{P}^1$ to transform $\overline{V}_{0,r}$ into the biholomorphically equivalent family of r-pointed spheres over the base $T_{0,r}$ where each fibre $\mathbb{P}^1$ has $s_1 = 0$, $s_2 = 1$, $s_r = \infty$. Let x_t denote the normalised co-ordinate function obtained from the map Φ by restriction to the fibre over t: here we emphasize again the relationship

---

[2] although $V(\Gamma)$ and $V_g(H_0)$ are different

with the (§3) elliptic curve case, where $x = x_\tau$ is the affine variable for the Legendre normal form

$$y^2 = x(x-1)(x-\lambda), \qquad \lambda = \lambda(\tau).$$

We can now introduce the *higher genus* $\underline{\lambda}$*-function* associated to (S_0, H_0).

DEFINITION. The functions $\lambda_j : T_g(H_0) \to \mathbb{C} - \{0,1\}$ are given by the rule

$$\lambda_j(t) = x_t(s_{r-j}(t)), \qquad j = 1, \ldots, r-3.$$

They determine a holomorphic function $\underline{\lambda}$ whose image lies in Ω^{r-3}, the complement in $\mathbb{C}^{r-3}$ of the normalised diagonal subset

$$\Delta_* = \{\lambda_j = 0 \text{ or } 1, \text{ for some } j\} \cup \{\lambda_j = \lambda_i,\ i \neq j\}.$$

In fact, the above construction of the map Φ has a further implication.

PROPOSITION 3. *The function* $\underline{\lambda}$ *is surjective.*

PROOF. Given a point $\underline{\omega} \in \mathbb{C}^{r-3}$ with $\omega_i \neq 0, 1$ and $\omega_i \neq \omega_j$ let $\mathbf{P}_\omega = \mathbb{C} - \{0, 1, \omega_1, \ldots, \omega_{r-3}\}$ and let $S = S_\omega$ be the Riemann surface obtained as the covering of $\mathbf{P}_1$ smooth over $\mathbf{P}_\omega$ and ramified at each excluded point according to the data of the covering $\phi : S_0 \to S_0/H_0 \cong \mathbf{P}^1$ together with the bijection (implicit in the discussion above) between the excluded set and the ramification subset $\boldsymbol{b}$ of S_0/H_0. Then there is a homeomorphism $\theta : S_0 \to S$ which furnishes a marking and determines a point $t = [S, \theta] \in T_g(H_0)$ whose $\underline{\lambda}$-image is ω. ∎

Recalling the definition (§1) of the relative modular group $Mod_g(H_0)$, we apply the isomorphism described in ([20],§4) to identify this group, in the context of (4.1), with a subgroup of the *algebraic modular group* $Out\,\Gamma$. Each H_0-symmetric mapping class determines a class of homeomorphisms of U, compatible with the group Γ, and inducing an automorphism of Γ which preserves the subgroup Γ_0 defining the surface S_0. The group Γ contains all the lifts of elements of H_0 to U, the universal covering of S_0, and there is, analogous to the sequence of groups in genus 1

$$1 \longrightarrow \langle J \rangle \hookrightarrow SL_2(\mathbb{Z}) \longrightarrow PSL_2(\mathbb{Z}) \longrightarrow 1,$$

an exact sequence for $Mod(\Gamma, \Gamma_0)$, the *effective* relative modular group,

$$1 \longrightarrow H_0 \hookrightarrow Mod_g(H_0) \longrightarrow Mod(\Gamma, \Gamma_0) \longrightarrow 1. \tag{4.3}$$

which is operating on $T(\Gamma)$. Thus $Mod(\Gamma, \Gamma_0) \subseteq Mod\,\Gamma$.

Regarding activity in the (isomorphic) space $T_{0,r}$ we let $P_{0,r}$ denote the subgroup of *pure mapping-classes* in $Mod_{0,r}$, which are those fixing each of the r distinguished points of $\mathbb{P}^1$. This group, which forms the analogue of the group $P\Gamma(2) = \Gamma(2)/\langle J\rangle$ from genus 1, regulates construction of modular families of coverings of $\mathbb{P}^1$ ramified over r points.

PROPOSITION 4. *$P_{0,r}$ is isomorphic, via the identification of Proposition 2, to a subgroup $P_g(H_0)$ of $Mod(\Gamma, \Gamma_0)$.*

PROOF. By covering surface theory, each mapping class in $P_{0,r}$ can be lifted to a mapping class on the surface S; any two liftings are related by composition with the Galois group H_0 of $S_0 \to \mathbb{P}^1$. A lifted homeomorphism is homotopic to an element of H_0 if and only if the original map is the identity on $\mathbb{P}^1$. The result now follows from the exact sequence (4.3). ∎

We are ready to establish the crucial property of the $\underline{\lambda}$-function.

THEOREM 3. *The function $\underline{\lambda}$ is automorphic with respect to the group $P_g(H_0)$ and induces a holomorphic isomorphism between $T_g(H_0)/P_g(H_0)$ and Ω^{r-3}.*

PROOF. The space $T_{0,r}$ is a cell and the function $\underline{\lambda} : T_g(H_0) \to \Omega^{r-3}$ is surjective, holomorphic and non-degenerate. It follows that $\underline{\lambda}$ is the universal covering projection, and tracing through the definitions one finds that the group $P_g(H_0)$ is the Poincaré representation of $\pi_1(\Omega^{r-3})$ acting on its universal covering. ∎

NOTES. 1) An alternative proof follows from that of Theorem 5.6 in [10].

2) Closer study reveals a connection between $P_g(H_0)$ and the *braid groups* $B_r(\mathbb{P}^1)$ and $B_r(\mathbb{C})$. In fact there is a surjective homomorphism from Artin's braid group $B_r = \pi_1(\mathbb{C}^r - \Delta)$ to $P_g(H_0)$, which factors through the well-known homomorphism onto $B_r(\mathbb{P}^1)$.

Next we construct a holomorphic family of Riemann surfaces, with base the finite covering $\widetilde{\mathcal{M}}_g^{pure}(H_0)$ of $\widetilde{\mathcal{M}_g}(H_0)$ associated with the pure modular subgroup $P_g(H_0)$, whose fibres exhibit the ramification characterised by the model covering $S_0 \to S_0/H_0$. This family will therefore carry the distinguishing mark of a *coarse modular family* for H_0 actions: every Riemann surface with symmetry group topologically conjugate to H_0 is represented at least once and at most $m = m(H_0)$ times, where m is the index of $P_g(H_0)$in

$Mod_g(H_0)$. As a preliminary, denote by $\mathcal{C}_{0,r}$ the family of r-pointed projective lines $\mathbb{P}^1$ over Ω^{r-3}.

THEOREM 4. *There is a holomorphic family $\mathcal{C}_g(H_0)$ over the manifold $\widetilde{\mathcal{M}}_g^{pure}(H_0)$ and a morphism of holomorphic families $\mathcal{C}_g(H_0) \to \mathcal{C}_{0,r}$ with the property that, over each point $[S_t, H_t]$ of $\widetilde{\mathcal{M}}_g^{pure}(H_0)$, the fibre of the family $\mathcal{C}_g(H_0)$ is a Riemann surface isomorphic to a Galois covering of $\mathbb{P}^1$ with group H_0 and unramified over $\mathbb{P}_\omega$, where $\omega = \underline{\lambda}(t)$, in the fibre $\mathbb{P}^1$ of $\mathcal{C}_{0,r}$.*

PROOF. We operate within the extended action of $mod\,\Gamma_0$ on $F(\Gamma_0)$. The total space $\mathcal{C}_g(H_0)$ will be achieved as the quotient of the fibre space $V_g(H_0)$ by the action of $P_g(H_0)$, obtained by lifting from the base $T_g(H_0)$; this is seen more clearly as the result of first lifting $P_g(H_0)$ to a subgroup of $mod\,\Gamma_0$ acting on $F_g(H_0)$ and then passing to the quotient $V_g(H_0)$ via the action of Γ_0. Therefore, parallel with the sequence (4.2), we have a sequence

$$1 \longrightarrow \Gamma_0 \longrightarrow \widetilde{P}_g(H_0) \longrightarrow P_g(H_0) \longrightarrow 1 \tag{4.4}$$

and the middle group acts discretely and without fixed points on $F_g(H_0)$, to produce a quotient manifold $\mathcal{C}_g(H_0)$. This is fibred over $\widetilde{\mathcal{M}}_g^{pure}(H_0) \simeq \Omega^{r-3}$, with fibres the compact genus g surfaces $S_t \cong U_t/\Gamma_t$. The mapping Φ of Proposition 2 furnishes the rest of the statement. ∎

NOTES. 1) From this viewpoint, the finite group H_0 acts holomorphically fibrewise on the complex manifold $\mathcal{C}_g(H_0)$ with fixed point sets given by the H_0-orbits of the r disjoint codimension-1 complex subvarieties, each isomorphic to the base manifold $\widetilde{\mathcal{M}}_g^{pure}(H_0)$ or a finite cover.

2) If instead of $P_g(H_0)$ one employed the full relative modular group, $Mod_g(H_0)$, the result would no longer be a family of genus g surfaces over the corresponding base variety $\mathcal{M}_g(H_0)$. This situation occurs already in genus 1 [26].

The particular case in which $H_0 \cong \mathbb{Z}_p$ is of prime order was thoroughly analysed in [10] using the theory of the Riemann theta function. That approach works equally in the present situation with certain modifications. We describe this alternative, more explicit, construction of the $\underline{\lambda}$ function briefly, as a concrete illustration of the relationship between special values of the Riemann theta function and the uniformisation of this particular type of Riemann surface.

By the Bers construction of the holomorphic bundle over T_g of abelian differentials, it is possible to give a Teichmüller space formulation of the *period mapping*, which associates to a marked Riemann surface its period matrix in the standard normalisation, as a mapping into the Siegel space $\mathfrak{S}_g$ equivariant with respect to the actions of Mod_g and the symplectic modular group $Sp(2g, \mathbb{Z})$. A fundamental result of Earle [6] exploits the notion of Riemann constant to produce a holomorphic embedding of the Teichmüller-Bers universal curve V_g into the holomorphic fibre space $\mathcal{J}(V_g)$ of Jacobi varieties which results from pulling back to T_g the fibre bundle of complex tori defined over $\mathfrak{S}_g$. Each projection $\phi_t : S_t \to \mathbf{P}^1$, $t \in T_g(H_0)$, determines two ramification divisors, $\mathfrak{b}_t$ on $\mathbf{P}^1$, $\mathfrak{B}_t$ on S_t, such that $\mathfrak{B}_t = \phi_t^{-1}(\mathfrak{b}_t)$, with every point $s_i(t) \in \mathfrak{b}_t$ yielding an inverse image set $\phi_t^{-1}(s_i(t)) = \{B_{ik}(t),\ k = 1, \ldots, \ell_i\}$ where ℓ_i divides the order of H_0 and each point $B_{ik}(t)$ occurs with multiplicity $n_i = n/\ell_i$. Thus we have a prime decomposition, suppressing dependence of the B_{ik} on t,

$$\mathfrak{B}_t = \mathfrak{B}_1^{n_1} \cdot \mathfrak{B}_2^{n_2} \cdots \mathfrak{B}_r^{n_r} \qquad \text{with} \quad \mathfrak{B}_i = \prod_{k=1}^{\ell_i} B_{ik}.$$

Now it is possible to choose (see [10] for an explicit procedure using the ramification set if $H_0 = \mathbb{Z}_p$) a degree $(g-1)$ divisor D_t on each surface $[S_t] \in T_g(H)$ so that all the products $B_{ik} \cdot D_t$, $i = 1, \ldots, r$, are non-special on every S_t. Note that the divisor D_t may *not* in general be supported in the ramification set of the group H_t, as in [10], proposition 2-1, but a holomorphic choice of $B_{ik} \cdot D_t$ *can* be made, since the (degree g) special divisors on S are a subvariety of the symmetric product $S^{(g)}$ of codimension at least 2 ([21], p. 155). This leads, *via* the embedding in $\mathcal{J}(V_g)$ of these product divisors and the vanishing properties of the corresponding Riemann theta functions, to the following alternative version of Proposition 2, which provides a definition of the basic x-co-ordinate function on the family of surfaces $V_g(H_0)$.

PROPOSITION 2′. *There is a collection of rational theta characteristics* $[a_{ik}, a'_{ik}], [b_{ik}, b'_{ik}]$, *one for each* B_{ik} *in the ramification divisor* $\mathfrak{B}_i$, *such that the meromorphic functions* $\phi_i : V_g(H_0) \to \mathbf{P}^1$ *defined by*

$$\phi_i(t,z) = \prod_{k=1}^{\ell_i} \left\{ \vartheta \begin{bmatrix} a_{ik} \\ a'_{ik} \end{bmatrix} (t,z) \Big/ \vartheta \begin{bmatrix} b_{ik} \\ b'_{ik} \end{bmatrix} (t,z) \right\} \qquad i = 1, \ldots, r-1$$

determine on each fibre S_t *a function whose divisor is* $\mathfrak{B}_i(t)/\mathfrak{B}_r(t)$. *The normalised* x*-coordinate function on* $V_g(H_0)$ *is then* $\left[\phi_1/\phi_1(s_2(t))\right]^{n_1}$.

In the same way as before, one may then produce a $\underline{\lambda}$-function on $V_g(H_0)$, which is given by quotients of powers of special values of Riemann's theta. It is more problematical, though theoretically possible, to produce a precise algebraic equation for $V_g(H_0)$ as a family of plane curves; for this one needs to construct a companion y-co-ordinate function with coefficients rational in the $\underline{\lambda}$-variable. As a final comment on this approach, the matter of deciding purely from the value of the period mapping what the corresponding surface is (the "Torelli problem" in a strong form) seems to be still open in general; if $H_0 = \mathbb{Z}_p$, then the more explicit form of Proposition 2′ in [10] solves it.

Returning to our main theme, we derive as a consequence of Theorem 3 the virtual holomorphic rigidity of the domain $\mathbb{C}^{r-3} - \Delta_* = \Omega^{r-3}$. First we define a natural action of the symmetric group Σ_r on Ω^{r-3} by projecting the permutation action of Σ_r on the components of $(\mathbb{P}^1)^r$:

$$\mathbb{C}^r - \Delta \hookrightarrow (\mathbb{P}^1)^r \longrightarrow (\mathbb{P}^1)^r/\Sigma_r .$$

This involves the Σ_r-equivariant action of the group $PGL_2(\mathbb{C})$ as Möbius transformations on each component $\mathbb{P}^1$. The quotient of $\mathbb{C}^r - \Delta$ by this action is just Ω^{r-3}. The combined result is an action of the Lie group $G = PGL_2(\mathbb{C}) \wr \Sigma_r$ on $\mathbb{C}^r - \Delta$ which projects to an action of Σ_r on Ω^{r-3} as biholomorphic isomorphisms, free of fixed points.

In the commutative diagram below, the vertical arrows denote quotient by the action of $PGL_2(\mathbb{C})$ on each component. The unramified part of the fibration F is precisely the left hand part of the diagram.

$$\begin{array}{ccc} (\mathbb{C}^r - \Delta)/\Sigma_r & \hookrightarrow & (\mathbb{P}^1)^r/\Sigma_r \\ \downarrow & & \downarrow F \\ \Omega^{r-3}/\Sigma_r & \hookrightarrow & W \end{array}$$

NOTATION. For a domain $\Omega \subset \mathbb{C}^n$, write $Aut(\Omega)$ for the group of biholomorphic self mappings of Ω.

THEOREM 5. $$Aut(\Omega^{r-3}) \cong \Sigma_r .$$

PROOF. Since $T_{0,r}$ is the universal cover of Ω^{r-3}, the familiar theory of covering spaces coupled with local geometric structure (see for instance [19]) implies that

$$Aut(\Omega^{r-3}) \cong Aut(T_{0,r})/P_{0,r} \cdot$$

But by Royden's rigidity theorem (see for instance [9] or [22]), $Aut(T_{0,r})$ is just $Mod_{0,r}$ and the quotient $Mod_{0,r}/P_{0,r}$ is the permutation group of the punctures acting in the manner described above. ∎

In the case of our Galois H_0-covering S_0 of $\mathbf{P}^1$, there is a subgroup of Σ_r which consists of those permutations $\sigma \in \Sigma_r$ which extend the pure modular H_0 action in the sense that there is an element of $Mod_g(H_0)$ which performs the permutation σ on the ramification points of the H_0-action. This group, denoted $\mathcal{G}(H_0)$, is isomorphic to the quotient group $Mod_g(H_0)/P_g(H_0)$. From the present discussion, in view of Theorem 1, we may infer the final result of this paper.

THEOREM 6. $$\widetilde{\mathcal{M}_g}(H_0) \cong \Omega^{r-3}/\mathcal{G}(H_0).$$
Hence both $\widetilde{\mathcal{M}_g}(H_0)$ and $\mathcal{M}_g(H_0)$ are unirational and quasi- projective.

PROOF. The first statement is already clear from the definitions. A unirational variety is just the rational image of a projective space under a rational map so this follows immediately. The quasi-projectivity is a consequence of the general fact that any quotient of a quasi-projective variety by a finite group is again quasi-projective. ∎

5. Final Comments.

1. The permutation group $\mathcal{G}(H_0)$ associated with a finite group of mapping classes has a simple characterisation in terms of lifting properties for homeomorphisms of the quotient space S_0/H_0 ([10], [15]).

Using this it is elementary to produce examples where $\mathcal{G}(H_0)$ is trivial and others for which $\mathcal{G}(H_0) \cong \Sigma_r$; the hyperelliptic involutions discussed previously provide examples in every genus where $\mathcal{G}(H_0) \cong \Sigma_r$ with $r = 2g+2$.

2. It appears to be a difficult problem to understand what an appropriate analogue of the j-function should be; indeed, the question whether such functions exist even in principle is part of the classic Lüroth problem which asks when a unirational variety is in fact rational.

3. If the quotient surface $R_0 = S_0/H_0$ has positive genus g_1, then no picture of the corresponding modular variety $\widetilde{\mathcal{M}_g}(H_0)$ has the same clear pattern. Instead, one must contemplate the (at present inaccessible) nature of the varieties of positive divisors on (variable) surfaces of genus g_1. Some progress may be possible if $g_1 = 1$, however.

4. In this paper we have developed a coarse moduli theory for surfaces with a specified automorphism group. In the longer term, one hopes to include this in a fine moduli theory for algebraic curves defined over $\overline{\mathbf{Q}}$, where it will be necessary to incorporate these ideas with the growing body of work on Galois fields over $\mathbf{Q}$ and the representation theory of braid groups. For a survey of this work, see [24], [25] and articles in the MSRI Volume 16 *Galois groups over* $\mathbf{Q}$. As a central fact which highlights the interaction between Galois groups over $\mathbf{Q}$ and Galois coverings of $\mathbf{P}^1$, the theorem of Belyi [2] may provide a suitable epilogue:

An algebraic curve is defined over some number field (Galois extension over $\mathbf{Q}$*) if and only if it is a covering of* $\mathbf{P}^1$ *with three ramification points.*

References

[1] W. L. BAILY. On the theory of θ-functions, the moduli of abelian varieties, and the moduli of curves. *Ann. of Math.* **75** (1962) 342-381.

[2] G. V. BELYI. Galois extensions of a maximal cyclotomic field. *Math. of U.S.S.R. Izv.* **14** (1980) 247-256.

[3] L. BERS. Uniformization, moduli and Kleinian groups. *Bull. London Math. Soc.* **4** (1972) 250-300.

[4] H. CARTAN Quotient d'un espace analytique par un groupe d'automorphismes. *Algebraic Geometry and Topology,* (Princeton U. Press, 1957) 90-102 or *Collected Works Vol II* (Springer-Verlag, 1979).

[5] C. H. CLEMENS. *A Scrapbook of Complex Curve Theory.* (Plenum Press, N. Y., 1980).

[6] C. J. EARLE. Families of Riemann surfaces and Jacobi varieties. *Ann. of Math.* (1978) 225-286.

[7] C. J. EARLE and I. KRA. On sections of some holomorphic families of closed Riemann surfaces. *Acta Math.* (1976) 49-79.

[8] H. M. FARKAS and I. KRA. *Riemann Surfaces.* (Springer, Berlin, 1980).

[9] F. P. GARDINER. *Teichmüller Theory and Quadratic Differentials.* (John Wiley & Sons, N. Y., 1987).

[10] G. GONZALEZ-DÍEZ. Loci of curves which are prime Galois coverings of $\mathbf{P}^1$. *Proc. London Math. Soc.* **62** (1991) 469-489.

[11] H. GRAUERT and R. REMMERT. *Coherent Analytic Sheaves.* (Springer, Berlin, 1980).

[12] L. GREENBERG. Maximal Fuchsian groups. *Bull. Amer. Math. Soc.* 4 (1963) 569-573.

[13] R. HARTSHORNE. *Algebraic Geometry.* (Springer, Berlin, 1977).

[14] W. J. HARVEY. On branch loci in Teichmüller space. *Trans. Amer. Math. Soc.* **153** (1971) 387-399.

[15] W. J. HARVEY. A lifting criterion for surface homeomorphisms. (Preprint)

[16] G. A. JONES and D. SINGERMAN. *Complex Functions.* (Cambridge Univ. Press, 1987).

[17] S. KERCKHOFF. The Nielsen realisation problem. *Ann. of Math.* **117** (1983) 235-265.

[18] S. KRAVETZ. On the geometry of Teichmüller spaces and the structure of their modular groups. *Ann. Acad. Sci. Fenn.* (1959) 1-35.

[19] A. M. MACBEATH. On a theorem of Hurwitz. *Proc. Glasgow Math. Assoc.* **5** (1961) 90-96.

[20] C. MACLACHLAN and W. J. HARVEY. On mapping-class groups and Teichmüller spaces. *Proc. London Math. Soc. (3)* **30** (1975) 496-512.

[21] D. MUMFORD. *Tata Lectures on theta I.* Progress in Maths. **28** (Birkhaüser, Boston, 1983).

[22] S. NAG. *The Complex Analytic Theory of Teichmüller Spaces.* (Wiley-Interscience, N. Y., 1988).

[23] H. E. RAUCH. A transcendental view of the space of algebraic Riemann surfaces. *Bull. Amer. Math. Soc.* **71** (1965) 1-39.

[24] J-P. SERRE. Groupes de Galois sur **Q**. *Séminaire Bourbaki, Exposé* 689 (November, 1987).

[25] H. VÖLKLEIN. $PSL_2(q)$ and extensions of $\mathbf{Q}(x)$. *Bull. Amer. Math. Soc.* **24** (1991) 145-154.

[26] S. A. WOLPERT. Homology of the moduli space of stable curves. *Ann. of Math.* **118** (1983) 491-522.

Departamento de Matemáticas
Universidad Autónoma de Madrid
28049 Madrid, Spain

Department of Mathematics
King's College London
London WC2R 2LS
Email: udah055@uk.ac.kcl.cc.oak

Modular groups - geometry and physics

W. J. Harvey

To Murray Macbeath on the occasion of his retirement

Introduction

Recent developments in the study of discrete groups associated to Riemann surfaces serve to emphasize once again the importance of geometric ideas in understanding structural properties of the most abstract kind. We consider here two instances of this maxim applied to the mapping-class groups $\mathrm{Mod}_{g,n}$; each occurs within the action as the modular group on the corresponding Teichmüller space, $T_{g,n}$, which classifies marked complex structures on an n-pointed genus g surface. This space carries intrinsically a structure of complex manifold with dimension $3g + n - 3$ and a complete global metric d defined in terms of the least (logarithmic) distortion necessary in deforming one complex structure to another. The modular group action is isometric and serves to identify points of $T_{g,n}$ that represent holomorphically equivalent structures.

The first question we discuss concerns geometric structures of a more concrete type within this metric framework, defined by a particular kind of deformation which one now calls a Teichmüller geodesic disc, a natural class of complex submanifold of $T_{g,n}$ isometric to the Poincaré metric model of the hyperbolic plane; they exist in profusion (through an arbitrary point in any given direction). We present a construction via hyperbolic geometry of some examples of Teichmüller discs which project to finite volume Riemann surfaces immersed in $\mathcal{M}_g$; on suitable finite coverings they form smooth totally geodesic submanifolds, isomorphic to affine plane models of certain familiar algebraic curves.

Our second topic involves a more formal geometric aspect of the modular action, relating to the combinatorial picture of how it extends to the boundary of $T_{g,n}$. This is embodied in a certain simplicial complex $\mathcal{T}_{g,n}$, whose chambers will be shown to define partial presentations for $\mathrm{Mod}_{g,n}$ in terms of braid subgroups. The structural pattern which emerges is closely related to contemporary work in theoretical physics on conformally invariant quantum field theories on $\mathbf{CP}^1$ and projective representations of braid and modular groups.

Many of the results described here are already to be found in the literature; partly for this reason, I have in places chosen to neglect the strict frontiers imposed by rigour in pursuit of the muse of geometry.

1. Geodesic surfaces in Teichmüller space

1.1 Let X_0 be a fixed compact Riemann surface of genus $g \geq 2$; a marking on a genus g surface is a homotopy class of homeomorphisms $f : X_0 \to X$, viewed modulo composition with a conformal map of X. We shall operate in the setting given by Bers [2] which produces models of all the Teichmüller spaces of marked surfaces in the wider context of deformations of Fuchsian groups G within the upper half plane $\mathcal{U}$ by means of the space Q of quasiconformal (q-c) self-homeomorphisms of $\mathcal{U}$ fixing the boundary points $0, 1$ and ∞. Let $Q(G)$ be the set of all $f \in Q$ such that fGf^{-1} is also Fuchsian. For the groups G that concern us, which are of finite type with limit set $\mathbf{R} \cup \infty$, the Teichmüller space $T(G)$ is just the quotient space $Q(G)$ modulo $Q(G) \cap Q_0$, where $Q_0 \subset Q$ is the normal subgroup of maps that fix $\mathbf{R} \cup \infty$.

The extended modular group $mod(G)$ of a Fuchsian group G is the quotient group $N(G)/N(G) \cap Q_0$, where $N(G)$ denotes the normaliser of G in the group $\tilde{Q}$ of *all* q-c maps of $\mathcal{U}$. By the Nielsen theorem on geometric realisation of automorphisms, $mod(G) \cong \mathrm{Aut}(G)$, the group of type-preserving automorphisms of G ; the quotient $mod(G)/G$ is the modular group $Mod(G)$, isomorphic to the mapping class group of the surface $\mathcal{U}/G$.

1.2 DEFINITION. A *Teichmüller disc* is a set of marked complex structures obtained in the following way. Fix a quadratic differential form $\phi \in \Omega^2(X_0)$, regarded as a holomorphic form of weight 4 for a suitable G on $\mathcal{U}$. We define a *Beltrami form* ν_ϕ for G by the rule $\nu_\phi(z) = \overline{\phi(z)}/|\phi(z)|$; this is an element of the Lebesgue space $L^\infty_{-1,1}(\mathcal{U}, G)$ of $(-1,1)$-forms for G, which represent complex dilatations on the surface X_0. For each ε with $|\varepsilon| < 1$, we solve in $\mathcal{U}$ the Beltrami equation $\bar{\partial}w = \varepsilon \cdot \nu_\phi \partial w$. Conjugating the group G by the family w_ε of normalised solutions fixing 0,1 and ∞ yields the requisite holomorphic family of Riemann surfaces over the unit disc Δ. The subset with $0 \leq \varepsilon < 1$ is called the *Teichmüller ray* in the ϕ direction.

Here is an alternative geometric description of this deformation. A type of parametrisation of X_0 by *affine charts* is determined by ϕ: away from the zeros of ϕ, write $\phi = dw^2$ to get local parameters w up to transition functions of the form $w \mapsto \pm w + c$. At a zero of order n, a singular local chart is defined by means of the branched $(n+2)$-fold covering map. Call such an atlas an *$\mathcal{F}$-structure* on X_0.

Now for each ε with $|\varepsilon| < 1$, define a new $\mathcal{F}$-structure on the underlying topological surface S by rotating non-singular charts through $\arg \varepsilon$ and applying the mapping

$$w = x + iy \longmapsto K_\varepsilon^{\frac{1}{2}} x + i K_\varepsilon^{-\frac{1}{2}} y \qquad \text{where} \quad K_\varepsilon = \frac{1 + |\varepsilon|}{1 - |\varepsilon|}.$$

This has the effect of expanding the real (horizontal) foliation of the w-plane while contracting the imaginary (vertical) one. The result of extending the maps to the zeros of ϕ determines the same disc at X_0 in the direction ϕ.

1.3 The Teichmüller deformation defines a holomorphic mapping λ_ϕ of Δ into $T(G)$ which is viewed as a model of $T(X_0)$.

PROPOSITION. *λ_ϕ is a proper holomorphic embedding.*

The mapping is proper because as $|\varepsilon| \to 1$, the L^∞ norm $\|\varepsilon \nu_\phi\| \to 1$ so $\lambda_\phi(\varepsilon)$ tends to the boundary $\partial T(X_0)$. It is injective by Teichmüller's uniqueness theorem, holomorphic because the complex structure on $T(X_0)$ is determined via projection from the complex Banach space $L^\infty_{-1,1}(\mathcal{U}, G)$.

In the next sections, we shall give a recipe for such structures which relates naturally to the hyperbolic geometry of surfaces given in explicit form over $\mathbf{P}^1$ as ramified coverings; the first examples were described from a different viewpoint by W. Veech [15].

1.4 The hyperelliptic curve X_n, where $n > 4$ is odd, is defined by the familiar affine equation

$$y^2 = 1 - x^n,$$

which determines a compact Riemann surface $\overline{X}_n$ of genus $g = (n-1)/2$ with a single point at ∞.

THEOREM 1. *The stabiliser in the modular group* Mod_g *of the Teichmüller disc determined by the differential* $q = \omega_1^2$ *on* X_n *is a Fuchsian triangle group*

$$G_n = \left\langle \begin{pmatrix} 1 & 2\cot(\frac{\pi}{n}) \\ 0 & 1 \end{pmatrix}, \begin{pmatrix} \cos(\frac{2\pi}{n}) & -\sin(\frac{2\pi}{n}) \\ \sin(\frac{2\pi}{n}) & \cos(\frac{2\pi}{n}) \end{pmatrix} \right\rangle = \left\langle \sigma, \beta \right\rangle$$

which is isomorphic to $\langle x_1^n = x_2^2 = 1 \rangle$, *where* $x_1^{g+1} x_2 = \sigma$, $x_1 = \beta$.

These groups G_n are conjugates of the Hecke groups, so named because of his work on them in connection with Dirichlet series having a functional equation. A fundamental polygon for the group G_n may be chosen in $\mathcal{U}$ whose sides are vertical line segments, $L = \{\text{Re}(w) = -\cot\frac{\pi}{n}\}$ and $\sigma(L)$, together with the hyperbolic perpendiculars to them from i. The element x_2 of order 2 fixes the foot of the perpendicular from i to the hyperbolic line L. To produce a copy of the surface X_n, one takes the quotient space of $\mathcal{U}$ by the commutator subgroup $C_n = [G_n, G_n]$, the kernel of a homomorphism from G_n onto the group $\mathbf{Z}_2 \oplus \mathbf{Z}_n$ representing the relevant automorphism group. The surface so obtained is identified as X_n by recognising the hyperelliptic involution: this is induced by conjugation with x_2 on $\mathcal{U}/C_n$.

The proof is completed by showing that if we view $\mathcal{U}$ as the Teichmüller disc through $\overline{X}_n$ given by the differential $q = {\omega_n}^2$, where $\omega_n = dx/y$, some power of the parabolic element σ determines a product of Dehn twists about loops in X_n defined by representative closed vertical trajectories of q, one for each of the $g = (n-1)/2$ cylinders of the $\mathcal{F}$-structure. Each loop shrinks in length as one follows the ray from i, which represents the Teichmüller point $[\overline{X}_n]$, to the cusp at ∞, according to our earlier alternative description of Teichmüller ray. Putting this together with the torsion element x_1 in G_n which fixes i, we see that the stabiliser in Mod_g of this complex disc is just G_n, since this group is maximal among Fuchsian groups.

COROLLARY. *The quotient in* $\mathcal{M}_g$ *of the Teichmüller disc through* $\overline{X}_n$ *corresponding to* q *is isomorphic to* $X_n/\text{Aut}(X_n) = \mathcal{U}/G_n$.

1.5 We may apply a similar method to the Fermat curve F_n,

$$u^n = 1 - v^n \qquad (n \geq 4)$$

which determines a closed Riemann surface $\overline{F}_n$ of genus $\gamma = (n-1)(n-2)/2$ with n points at ∞. It possesses a large symmetry group of order $6n^2$; the quotient surface is a projective line with three ramification points of orders 2,3 and $2n$. From the point of view of uniformisation, there exists a representation of the Fuchsian triangle group Γ_n with these periods onto the group Aut$\overline{F}_n$, with kernel K_n isomorphic to $\pi_1(\overline{F}_n)$.

There is a related tower of coverings (and subgroups) which reflects the relationship between F_n and a certain n-sheeted quotient surface $Y_n = \mathcal{U}/H_n$ which is hyperelliptic; this depends on the parity of n and for simplicity we again restrict to the case of odd n, for which one has

$$K_n \overset{n}{\triangleleft} H_n \overset{2n}{\triangleleft} \{2, n, 2n\} \overset{3}{\triangleleft} \Gamma_n.$$

The surface Y_n has genus $(n-1)/2$ by the Riemann-Hurwitz branching formula and, by virtue of its evident symmetries, coincides with the surface $\overline{X}_n$ of §1.4. By direct computation of a rational mapping between these surfaces one shows that the holomorphic 1-form ω_n on $\overline{X}_n$ pulls back to the form $\Omega_n = dv/u^{n-1}$ on $\overline{F}_n$, which has zeros of order $n-3$ at each of the n points at infinity. Now the trajectory structure of Ω_n on $\overline{F}_n$ is determined by that of ω_n on $\overline{X}_n$ and the monodromy of the smooth covering projection. Arguing as before, or using the inclusion of $T(Y_n)$ in $T(\overline{F}_n)$, we obtain an analogous result for the modular image of $\mathcal{U}$.

COROLLARY. *The quotient in $\mathcal{M}_\gamma$ of the Teichmüller disc through $\overline{F}_n$ corresponding to Ω_n^2 is isomorphic to $F_n/\mathrm{Aut}(F_n)$.*

2. Braids and modular groups

2.1 As one knows from topology, the modular group acts on a wide variety of spaces; these actions have appeared in various parts of theoretical physics, often as a foil for some involvement of Teichmüller theory. What seems rather more mysterious is the emergence of certain representations of modular groups in conjunction with braids in quantum physics and the Yang-Baxter equations, connected furthermore via the ideas of E. Witten to the construction of polynomial knot invariants by V. Jones and others. This latter link brings us back to familiar ground as there are well-trodden paths between mapping classes and 3-dimensional manifolds by way of mapping-tori or Heegard diagrams, but the new route via conformal field theory and quantum physics, which we discuss in §3, appears here as something of a Magical Mystery tour.

I shall describe a venerable technique, with origin in work of Poincaré on Fuchsian groups, for generating certain geometric presentations for modular groups, which exhibit a braid-like nature. These have occurred (albeit in more or less concealed form) in recent and earlier work (see [4], [6]), but appear most clearly from the action of modular groups at the boundary of their corresponding Teichmüller spaces.

2.2 We recall the simplicial method for group presentation employed by A. Weil and by A. M. Macbeath in the study of lattice subgroups of Lie groups; see [10], [13] for details.

Let X be a simply connected topological space on which a group Γ acts, and let U be an open set, connected and simply connected, such that X is covered by the union of all translates γU, $\gamma \in \Gamma$. Let Σ denote the set of those γ for which $\gamma U \cap U$ is non empty. If x_γ denote generating elements of the free group $F(\Sigma)$ and $E \subset \Sigma \times \Sigma$ is the subset comprising γ, δ such that $U \cap \gamma U \cap \delta U \neq \emptyset$, then by a theorem of Macbeath there is a natural surjection $\phi : F(\Sigma) \to \Gamma$ whose kernel is the normal closure of the subgroup generated by the elements $x_\gamma x_\delta x_{\gamma\delta}^{-1}$, $(\gamma, \delta) \in E$. Application of this result to the case where X is a simplicial complex on which the group Γ acts simplicially with finite quotient yields a presentation of Γ with a simple combinatorial description. Namely, let $W \subset X$ be a finite subcomplex which contains precisely one cell for each Γ-orbit of cells in X and let $\mathcal{V}(W)$, $\mathcal{E}(W)$ be the (finite) sets of vertices and edges of W. Denote by Γ_v $(v \in \mathcal{V})$ and Γ_e $(e \in \mathcal{E})$ the stabilisers in Γ of v and e respectively.

THEOREM 2. *Assume that X is simply connected and that the action of Γ preserves orientation on the edges $\mathcal{E}(W)$. Then Γ is isomorphic to the sum of the groups Γ_v amalgamated on the subgroups Γ_e.*

2.3 This result will be applied to the simplicial action of the modular group $\mathrm{Mod}_{g,n}$ on the complex $\mathcal{T}_{g,n}$ introduced in [5] and studied since by various authors, including J. Harer [4] and N. V. Ivanov [7].

The *complex of closed loops*, $\mathcal{T}_{g,n} = \mathcal{T}(S_{g,n})$ associated to a topological surface $S = S_{g,n}$ is the geometric realisation of the partially-ordered set $\mathcal{L}$ formed by all systems Λ of (homotopy classes of) simple closed loops in $S_{g,n}$, ordered by inclusion. A maximal set Λ in $\mathcal{L}$ contains $3g - 3 + n$ loops and determines a *decomposition graph* Υ for S. These trivalent graphs consist of a vertex for each connected component (3-holed sphere) of $S \setminus \Lambda$ with an edge for each connection, by a loop in the set Λ, to another (or the same) component. The barycentric subdivision $\mathcal{T}'_{g,n}$ has a grading on vertices by the number of loops in the corresponding (ordered) set Λ.

It is known that $\mathcal{T}_{g,n}$ is connected and simply connected if the dimension $3g - 4 + n$ is at least 2. Furthermore the subcomplex consisting of loop systems that do not disconnect S is an equivariant retract of $\mathcal{T}_{g,n}$ and there is a homotopy equivalence between the 2-skeleton of $\mathcal{T}'_{g,n}$ and the complex $\mathcal{X}$ used in [6] to describe a presentation of Mod_g. Prominent in the structure of $\mathcal{X}$ is the class of 2-cells (pentagons) which correspond to pentagonal bracelets; these are 5-cycles of loops, each intersecting the next in a single point, lying in a subsurface $S_{2,*} \subset S$. Following a path through a sequence of chambers in the complex $\mathcal{T}'_{g,n}$ involving such a cycle of *fusion* moves – deletion of a loop and insertion of its successor which intersects it in one point while remaining disjoint from all others in the loop system – defines a closed path in the complex which is spanned by a pentagon in $\mathcal{X}$ or simplicial 2-cycle in $\mathcal{T}'_{g,n}$. Together with less intricate bracelet types of shorter length, these cells make up the whole complex $\mathcal{X}$.

The modular group permutes the set $\mathcal{L}$ and induces actions on all these complexes. A finite fundamental subcomplex exists in each case (after barycentric subdivision). In view of theorem 2 above, the structure of the group $\mathrm{Mod}_{g,n}$ is determined by amalgamating the vertex stability groups Γ_v over the edge groups Γ_e as v, e run through the finite sets of top-graded vertices and edges corresponding to fusions.

2.4 By an induction argument on grading it is possible to deduce that the structure of $\mathrm{Mod}_{g,n}$ is built up from that of certain subgroups of the classical modular group by iterated amalgamations. Rather than study that pattern, we outline a different aspect which emerges from analysing the subcomplex of $\mathcal{T}_{g,n}$ comprising chambers with a fixed type of decomposition graph Υ. This is a connected, simply connected subcomplex on which $\mathrm{Mod}_{g,n}$ acts, with an interesting structure in its own right.

The following result is indicative of the underlying braid-like character of modular groups.

THEOREM 3. *Let $\sigma = \sigma_\Lambda$ be a chamber of $\mathcal{T}_{g,n}$. There is an associated presentation of* $\mathrm{Mod}_{g,n}$ *which is given by a surjection $\Phi : B(\sigma) \to \mathrm{Mod}_{g,n}$, where the group $B(\sigma)$ is the quotient of the free group $\mathcal{F}_\Lambda$ generated by symbols $\{s_\ell, t_\ell;\ \ell \in \Lambda\}$, with braid relations*

$$s_\ell t_\ell s_\ell = t_\ell s_\ell t_\ell \quad \text{for every non-separating (n-s) } \ell \in \Lambda,$$
$$t_\ell, s_\ell \text{ commute with } t_m, s_m \quad \text{for every pair } \ell, m \in \Lambda.$$

The proof rests on elementary facts about Dehn twists. We show that the set

$$\Sigma(\sigma_\Lambda) = \{\gamma \in \mathrm{Mod}_{g,n}; \sigma_\Lambda \cap \gamma\sigma_\Lambda \neq \emptyset\}$$

contains a subset of the requisite type. Observe that any n-s loop ℓ in Λ has a companion loop $\ell' \in \Sigma$ with the property that $\ell' \cap \ell = \ell' \cap \Lambda$ is a single point. The corresponding twists t_ℓ and s_ℓ are known to satisfy the above relations in $\mathrm{Mod}_{g,n}$. Furthermore, this set of $6g-6+2n$ twists is sufficient to generate $\mathrm{Mod}_{g,n}$ by a modification of standard results on generating sets, due to Dehn and Lickorish and refined by Humphries.

The kernel of Φ is dictated by the 2-cells of $\mathcal{X}$.

3. Geometry and physics

In this final section we comment briefly on some of the implications and further developments of these results.

3.1 A harbinger of the current interaction between Riemann surfaces and theoretical physics was the picture developed during the 1960's by particle physicists (including Koba, Nielsen, Alessandrini and Mandelstam) of light cone diagrams for a moving string, which sweeps out a closed surface S in space-time, generating non-trivial topology by dividing and/or rejoining. The built-in singular foliation of S by level curves of time represents a closed 1-form, which can be rendered holomorphic by suitable choice of complex structure for S. In this connection, the particular examples described in §1 correspond to a special part of some decomposition of the Teichmüller space into cells representing the various choices of combinatorial pattern possible for such string diagrams on a genus g surface. For further discussion of this theory see for instance [3].

3.2 A further application of the Teichmüller deformation described in §1 lies in the connection with Euclidean polygons and their dynamical properties, described in [8], [15]; here it is worth mentioning that Veech's work demonstrates a close relationship between the analytic theory of Eisenstein series for Hecke groups and the asymptotic behaviour of the length spectrum for periodic billiard trajectories in certain isosceles triangles and regular n-gons. For instance he shows that the zeta function of the simple length spectrum for the affine q-structure on X_n is a weighted sum of Eisenstein series, whence by a Tauberian theorem the growth rate is quadratic.

3.3 Finally, we indicate the relationship between §2 and quantum field theory. A further complex $\mathcal{Y} = \mathcal{Y}_{g,n}$ on which the modular group $\mathrm{Mod}_{g,n}$ acts was introduced by Moore and Seiberg [11] in their work on the construction of conformal field theories (see also [9] and [14]); this complex is closely related to $\mathcal{X}$. A method is given for constructing projective representations of $\mathrm{Mod}_{g,n}$ by means of the standard spin representations of the Lie algebra $sl_2(\mathbf{C})$, which are used as labels on the edges of the decomposition graphs

Υ. These labellings are the index sets for the basis elements of finite dimensional vector spaces $V(\Upsilon)$ (known as *conformal blocks*) on which $\text{Mod}_{g,n}$ acts by combining its permutation action on the simple loops and boundary components of $S_{g,n}$ with the various label representations. We note two important implications for this theory of results mentioned in §2.

Firstly, as a consequence of the connectedness of $\mathcal{Y}$, which follows from that of $\mathcal{X}$, the vector spaces $V(\Upsilon)$ constructed (though not the representations) are independent of the graph Υ employed, and one obtains in this way projective representations of modular (and braid) groups, because there are natural isomorphisms $\Phi_{\Upsilon\Upsilon'} : V(\Upsilon) \to V(\Upsilon')$ for each Υ, Υ' implied by the simple connectedness of $\mathcal{Y}$.

Secondly, there is a deep implicit connection with the study of polynomial invariants for knots and 3-manifolds; this is explained in particular by Kohno in a recent Nagoya preprint via Heegard diagrams. The Lincei lectures by Atiyah [1] describe Witten's approach to this in some detail; also discussed is a further important step (due to G. B. Segal) towards a mathematical framework for QFT that involves assembly of all the requisite Hilbert spaces and unitary operators (one for every Riemann surface) into a *modular functor*. The natural relationships necessary between the various entities are synthesized into a categorical formulation which is then shown [12] to produce precisely the projective bundles over (all) moduli spaces of Riemann surfaces constructed more painfully by other authors.

References

[1] M. F. Atiyah. *Geometry and Physics of Knots.* (Cambridge University Press 1990).

[2] L. Bers. Fiber spaces over Teichmüller spaces. *Acta Math.* **130** (1973) 89-126.

[3] S. B. Giddings, S. A. Wolpert. A triangulation of moduli space from light-cone string theory. *Commun. Math. Phys.* **109** (1987) 177-190.

[4] J. Harer. The cohomology of the moduli space of curves. *Lecture Notes in Math.* **1337** (Springer 1988) 138-221.

[5] W. J. Harvey Boundary structure of the modular group. *Ann. of Math. Studies* **97** (Princeton Univ. Press 1981) 245-251.

[6] A. Hatcher and W. P. Thurston. A presentation for the mapping class group of a closed orientable surface. *Topology* **19** (1980) 221-237.

[7] N. V. Ivanov. Complexes of curves and the Teichmüller modular group. *Russian Math. Surveys* **42** (1987) 55- 107.

[8] S. KERCKHOFF, H. MASUR, J. SMILLIE. Ergodicity of billiard flows and quadratic differentials. *Ann. of Math.* **124** (1986) 293-311.

[9] T. KOHNO. Monodromy representations of braid groups and Yang-Baxter equations. *Ann. Inst. Fourier* **37** (1987) 139-160.

[10] A. M. MACBEATH. Groups of homeomorphisms of a simply connected space. *Ann. of Math.* **79** (1964) 473-488.

[11] G. MOORE and N. SEIBERG Classical and quantum conformal field theory. *Comm. Math. Physics* **123** (1989) 177-254.

[12] G. B. SEGAL. Conformal field theories and modular functors. *Proc. IXth Int. Cong. of Math. Phys.* (IOP Publ. 1989) 22-37.

[13] C. SOULÉ. Groupes opérant sur un complexe simplicial. *C. R. Acad. Sci. Paris* (A) **276** (1973) 607-609.

[14] A. TSUCHIYA, K. UENO, Y. YAMADA. Conformal field theory on universal family of stable curves. *Adv. Stud. Pure Math.* **19** (1989) 459-566.

[15] W. VEECH. Teichmüller curves in moduli space,.... *Invent. Math.* **97** (1989) 553-583.

Department of Mathematics
King's College London
Strand, London WC2R 2LS
Email: udah055@uk.ac.kcl.cc.oak

On automorphisms of free products

A. H. M. Hoare and M. K. F. Lai

for Murray Macbeath

1. Introduction

Whitehead [8] represented elements of a free group by embedding surfaces in an orientable 3-manifold and represented the action of an automorphism by changing the surfaces. Combinatorial versions of his results were given by Rapaport [7] and by Higgins and Lyndon [3]. These were further developed by McCool [6] and a graphical version of his results is given in [4]. More recently Whitehead's work was extended by Gersten [2] to subgroups of a free group and by Collins and Zieschang [1] to elements of a free product.

In this paper we prove the Peak Reduction theorem of [1] for subgroups of a free product. The notation used differs slightly from that in [1] for reasons that will be apparent. We use a concept from [2] and an action by automorphisms on coset graphs similar to that in [5] and dual to that in [4]. This action is also similar to some ideas of Wicks [9]. The authors are grateful to Don Collins for useful comments and criticisms.

2. Preliminaries.

Let G be a group with identity e. Let X be a set of generators for G closed under taking inverses. If H is a subgroup of G then the based coset graph (Γ, H) has for vertices the right cosets of H, with H itself being the base vertex, and has for each x in X a directed edge labelled x joining K to Kx for each coset K. The inverse of each such edge is the edge labelled x^{-1} from Kx to K. The labelling, λ say, of the edges induces for each vertex K a one to one correspondence between paths p starting at K and words in X. We shall use $\lambda(p)$ to denote the word in X or the corresponding element of G according to the context. If p is a path with initial vertex K then its terminal vertex is $K\lambda(p)$. In particular a word w in X defines an element of H if and only if the path p starting at H with label w ends at H. A shift of base point to K gives the coset graph of $K^{-1}HK$, so an unbased coset graph uniquely represents a conjugacy class of subgroups of G.

DEFINITION. If v is any vertex of a directed graph then *star* v is the set of edges with initial vertex v. Each such edge is called an *exit* from v. [The latter useful terminology is due to Collins.]

Now suppose that G is expressed as a free product $F(S) * (*_{j \in J} G_j)$, where $F(S)$ is the free group on the set S and each G_j is indecomposable and not infinite cyclic. Let X consist of the generators S and their inverses together with all elements of the factors G_j excluding the identity. A word $w = x_{i_1} x_{i_2} \ldots x_{i_n}$ in X is *reduced* if no $x_{i_r} x_{i_{r+1}}$ is equal in G to an element of X or to the identity e. Each element of G is given uniquely by a reduced word.

DEFINITION. A path p in (Γ, H) is *reduced* if $\lambda(p)$ is reduced.

NOTE. We consider the edge labelled x from K to Kx to be identified topologically with its inverse. This means that in what follows when we delete an edge we also delete its inverse and when we insert an edge with a given label we also insert the inverse edge with the inverse label.

3. S-Whitehead automorphisms.

Let $A = \{a, a^{-1}\}$ where a is an element of $S^{\pm 1}$ and let A_0, A_1 be a partition of X such that $a \in A_1$ and $a^{-1} \in A_0$ and, for each $j \in J$, all the non-identity elements of G_j are contained in the same A_i. Let $a_0 = e$ and $a_1 = a$. The corresponding S-Whitehead automorphism σ acts on reduced words in X by

$$x \mapsto x \text{ if } x \in A \text{ and}$$

$$x \mapsto a_i x a_j^{-1} \text{ otherwise, where } x \in A_i, x^{-1} \in A_j,$$

and then cancelling any e or $a^{-1}a$. We denote this Whitehead automorphism by (A_1, a_1).

EXAMPLE. Let $G = \langle a, b, s, t; a^3 = b^2 = 1\rangle$ then $G = \langle a\rangle * \langle b\rangle * \langle s\rangle * \langle t\rangle$ and $X = \{a, a^{-1}, b, s, s^{-1}, t, t^{-1}\}$. Let $\sigma = (S_1, s_1)$ where $s_1 = s$ and $S_1 = \{t^{-1}, s, b = b^{-1}\}$. Then σ takes a to a, b to sbs^{-1}, s to s, and t to ts^{-1} so for example

$$\sigma : ts^{-1} \mapsto ts^{-1}s^{-1}$$

$$\sigma : tba^{-1} \mapsto tbs^{-1}a^{-1}$$

$$\sigma : taba^{-1}t \mapsto ts^{-1}asbs^{-1}a^{-1}ts^{-1}$$

We now define the action of σ on (Γ, H).

(i) Split each vertex K into two vertices K_1 and K_0 such that $\lambda(\text{star } K_i) = A_i$ and join K_0 to K_1 by a new directed edge labelled a from K_0 to K_1.

(ii) Delete all the old edges labelled a identifying their initial and terminal vertices.

EXAMPLE. Let G and σ be as above. Let H be a subgroup of G such that Figure 1 is part of (Γ, H). The result of applying step (i) of σ to this part of (Γ, H) is given in Figure 2(i) where the straight vertical edges are the new edges. The result of applying step (ii) is given in Figure 2(ii).

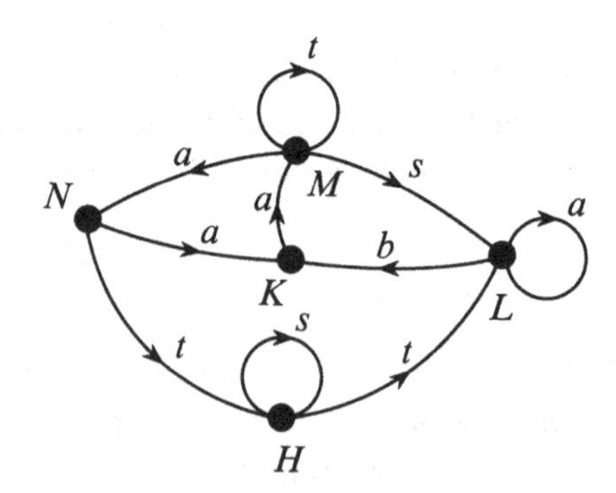

Figure 1

The new vertices are obtained by combining each vertex K_0 with L_1 where $La = K$. Denote the resulting vertex by K', and take H' as the new base vertex. Denote the resulting graph by $(\Gamma', H') = (\Gamma, H)\sigma$.

If p is a path in (Γ, H) from H to K then the action above takes p to a path p' in (Γ', H') from H' to K' as follows.

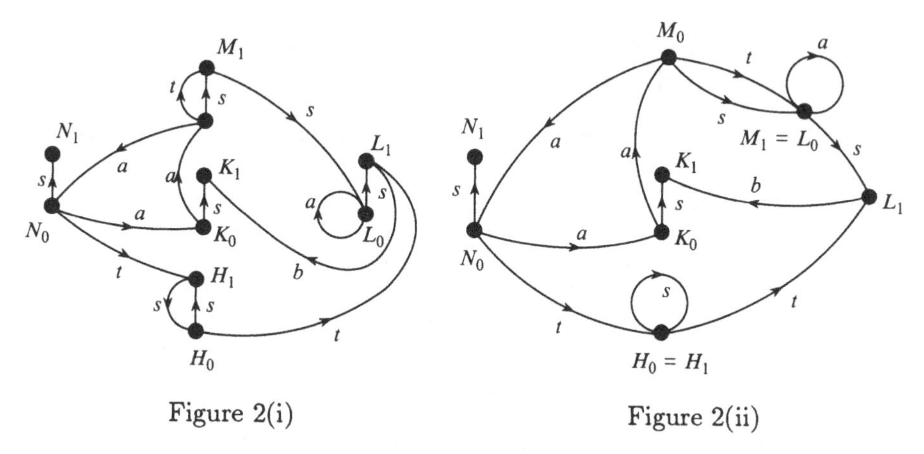

Figure 2(i) Figure 2(ii)

Firstly by inserting the new edge labelled a or a^{-1} whenever the path is split by (i) we get a path from H_0 to K_0. This path begins with the new edge labelled a from H_0 to H_1 if and only if p begins with a directed edge having label in A_1. Similarly it ends with the new edge labelled a^{-1} from K_1 to K_0 if and only if p ends with a directed edge whose inverse has label in A_1.

Secondly by deleting all the old edges in p labelled a or a^{-1} and identifying their initial and terminal vertices we get the path p' from H' to K' in (Γ', H').

If p is a reduced path then so is p'.

LEMMA 1. *Let p be a reduced path from H to K in Γ and let p' be the path in Γ' from H' to K' as defined above, then $\lambda(p') = \lambda(p)\sigma$.*

PROOF. Suppose p has label $\lambda(p) = x_1x_2\ldots x_n, x_i \in X^{\pm 1}$. Then, by the definition of $p', \lambda(p')$ is obtained from $\lambda(p)$ by putting a new a in front of x_1 if and only if $x_1 \in A_1$; putting a new a between x_i and x_{i+1} if and only if $x_i^{-1} \in A_0$ and $x_{i+1} \in A_1$; putting a new a^{-1} between x_j and x_{j+1} if and only if $x_j^{-1} \in A_1$ and $x_{j+1} \in A_0$; putting a new a^{-1} after x_n if and only if $x_n^{-1} \in A_1$; and finally deleting any $x_k = a^{\pm 1}$. It is easily checked that this is the same as the result of σ acting on the reduced word $\lambda(p)$. ∎

EXAMPLE. The paths in Figure 1 starting from H with labels ts^{-1}, tba^{-1} and $taba^{-1}t$ ending at M, N and H go to the paths labelled $ts^{-1}s^{-1} = (ts^{-1})\sigma$, $tbs^{-1}a^{-1} = (tba^{-1})\sigma$ and $ts^{-1}asbs^{-1}a^{-1}ts^{-1} = (taba^{-1}t)\sigma$ from

H' to M', N' and H' respectively, that is M_0, N_0 and H_0 in Figure 2(ii).

COROLLARY 1. *The graph* (Γ', H') *is the coset graph of* $H\sigma$.

PROOF. (Γ, H) has unique exits with each label in X and the paths p' from H' to H' are precisely those with label $\lambda(p')$ in H'. ∎

DEFINITION. (see [2]) The *core* of (Γ, H), denoted by Core(Γ, H), is the smallest connected subgraph of (Γ, H) containing all reduced circuits. The vertex H itself need not be in Core(Γ, H).

We now consider the action of σ restricted to Core(Γ, H).

(i) Split K into vertices K_1 and K_0 as above so that K_i exists if $\lambda(\text{star } K) \cap A_i \neq \emptyset$ where star K now refers to the exits from K in Core(Γ, H). If both K_0 and K_1 exist join them by a new edge labelled a from K_0 to K_1.

(ii) Delete all old edges labelled a in Core(Γ, H) and identify each pair of end vertices.

COROLLARY 2. Core(Γ', H') *is given by the action of* σ *on* Core(Γ, H) *as defined above.*

PROOF. This follows from Corollary 1. The restriction on the splitting of vertices in (i) ensures that there are no end vertices in the resulting graph. ∎

EXAMPLE. If Figure 1 gives the whole of Core(Γ, H) then, apart from the vertex N_1, Figure 2(ii) gives Core(Γ', H').

REMARK 1. From the definition $\sigma^{-1} = (A - a + a^{-1}, a^{-1})$, moreover applying step (i) of σ^{-1} to Core(Γ', H') reverses step (ii) above in such a way that $K_0' = K_0$ for all K.

4. J-Whitehead automorphisms.

Let A be one of the free factors which is not infinite cyclic. Let $e = a_0, a_1, \ldots, a_n, \ldots$ be the elements of A and let $A_0, A_1, \ldots, A_n, \ldots$ be a partition of $X \backslash A$, where some A_i may be empty, such that all non-identity elements of each G_j are contained in the same A_i. The corresponding (multiple) J-Whitehead automorphism τ (see [1]) is defined on reduced words in X by

$$x \mapsto x \text{ if } x \in A$$

$$x \mapsto a_i x a_j^{-1} \text{ otherwise where } x \in A_i, x^{-1} \in A_j,$$

and then coalescing any product of elements of A and deleting e. We denote this automorphism by $\tau = \prod_{i\neq 0}(A_i, a_i)$.

The action of this automorphism on the coset graph (Γ, H) is defined by

(i) Delete all edges with label in A, then split each vertex K into $K_0, K_1, \ldots, K_n, \ldots$ such that $\lambda(\text{star } K_i) = A_i$. We assume at this point that K_i exists for all i even if A_i is empty.

(ii) Identify vertices equivalent under the relation $K_i \approx L_j$ if $Ka_i = La_j$. Let $[K_i]$ denote the vertex defined by the equivalence class of K_i. For each K and each $i \neq j$ join $[K_i]$ to $[K_j]$ by a new directed edge labelled $a_i^{-1}a_j$.

EXAMPLE. Let G and H be as above and let $A = \langle a \rangle$ with $a_0 = e, a_1 = a$ and $a_2 = a^{-1}$. Let $A_0 = \{s^{-1}, t\}, A_1 = \{b = b^{-1}, s\}$ and $A_2 = \{t^{-1}\}$. The corresponding τ takes a to a, b to aba^{-1}, s to as, and t to ta. The two steps of the action of τ on the part of (Γ, H) in Figure 1 are illustrated in Figure 3(i) and 3(ii).

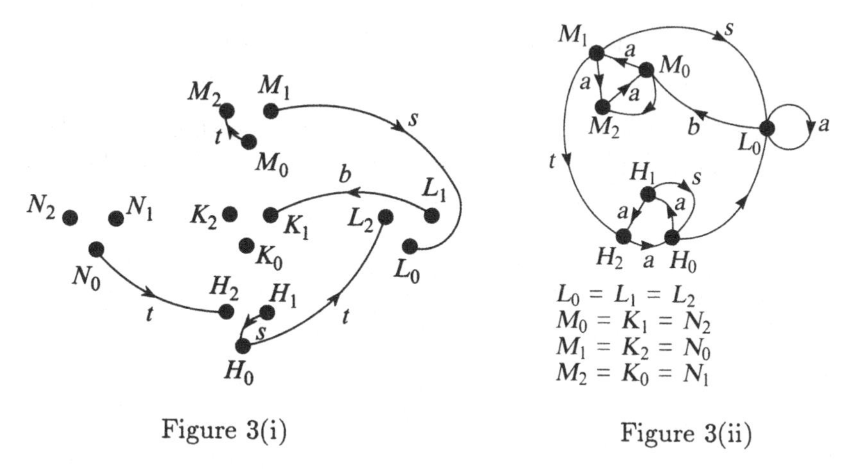

Figure 3(i) Figure 3(ii)

For each K let K' denote the vertex $[K_0]$. There is precisely one vertex with each suffix in each equivalence class under the relation in (ii). Thus K' has precisely one exit with each label in $X\backslash A$. For each $a_i \in A, a_i \neq e$, there is an edge labelled a_i joining K' to $[K_i] = L' = [L_0]$ where $Ka_i = L$. Thus K' has an exit with each label in A. Moreover if $[K_i] = [L_r]$ and if $a_i^{-1}a_j = a_r^{-1}a_s$ then $[K_j] = [L_s]$, so K' has precisely one exit with each label in A. Thus the new graph is a coset graph.

To prove the analogue of Lemma 1 and its corollaries consider the action of σ on reduced paths in Γ from H to K. Suppose that some such path p with label w has successive edges labelled x^{-1}, a and y with vertices M and N, not necessarily distinct, as below

$$\overset{x}{\longleftarrow} M \overset{a}{\longrightarrow} N \overset{y}{\longrightarrow}$$

where $a \in A, x \in A_i$ and $y \in A_j$. Then $Maa_j = Na_j$, so $[N_j] = [M_k]$ where $a_k = aa_j$, and, after operating on the coset graph, p is replaced by the path p' from $[H_0]$ to $[K_0]$ with successive edges

$$\overset{x}{\longleftarrow} [M_i] \overset{b}{\longrightarrow} [M_k] = [N_j] \overset{y}{\longrightarrow}$$

where $b = a_i^{-1}a_k = a_i^{-1}aa_j$ provided $a_i \neq aa_j$, or

$$\overset{x}{\longleftarrow} [M_i] = [N_j] \overset{y}{\longrightarrow}$$

if $a_i = aa_j$.

The end vertices and edges of p and any successive edges labelled x^{-1}, y with $x \in A_i$ and $y \in A_j$ are transformed in the obvious analagous manner. Comparing this with the action of τ on reduced words we see that as in Lemma 1 the path p' has label $w\tau$, and hence Corollary 1 holds.

EXAMPLE. Consider the path in Figure 1 from H to N with label $sts^{-1}ta$. Under the action of τ this becomes the path in Figure 3(ii) from $[H_0]$ to $[N_0]$ with label $astas^{-1}a^{-1}ta = (sts^{-1}ta)\tau$.

To define the action of τ on Core(Γ, H) we first observe that by the definition of the core every vertex has at least one adjacent edge with label not in A. The effect of τ on Core(Γ, H) is given by the following.

(i) Delete all edges with label in A. Split each vertex K into $\{K_i : \lambda(\text{star } K) \cap A_i \neq \emptyset\}$, star K as before being defined in Core(Γ, H), so K_i exists only if K has exits with label in A_i.

(ii) Identify K_i with L_j whenever K is joined to L in Core(Γ, H) by an edge labelled $a_ia_j^{-1}$. Join $[M_i]$ to $[N_j]$ by a new edge labelled $a_i^{-1}aa_j$ whenever M is joined in Core(Γ, H) to N by an edge labelled a, that is whenever $Ma = N$.

It is clear that this is the action of τ on the coset graph restricted to Core(Γ, H). Moreover as in Corollary 2 the resulting graph is Core(Γ', H').

NOTE. We use the fact that any reduced path in the coset graph (Γ, H) with end points in the core is itself in the core.

EXAMPLE. If H is as above then applying (i) of τ to Core(Γ, H) gives only the six edges labelled t, b and s with their adjacent vertices in Figure 3(i). Applying (ii) we have $N_0 \approx M_1$ and $[K_1]$ is joined to $[M_1]$ by an edge labelled $a_1^{-1}aa_1 = a$, etc. Figure 3(ii) gives the result.

REMARK 2. The J-Whitehead automorphism τ^{-1} is given by the elements $a_0^{-1} = e, a_1^{-1}, \ldots, a_n^{-1}, \ldots$ and the same partition $A_0, A_1, \ldots, A_n \ldots$; moreover applying (i) of τ^{-1} to Core(Γ', H') brings us back to the graph obtained by (i) of τ applied to Core(Γ, H). The two graphs are identified by putting $[K_i]_i = K_i$, in particular $K_0' = K_0$.

5. Combinatorial Lemmas.

Henceforth suppose that Core(Γ, H) has only finitely many vertices and let $\|H\|$, the *complexity* of H, be the number of such vertices (see [2]). Suppose also that, discounting edges with labels in the same non-cyclic free factor G_j, each vertex has only finitely many adjacent edges. This means that each vertex K splits under (i) into only finitely many K_i.

In this section we use the term *exit in* S, where S is a subset of X, to mean exit with label in S.

Let $\sigma = \prod_{i \neq 0}(A_i, a_i)$ and $\tau = \prod_{j \neq 0}(B_j, b_j)$ be S-Whitehead or J-Whitehead automorphisms with $A \neq B$. Let $\tilde{A} = \bigcup_{i \neq 0} A_i$ and $\tilde{B} = \bigcup_{j \neq 0} B_j$.

LEMMA 2. *If $A \cup \tilde{A} \subseteq B_0$ then $\sigma\tau = \tau\sigma$ and*

$$\|H\sigma\tau\| - \|H\sigma\| = \|H\tau\| - \|H\|.$$

PROOF. The first part is clear. To prove the second part we first consider the case when σ is S-Whitehead. Apply step (i) of σ to Core(Γ, H) and let the resulting graph be denoted by Δ. In Figure 4(i) we have a vertex K in Core(Γ, H) and in Figure 4(ii) the corresponding vertices K_1 and K_0

of Δ. In general not both K_i will exist. Here and in later figures, broad arrows represent all exits from the relevant vertex whose labels are in the given subset of X.

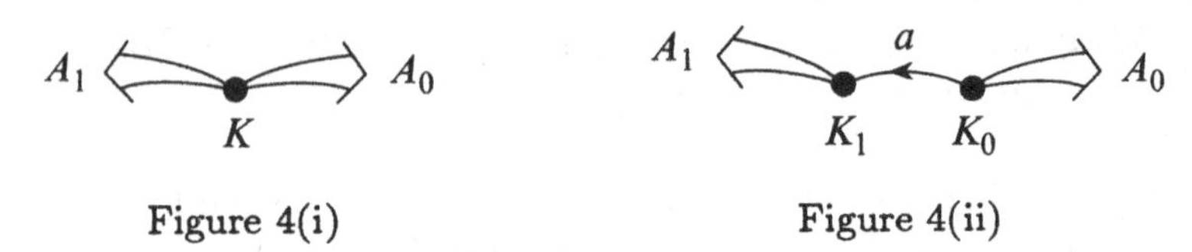

Figure 4(i) Figure 4(ii)

We now apply τ to the graph Δ and compare this with the action of τ on Core(Γ, H). Consider the two steps by which the action of τ on Δ is defined.

Step (i). If K_1 exists then it is not split since all its exits are in $A \cup A_1 \subseteq B_0$. Now consider any vertex K_0. Since $B_j \subseteq A_0$ for $j \neq 0, K_0$ exists and has exits in B_j if and only if K has such exits. Moreover because $a \in A \subseteq B_0, K_0$ also has exits in B_0 if and only if K has, i.e. either K_1 exists or K has exits in $A_0 \cap B_0$ (see Figure 4(ii)). Therefore the splitting of K_0 by τ mirrors the splitting of K in the sense that K splits if and only if K_0 exists and splits in the same way as K.

Step (ii). If τ is S-Whitehead then the number of edges labelled b in Core(Γ, H) is the same as the number labelled b in Δ and is in each case equal to the number of identified pairs of vertices. Thus

$$\|\Delta\tau\| - \|\Delta\| = \|H\tau\| - \|H\|$$

where $\|\Delta\|$ is the number of vertices of Δ.

If τ is J-Whitehead then since $B \subseteq A_0$ the split vertices of Δ are identified in step (ii) of τ if and only if the corresponding split vertices of Γ are identified. Thus in this case also

$$\|\Delta\tau\| - \|\Delta\| = \|H\tau\| - \|H\|.$$

We now consider the action of τ on Core(Γ', H'). By Remark 1 applying step (i) of σ^{-1} to Core(Γ', H') gives the same graph Δ, and by the same argument as above we also have

$$\|\Delta\tau\| - \|\Delta\| = \|H'\tau\| - \|H'\|$$

where $H' = H\sigma$. This completes the proof for the case when σ is S-Whitehead.

If σ is J-Whitehead the argument is essentially the same but we need to modify the splitting of Δ by τ as follows.

(i)$'$ If τ is J-Whitehead delete all edges with label in B. Split each vertex L of Δ into $\{L_j : \lambda(\text{star } L) \cap B_j \neq \emptyset\}$. Additionally add a vertex L_0 whenever $L = K_0$ and K has exits in B_0 but K_0 has not. If τ is S-Whitehead add a new edge labelled b from L_0 to L_1 whenever they both exist.

NOTE. Clearly (i)$'$ depends on Core(Γ, H). The extra vertex L_0 is added whenever $L = K_0$ and K has exits in A_0 and in B_0 but none in $A_0 \cap B_0$.

As before any vertex $K_i, i \neq 0$, in Δ is not split since $A_i \subseteq B_0$. Also as before K_0 exists and has exits in B_j if and only if K has. Moreover, with the modification, L_0 exists if and only if K has exits in B_0. Thus the splitting of K_0 mirrors the splitting of K. This is illustrated in Figures 5,6 and 7, when K has exits in A_0 and B_0 but none in $A_0 \cap B_0$.

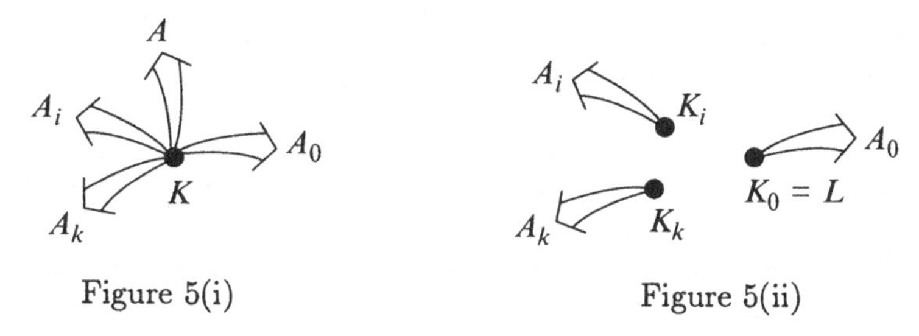

Figure 5(i) Figure 5(ii)

Figure 5(i) gives the vertex K in Core(Γ, H). Figure 5(ii) gives the part of Δ arising from the vertex K.

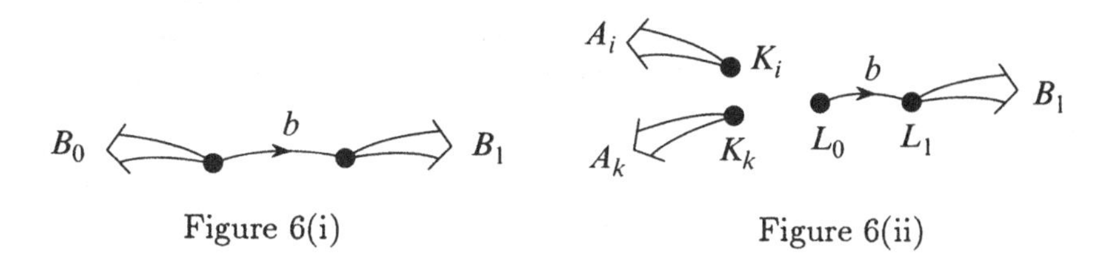

Figure 6(i) Figure 6(ii)

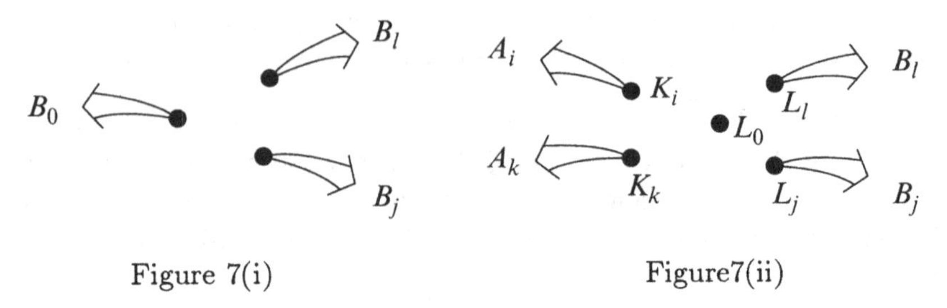

Figure 7(i) Figure7(ii)

Figures 6(i) and 7(i) give the splitting of K by step (i) of τ when τ is S-Whitehead and J-Whitehead respectively. Figures 6(ii) and 7(ii) give the corresponding modified splitting of the relevant part of Δ, i.e. of Figure 5(ii).

Applying step (ii) of τ as above we get

$$\|\Delta\tau\| - \|\Delta\| = \|H\tau\| - \|H\|.$$

By Remark 2 applying step (i) of σ^{-1} to Core(Γ', H') gives the same graph Δ with K_i corresponding to $[K_i]_i$. Thus to complete the proof we need to show that the two modified actions of τ on Δ depending on Core(Γ, H) and Core(Γ', H') are the same.

Suppose that K has exits in A_0 and in B_0 but none in $A_0 \cap B_0$. Then K' has the same exits in A_0 as K has. Thus we need to show that K' has exits in B_0, for then K' will have exits in A_0 and in B_0 but none in $A_0 \cap B_0$. By the condition on the exits from K and the hypothesis of the Lemma, K has an exit in $A \cup \tilde{A}$. Suppose K has an exit in A_i where $i \neq 0$ then K_i is a vertex of Δ and so $K' = [K_0]$ has an exit with label $a_i \in B_0$ joining it to $[K_i]$. On the other hand suppose K has an exit with label a in A joining it to M say, then M has an exit in some A_j where j could be zero. If $a = a_j^{-1}$ then $K' = [K_0] = [M_j]$ has an exit in A_j otherwise $K' = [K_0]$ is joined to $[M_j]$ by an edge labelled $aa_j \in B_0$. Thus in either case K' has an exit in B_0. Interchanging K and K' we deduce that the two modified actions of τ on Δ are the same. Thus we have

$$\|\Delta\tau\| - \|\Delta\| = \|H\sigma\tau\| - \|H\sigma\|$$

when τ is J-Whitehead. This completes the proof of the Lemma. ∎

LEMMA 3. *If A is not equal to B but $A \cup A_0 = B \cup B_0$ then*

$$\|H\sigma\| + \|H\tau\| \geq 2\|H\|$$

PROOF. We show that each vertex K of Core(Γ, H) contributes at least two to the L.H.S. To distinguish the split vertices we use the notation K_{A_i} and K_{B_j}. By the definitions no K_{A_0} is identified with any L_{A_0} under step (ii) of σ and similarly with K_{B_0}. So we count $[K_{A_0}]$ and $[K_{B_0}]$ if they exist each as contributing one to the L.H.S. Suppose there is no K_{A_0} given by step (i) of σ then, whether σ is S-Whitehead or J-Whitehead, K has at least one exit with label x not in A_0 or A. By the hypothesis x is in some $B_j, j \neq 0$, so K_{B_j} is created by step (i) of τ. Moreover, since there is no K_{A_0}, K has no exit in B so K_{B_j} is not identified with any L_{B_k} under step (ii) of τ and we count this $[K_{B_j}]$ as contributing one to the L.H.S. Similarly if there is no K_{B_0} then there is some K_{A_i} which is not identified with any L_{A_k}. So in all cases K contributes at least two to the L.H.S. ∎

Now suppose that $A \cup B \subseteq B_0 \cup A_0$ and $A \neq B$. If σ is J-Whitehead let σ_0 and σ_1 be the J-Whitehead automorphisms defined by the elements $e = a_0, a_1, \ldots, a_n, \ldots$ and the partitions

$$A_0 \cup (\tilde{A} \backslash B_0), A_1 \cap B_0, A_2 \cap B_0, \ldots, A_n \cap B_0, \ldots$$

and

$$A_0^\sharp = A_0 \cup (\tilde{A} \cap B_0), A_1 \backslash B_0, A_2 \backslash B_0, \ldots, A_n \backslash B_0, \ldots$$

respectively.

If σ is S-Whitehead let σ_0 and σ_1 be the S-Whitehead automorphisms defined by the elements a, and a^{-1} and the sets

$$A_0 \cup (A_1 \backslash B_0), A_1 \cap B_0,$$

and

$$A_0^\sharp = A_0 \cup (A_1 \cap B_0 - a), A_1 \backslash (B_0 - a),$$

respectively.

Let τ_0 and τ_1 be defined similarly interchanging A and B.

COROLLARY 2. *Suppose $A \cup B \subseteq B_0 \cup A_0, A \neq B$, and suppose $a^{-1} \in B_0$ whenever σ is S-Whitehead, then*

$$\|H\sigma\| + \|H\tau\| \geq \|H\sigma_0\| + \|H\tau_0\|.$$

PROOF. The hypothesis implies that $A \cup A_0^\sharp = B \cup B_0^\sharp$ so Lemma 3 can be applied to σ_1 and τ_1 acting on $H\sigma_0\tau_0$ to give

$$\|H\sigma_0\tau_0\tau_1\| + \|H\sigma_0\tau_0\sigma_1\| \geq 2\|H\sigma_0\tau_0\|.$$

By applying Lemma 2 three times we have

$$\|H\sigma_0\tau\| - \|H\sigma_0\| = \|H\tau\| - \|H\|$$

since $A \cup (\tilde{A} \cap B_0) \subseteq B_0$,

$$\|H\sigma\tau_0\| - \|H\sigma\| = \|H\tau_0\| - \|H\|$$

since $A \cup \tilde{A} \subseteq B_0 \cup (\tilde{B} \backslash A_0)$, and

$$\|H\sigma_0\tau_0\| - \|H\sigma_0\| = \|H\tau_0\| - \|H\|$$

since $A \cup (\tilde{A} \cap B_0) \subseteq B_0 \cup (\tilde{B} \backslash A_0)$.
Since $\tau = \tau_0\tau_1$ and $\sigma\tau_0 = \sigma_0\sigma_1\tau_0 = \sigma_0\tau_0\sigma_1$ the result follows. ∎

We need two further lemmas to deal with the cases to which Corollary 2 does not apply.

Suppose σ and τ are the S-Whitehead automorphisms (A_1, a) and (B_1, b) respectively. Suppose also that the elements a, a^{-1}, b and b^{-1} are in $A_1 \cap B_1, A_0 \cap B_0, A_0 \cap B_1$ and $A_1 \cap B_0$ respectively. Let $\rho_1 = (A_1 \cap B_1, a), \rho_2 = (A_0 \cap B_0, a^{-1}), \rho_3 = (A_0 \cap B_1, b)$ and $\rho_4 = (A_1 \cap B_0, b^{-1})$.

LEMMA 4. *Under these conditions*

$$2\|H\sigma\| + 2\|H\tau\| \geq \|H\rho_1\| + \|H\rho_2\| + \|H\rho_3\| + \|H\rho_4\|$$

PROOF. We first consider the splitting of Core(Γ, H) by the various Whitehead automorphisms. By considering the labels of the exits we see that if a vertex K is split by ρ_1 or ρ_2 then it is also split by σ or τ. Moreover if it is split both by ρ_1 and by ρ_2 then it is also split both by σ and by τ. Thus the total number of vertices split by σ plus the number split by τ is greater than or equal to the number split by ρ_1 plus the number split by ρ_2. The same holds for ρ_3 and ρ_4.

Now the number of vertices lost in step (ii) of ρ_1, ρ_2 and σ are all equal to the number edges of Core(Γ, H) with label a. Similarly the number lost in step (ii) of ρ_3, ρ_4 and τ are all equal to the number of edges with label b. The result follows by counting gains and losses. ∎

Suppose again that σ and τ are S-Whitehead. Suppose now that b^{-1} and a are in $B_0 \cap A_1$ and a^{-1} and b are in $A_0 \cap B_1$, where a and b^{-1} are not necessarily distinct. Let $\rho_1 = (B_0 \cap A_1, a)$ and $\rho_2 = (A_0 \cap B_1, b)$.

LEMMA 5. *Under these conditions*

$$\|H\sigma\| + \|H\tau\| \geq \|H\rho_1\| + \|H\rho_2\|$$

PROOF. As above the total number of vertices split by σ plus the number split by τ is greater than or equal to the number split by ρ_1 plus the number split by ρ_2. Moreover as above the number of vertices lost in step (ii) of ρ_1, and σ are both equal to the number edges of Core(Γ, H) with label a and the number lost in step (ii) of ρ_2 and τ are equal to the number of edges with label b. The result follows as above. ∎

6. Peak/Plateau reduction.

Let Ω consist of all Whitehead automorphisms, all permutation automorphisms, i.e. permutations of isomorphic free factors, and all factor automorphisms, i.e. automorphisms of the free factors, see [1]. A peak/plateau consists of a subgroup H and two automorphisms σ and τ in Ω such that

$$\|H\sigma\| \leq \|H\| > \|H\tau\|.$$

A *reduction* of this peak/plateau consists of automorphisms $\tau_1, \ldots, \tau_m$ in Ω such that $\sigma\tau_1 \ldots \tau_m = \tau$ and $\|H\sigma\tau_1 \ldots \tau_q\| < \|H\|$ for $q = 1, \ldots, m-1$, (see Figure 8 in which height represents complexity).

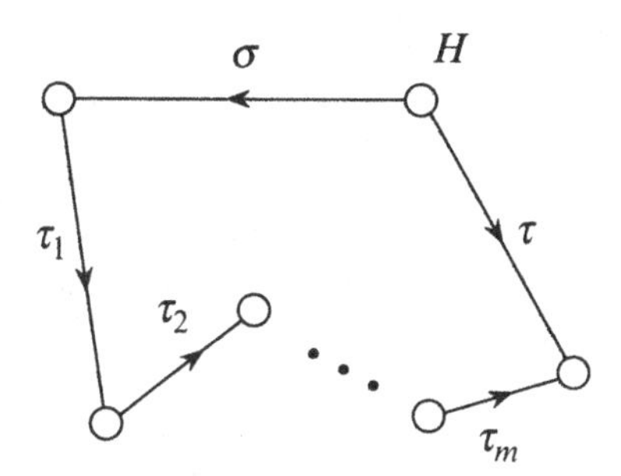

Figure 8

THEOREM. *Every peak/plateau has a reduction.*

PROOF. The argument is essentially the same as in [3], [6] and [1]. Suppose σ is a permutation or a factor automorphism then σ changes the labels of edges of Core(Γ, H) and does not change the complexity. Therefore $\|H\tau\sigma\| = \|H\tau\| < \|H\|$. Moreover $\tau^\sigma = \sigma^{-1}\tau\sigma$ is in Ω, so τ^σ, σ^{-1} is a reduction.

Now suppose that σ is a Whitehead automorphism. We observe that inner automorphisms are products of S-Whitehead automorphisms of the form $(X - a, a)$, with a in S, and J-Whitehead automorphisms $(X - A, a)$, with a not in S. Moreover inner automorphisms together with permutations and factor automorphisms may be collected at the end without affecting the theorem. Thus without loss we shall work modulo such automorphisms. In particular we may replace the S-Whitehead automorphism (A_1, a) by (A_0, a^{-1}) and the J-Whitehead automorphism $\prod_{i\neq 0}(A_i, a_i)$ by $\prod_{a\neq a_i}(A_i, a^{-1}a_i)$ where $a \in A$.

Suppose σ and τ are both S-Whitehead. If a, a^{-1}, b, and b^{-1} are all in different sets $A_i \cap B_j$ then without loss of generality the conditions of Lemma 4 hold. Moreover $2\|H\| > \|H\sigma\| + \|H\tau\|$ so $\|H\rho_i\| < \|H\|$ for some i. Suppose $\|H\rho_4\| < \|H\|$. Then putting ρ for $\rho_4, (A_0, a^{-1})$ for σ and (B_0, b^{-1}) for τ we have $A \cup A_0 \subseteq A_0 \cup B_1$ so by Lemma 2

$$\|H\sigma\rho\| - \|H\sigma\| = \|H\rho\| - \|H\|.$$

Moreover modulo an inner automorphism $\rho\sigma = \sigma\rho$. So $\|H\sigma\rho\| < \|H\|$ and $\rho, \sigma^{-1}, \rho^{-1}\tau = (B_0 \backslash A_1, b^{-1})$ is a reduction. The proof is similar when $\|H\rho_i\| < \|H\|$ for i = 1,2, or 3.

If σ and τ are S-Whitehead with a, b, a^{-1} and b^{-1} contained in two of the sets $A_i \cap B_j$ then we may apply Lemma 5 to find $\|H\rho_i\| < \|H\|$ and proceed as above. Note that this case includes $A = B$.

If σ and τ are S-Whitehead with a, b, a^{-1} and b^{-1} contained in three of the sets $A_i \cap B_j$ then, after putting (A_0, a^{-1}) for σ and (B_0, b^{-1}) for τ if necessary and possibly interchanging σ and τ, the conditions of Corollary 2 hold. Therefore either $\|H\sigma_0\| < \|H\|$ or $\|H\tau_0\| < \|H\|$ and we can proceed as before. Thus we have covered all the cases in which σ and τ are S-Whitehead.

Suppose now that τ is J-Whitehead. If $A = B$ then as shown in [1] $\sigma^{-1}\tau$ is also J-Whitehead and so forms a reduction. If $A \neq B$ then working modulo inner automorphisms etc. we can ensure that $B \subset A_0$ and then that $a \in B_0$ when σ is S-Whitehead, and $A \subset B_0$ when σ is J-Whitehead. Thus the hypotheses of Corollary 2 hold with σ and τ interchanged and we proceed as above.

The proof holds equally well if σ is J-Whitehead interchanging σ and τ. This completes the proof of the theorem. ∎

References

[1] D. J. Collins and H. Zieschang. Rescuing the Whitehead method for free products, I: Peak reduction. *Math. Z.* **185** (1984) 487-504.

[2] S. M. Gersten. On Whitehead's algorithm. *Bull. Amer. Math. Soc.* **10** (1984) 281-284.

[3] P. J. Higgins and R. C. Lyndon. Equivalence of elements under automorphipsms of a free group. *J. London Math. Soc., Second Series* **8** (1974) 254-258.

[4] A. H. M. Hoare. Coinitial graphs and Whitehead Automorphisms. *Canadian J. of Math.* **31** (1979) 112-123.

[5] A. H. M. Hoare. On automorphisms of free groups II. *J. London Math. Soc., Second Series* **42** (1990) 226-236.

[6] J. McCool. A presentation for the automorphism group of a free group of finite rank. *J. London Math. Soc., Second Series* **8** (1974) 259-266.

[7] E. S. Rapaport. On free groups and their automorphisms. *Acta Math.* **90** (1958) 139-163.

[8] J. H. C. Whitehead. On equivalent sets of elements in a free group. *Ann. of Math.* **37** (1936) 782-800.

[9] M. J. Wicks. Private communication (1977).

School of Mathematics and Statistics
University of Birmingham
P O Box 363
Edgbaston. B15 2TT.

Communications Division
Electronic Research Laboratory
Building 806 TSAS
P O Box 1600
Salisbury, South Australia 5108

The growth series of the Gieseking group

D. L. Johnson and H.-J. Song

To Murray Macbeath on the occasion of his retirement

The question.

The *growth series* of a group G with respect to a finite generating set X is a power series

$$G_X(t) = \sum_{n \geq 0} c_n t^n$$

in an indeterminate t with integer coefficients, where the c_n are the *growth coefficients* in the sense of Milnor [6]: c_n is the number of elements of G that can be written as a product of n elements of X and their inverses, and no fewer. For a brief history of the study of growth of groups, we refer to [3] and the references cited there.

The *Gieseking group* is the fundamental group of the Gieseking manifold and is given by the presentation

$$G = \langle\ x, y \mid x^2 y^2 = yx\ \rangle.$$

We write $G(t)$ for the growth series of this group with respect to the generating set $X = \{x, y\}$ and compute it explicitly below. It turns out to be a rational function of t whose coefficients grow exponentially.

The answer.

The group G is associated with a certain regular tesselation $\mathcal{T}$ of hyperbolic 3-space $\mathbf{H}^3$ by ideal tetrahedra, which may be derived either from the geometrical construction in [5], pp. 153-156, or from the Cayley diagram Γ of G in the following way.

We can picture Γ as being made up of hexagonal cells with oriented edges labelled by letters of the (cyclic) relator $r := x^2y^2x^{-1}y^{-1}$. Since the 12 subwords of length 2 in $r^{\pm 1}$ are all distinct (and so coincide with the 12 reduced words of length 2 in the free group $F(x, y)$), these hexagonal cells meet

in threes at each edge, and
in sixes at each vertex, since
there are 4 edges (labelled $x^{\pm 1}, y^{\pm 1}$) at each vertex of Γ.

As will be seen directly, these rules of incidence ensure that Γ embeds in $\mathbf{H}^3$, which we think of as the interior of an open Euclidean 3-ball whose geodesics are arcs of circles orthogonal to its boundary S^2.

The tesselation $\mathcal{T}$ is the dual of the 1-skeleton of Γ regularly embedded in $\mathbf{H}^3$ as follows. The vertices of a regular Euclidean tetrahedron inscribed in $S^2 = \delta\mathbf{H}^3$ define an ideal hyperbolic tetrahedron T in $\mathbf{H}^3$ whose faces turn out to meet in pairs at an angle of $\pi/3$. Repeated application of the reflections a, b, c, d in these four faces yields $\mathcal{T}$, made up of ideal tetrahedra meeting in sixes at each edge.

Now the growth coefficient c_n of G is just the number of vertices of Γ distant n, in the usual graph metric, from a preferred one. But this is the same as the number of tetrahedra in $\mathcal{T}$ got by a minimum of n reflections from a fundamental region, T say, and this in turn coincides with the nth growth coefficient of the corresponding hyperbolic Coxeter group (see [2], p. 142, foot of second column):

$$W = \langle a, b, c, d \mid a^2 = b^2 = c^2 = d^2 = 1, (ab)^3 = (bc)^3 = (cd)^3 = (da)^3 = (ca)^3 = (db)^3 = 1 \rangle.$$

The groups G and W are thus isometric, so that the growth series can be found from the general formulae in Exercise 26 of [1], p. 45 (or [2], p. 123):

$$\sum_{Y \subseteq S} \frac{(-1)^{|Y|}}{W_Y(t)} = \begin{cases} t^m/W(t) & \text{if } |W| < \infty, \\ 0 & \text{if } |W| = \infty, \end{cases}$$

where W is any Coxeter group with a finite set S of generators and $m = \max\{l(w)|w \in W\}$ in the finite case.

The subgroups W_Y of our W are enumerated as follows.

$|Y| = 0$: one trivial subgroup with growth series $\gamma_0 = 1$.
$|Y| = 1$: four cyclic subroups of order 2 with growth series $\gamma_1 = 1 + t$.
$|Y| = 2$: six dihedral subgroups of order 6 with growth series

$$\gamma_2 = (1 + t)(1 + t + t^2).$$

$|Y| = 3$: four affine subgroups with growth series given by

$$\frac{1}{\gamma_0} - \frac{3}{\gamma_1} + \frac{3}{\gamma_2} - \frac{1}{\gamma_3} = 0,$$

that is,

$$\gamma_3 = \frac{1 + t + t^2}{(1 - t)^2}.$$

$|Y| = 4$: the whole group with growth series γ_4 given by

$$\frac{1}{\gamma_0} - \frac{4}{\gamma_1} + \frac{6}{\gamma_2} - \frac{4}{\gamma_3} + \frac{1}{\gamma_4} = 0,$$

that is,

$$\gamma_4 = \frac{(1 + t)(1 + t + t^2)}{(1 - t)(1 - t - 3t^2)}.$$

PROPOSITION.

$$G(t) = \frac{(1 + t)(1 + t + t^2)}{(1 - t)(1 - t - 3t^2)}.$$

Remarks.

1. Because the denominator here has roots $(1 \pm \sqrt{13})/6$ inside the unit circle, the growth rate of the coefficients c_n is exponential. A simple calculation shows that they are given by the linear recurrence

$$c_0 = 1, \quad c_1 = 4, \quad c_2 = 12, \quad c_3 = 30, \quad c_n = 2c_{n-1} + 2c_{n-2} - 3c_{n-3}, \quad n \geq 4.$$

2. It would be interesting to have other non-trivial examples of such isometries between finitely-generated groups. For example, we can get one such

very cheaply as follows. It is a simple matter to check that any 2-generator one-relator group with the crucial property

> **P:** the 2-letter subwords of the cyclic relator and its inverse are all distinct and exhaust the set of words of length 2 in the corresponding free group,

is isomorphic either to G or to the group

$$G^{-} = \langle\ x, y \mid x^2 y^2 = y x^{-1}\ \rangle.$$

> M. Edjvet has pointed out that G and G^{-} have distinct Alexander polynomials, and so are not isomorphic.

3. Another crucial feature of our argument is the fact that the faces of T meet at an angle of $\pi/3$. The corresponding angles for the cube and regular octahedron are easily computed as $\pi/3$ and $\pi/2$, respectively.

4. The subgroup of index 2 in G is the group

$$L = \langle\ x, y \mid yxy^{-1}xyx^{-1}y^{-1}xy^{-1}x^{-1}\ \rangle$$

of the figure-eight knot. We have not yet computed the growth series of L.

Acknowledgements

The first author is grateful to Murray himself for the insight into hyperbolic space provided by his lectures at Birmingham in the Spring of 1964 and to J. Neubüser and his staff at Lehrstuhl D für Mathematik of the RWTH in Aachen for their hospitality while this article was being written. We both gratefully acknowledge the support of the Royal Society, the Korean Science and Engineering Foundation, the British Council and the Korean Research Foundation, without which this collaboration would not have been possible.

References

[1] N. BOURBAKI. *Groupes et algèbres de Lie.* Chapitres 4, 5, 6. (Hermann, Paris, 1968).

[2] J. E. HUMPHREYS. *Reflection Groups and Coxeter Groups.* (Cambridge University Press, Cambridge, 1990).

[3] D. L. JOHNSON. Rational growth of wreath products, in *Groups St. Andrews 89*, eds. C. M. Campbell and E. F. Robertson, (Cambridge University Press, Cambridge, 1991) 309–315.

[4] D. L. JOHNSON AND H.-J. SONG. The growth of torus knot groups, to appear.

[5] W. MAGNUS. *Noneuclidean Tesselations and Their Groups.* (Academic Press, New York – London, 1974).

[6] J. MILNOR. A note on curvature and the fundamental group. *J. Differential Geometry* **2** (1968) 1–7.

Mathematics Department
University Park
Nottingham, U. K.

Department of Applied Mathematics
National Fisheries University
Pusan, South Korea

Exceptional representations of $\mathrm{PSL}_2(q)$ of monodromy genus zero

Gareth A. Jones

Dedicated to Murray Macbeath on the occasion of his retirement

Abstract. We determine the values of q for which $\mathrm{PSL}_2(q)$, acting on the cosets of a subgroup isomorphic to A_4, S_4 or A_5, has monodromy genus 0.

1. Introduction

The *monodromy genus* $\gamma(G;\Omega)$ of a finite transitive permutation group (G,Ω) is the least genus of any branched covering S of the Riemann sphere with monodromy group (G,Ω); the *monodromy genus* $\gamma(G)$ of a finite group G is the minimum of $\gamma(G;\Omega)$ as Ω ranges over all faithful transitive representations of G, that is, the least genus of any S with monodromy group isomorphic to G.

It is well-known [7] that cyclic, dihedral, alternating and symmetric groups all have genus 0. On the other hand, Guralnick and Thompson [3] have conjectured that at most finitely many simple groups of Lie type can be composition factors of monodromy groups of any given genus; Liebeck and Saxl [8] have verified this for simple groups of bounded Lie rank, but the general problem remains open.

In [5] I determined those q for which the natural representation of $G = \mathrm{PSL}_2(q)$ on the projective line $\Omega = PG_1(q)$ has genus 0 (all the prime-powers $q \leq 43$ except 23, 27, 31 and 32). It is hoped eventually to determine all q for which $\mathrm{PSL}_2(q)$ has genus 0 (in any representation), and to do this it is sufficient to consider the primitive representations, those for which the stabilisers $H = G_\alpha (\alpha \in \Omega)$ are maximal subgroups. Apart from the natural point-stabilisers, of index $q+1$, the maximal subgroups of G are dihedral groups, projective groups $\mathrm{PSL}_2(q')$ or $\mathrm{PGL}_2(q')$ for certain q' dividing q, and exceptional subgroups $H \cong A_4$, S_4 or A_5 arising for various q. My aim here is to consider these last three *exceptional representations* of G (whether primitive or not), and to prove the following

THEOREM. *Let $G = \mathrm{PSL}_2(q)$ and let $\Omega = G/H$, where $G > H \cong A_4$, S_4 or A_5. Then $\gamma(G;\Omega) = 0$ if and only if*

(i) $q = 4, 5, 7, 9$ *or* 13 *for* $H \cong A_4$;

(ii) $q = 7, 9$ *or* 17 *for* $H \cong S_4$;

(iii) $q = 9, 11, 19$ *or* 29 *for* $H \cong A_5$.

(These representations are all primitive, except those for $q = 7$ and 9 in case (i)).

The proof follows that in [5] for the natural representation of G, with a major role being played by Macbeath's analysis [9] of generating triples for $\mathrm{PSL}_2(q)$.

2. Monodromy groups. ([3], [5], [6], [7], [10])

Let $\psi : S \to \Sigma$ be a connected n-sheeted covering of the Riemann sphere $\Sigma = \mathbb{C} \cup \{\infty\}$, branched at $P = \{p_1, \ldots, p_r\} \subset \Sigma$. If $p_0 \in \Sigma - P$, then the fundamental group $\Pi = \pi_1(\Sigma - P; p_0)$ induces, by unique path-lifting, a transitive group G of permutations of the fibre $\Omega = \psi^{-1}(p_0) \subset S$, called the *monodromy group* of S (or strictly, of ψ). Since Π has presentation

$$\Pi = \langle s_1, \ldots, s_r \mid s_1 \ldots s_r = 1 \rangle,$$

where s_i is the homotopy class of a suitable loop around p_i, G is generated by permutations g_i $(1 \leq i \leq r)$ induced by s_i, satisfying

$$g_1 \ldots g_r = 1. \tag{2.1}$$

Conversely, every transitive finite permutation group, with generators $g_1, \ldots, g_r$ satisfying (2.1), arises in this way for some $\psi : S \to \Sigma$.

The genus γ of S is given by the Riemann-Hurwitz formula

$$\gamma = 1 - n + \frac{1}{2}B, \tag{2.2}$$

where B is the total order of branching of ψ. Now the order of branching at p_i is $n - |\psi^{-1}(p_i)| = n - \nu(g_i)$, where $\nu(g_i)$ is the number of cycles of g_i on Ω , so

$$\gamma = 1 - n + \frac{1}{2}\sum_{i=1}^{r} \beta(g_i), \tag{2.3}$$

where

$$\beta(g) = n - \nu(g) = |\Omega| - |\Omega/\langle g \rangle|. \tag{2.4}$$

The monodromy genus $\gamma(G;\Omega)$ of a transitive finite permutation group (G,Ω) is the least genus γ of any S with monodromy group (G,Ω); it is obtained by minimising (2.3) over all generating r-tuples $\mathbf{g} = (g_1, \ldots, g_r)$ in G which satisfy $g_1 \ldots g_r = 1$. It is convenient to denote by K the multiset $\{k_1, \ldots, k_r\}$, where g_i has order k_i, and to let $\Delta(K)$ be the group with presentation

$$\Delta(K) = \langle s_1, \ldots, s_r \mid s_1^{k_1} = \ldots = s_r^{k_r} = s_1 \ldots s_r = 1 \rangle,$$

so that $\Delta(K)$ maps onto G in the obvious way.

In addition to proving the Theorem, I shall also determine the multisets K arising in each case. They are given in Table 1, where semi-colons separate different multisets, and $2^{(3)}$ denotes 2, 2, 2, etc.

H	q	n	K
A_4	$4,5$	5	$2^{(3)},3;\ 2^{(2)},3^{(2)};\ 2,3^{(3)};\ 3^{(4)};\ 2,3,5;\ 3,3,5$
	7	14	$2,3,7;\ 3,3,4$
	9	30	$3,3,4$
	13	91	$2,3,7$
S_4	7	7	$2^{(6)};\ 2^{(4)},3;\ 2^{(4)},4;\ 2^{(3)},7;\ 2^{(2)},3^{(2)};\ 2^{(2)},3,4;\ 2^{(2)},4^{(2)};\ 2,3,7;\ 2,4,7;\ 3,3,4;\ 3,4,4;\ 4^{(3)}$
	9	15	$2^{(3)},4;\ 2,4,5;\ 3,3,4$
	17	102	$2,3,8$
A_5	9	6	$2^{(5)};\ 2^{(4)},3;\ 2^{(3)},3^{(2)};\ 2^{(2)},3^{(3)};\ 2,3^{(4)};\ 3^{(5)};\ 2^{(3)},4;\ 2^{(3)},5;\ 2^{(2)},3^{(2)};\ 2^{(2)},3,4;\ 2^{(2)},3,5;\ 2,3^{(3)};\ 2,3^{(2)},4;\ 2,3^{(2)},5;\ 3^{(4)};\ 3^{(3)},4;\ 3^{(3)},5;\ 2,4,5;\ 2,5,5;\ 3,3,4;\ 3,3,5;\ 3,4,5;\ 3,5,5$
	11	11	$2^{(5)};\ 2^{(3)},5;\ 2^{(3)},6;\ 2^{(2)},3^{(2)};\ 2,3,11;\ 2,5,5;\ 2,5,6;\ 2,6,6;\ 3,3,5;\ 3,3,6$
	19	57	$2,3,9;\ 2,3,10$
	29	203	$2,3,7$

Table 1

Instances in which $\mathrm{PSL}_2(q)$, acting with degree n on the cosets of a subgroup $H \cong A_4$, S_4 or A_5, has monodromy genus $\gamma = 0$. The multisets K indicate the orders of the generators g_i associated with the branch-points.

3. Existence and conjugacy of exceptional subgroups. ([2], [4], [11])

From now on, G denotes $\mathrm{PSL}_2(q)$, where $q = p^e$ and p is prime; we have $|G| = q(q^2-1)/d$ where $d = (q-1, 2)$. Table 2 gives the numbers of conjugacy classes in G of exceptional subgroups $H \cong A_4$, S_4 or A_5, the size of each class, and the cases where H is maximal in G. (Whenever there are two classes, they are conjugate in $\mathrm{PGL}_2(q)$.)

H	q	number of classes	size of class	maximality
A_4	$= 2^{2f+1}$	0	–	–
	$= 2^{2f}$	1	$\lvert G\rvert/12$	$q = 4$
	$\equiv \pm 3(8)$	1	$\lvert G\rvert/12$	$e = 1$ and $q \equiv \pm 2(5)$
	$\equiv \pm 1(8)$	2	$\lvert G\rvert/24$	no
S_4	$\not\equiv \pm 1(8)$	0	–	–
	$\equiv \pm 1(8)$	2	$\lvert G\rvert/24$	$q = 9$ or $e = 1$
A_5	$= 2^{2f+1}$	0	–	–
	$= 2^{2f}$	1	$\lvert G\rvert/60$	f odd prime
	$= 5^{2f+1}$	1	$\lvert G\rvert/60$	e odd prime
	$= 5^{2f}$	2	$\lvert G\rvert/120$	no
	$\equiv \pm 3(10)$	0	–	–
	$\equiv \pm 1(10)$	2	$\lvert G\rvert/60$	$e = 1$, or $e = 2$ and $p \equiv -1(10)$

Table 2

4. Conjugacy classes in G

To prove the Theorem, we find all solutions $\mathbf{g} = (g_1, \ldots, g_r)$ in G of the equation

$$2n - 2 = \sum_{i=1}^{r} \beta(g_i) \tag{4.1}$$

or equivalently

$$2\left(1 - \frac{1}{n}\right) = \sum_{i=1}^{r} \frac{\beta(g_i)}{n} \tag{4.1$'$}$$

obtained by putting $\gamma = 0$ in (2.3); we then determine which of these solutions generate G and satisfy $g_1 \ldots g_r = 1$.

In order to solve (4.1′), we must first evaluate $\beta(g)/n$ for each $g \in G$.

By the Cauchy-Frobenius Lemma ([4], V.13.4)

$$\nu(g) = \frac{1}{k}\sum_{j=1}^{k}\pi(g^j) \tag{4.2}$$

and so

$$\frac{\beta(g)}{n} = 1 - \frac{1}{k}\sum_{j=1}^{k}\frac{\pi(g^j)}{n} \tag{4.3}$$

where k is the order of g and $\pi(h) = |\{\alpha \in \Omega \mid \alpha h = \alpha\}|$. A simple argument [5], counting such pairs (α, h) in two ways, gives

$$\frac{\pi(h)}{n} = \frac{|C \cap H|}{|C|} \tag{4.4}$$

where C is the conjugacy class h^G of G containing h, and $H = G_\alpha$ $(\alpha \in \Omega)$.

If C is to have non-empty intersection with H, its elements must have order $\ell \leq 5$; Table 3 lists all such non-identity classes C, also giving information needed in §5 on whether C is invariant under inverting or squaring its elements.

ℓ	q	number of classes C	$\lvert C\rvert$	$C = C^{-1}$?	$C = C^2$?
2		1	$\lvert G\rvert/(q-\delta)$	yes	no
3	$= 3^e$	2	$\frac{1}{2}(q^2-1)$	iff $e = 2f$	iff $e = 2f$
	$\neq 3^e$	1	$q(q+\varepsilon)$	yes	yes
4	$\equiv \pm 1(8)$	1	$q(q+\delta)$	yes	no
	$\not\equiv \pm 1(8)$	0	–	–	–
5	$= 5^e$	2	$\frac{1}{2}(q^2-1)$	yes	iff $e = 2f$
	$\equiv \pm 1(5)$	2	$q(q+\zeta)$	yes	no
	$\equiv \pm 2(5)$	0	–	–	–

Table 3

In this table,

$$\delta = \begin{cases} 1 & \text{if } q \equiv 1(4), \\ -1 & \text{if } q \equiv -1(4), \\ 0 & \text{if } q \equiv 0(2); \end{cases}$$

similarly, $\varepsilon = \pm 1$ as $q \equiv \pm 1(3)$ and $\zeta = \pm 1$ as $q \equiv \pm 1(5)$.

5. Conjugacy classes meeting H

If non-empty, $C \cap H$ must be a union of conjugacy classes of H. These are well-known, and it is easily seen that $C \cap H$ must consist of all the elements of order ℓ in H, except possibly in the cases $\ell = 3$ for $H \cong A_4$, and $\ell = 5$ for $H \cong A_5$, which we now consider.

A_4 has two mutually inverse classes of four elements of order 3. If $p \neq 3$ the unique class C of elements of order 3 in G contains both H-classes, so $|C \cap H| = 8$. If $q = 3^{2f+1}$ then the two G-classes, being mutually inverse, each satisfy $|C \cap H| = 4$; if $q = 3^{2f}$, however, the two G-classes are self-inverse, so one has $|C \cap H| = 8$ and the other $|C \cap H| = 0$. Similar arguments apply to $H \cong A_5$, which has two self-inverse classes of twelve elements of order 5, each the square of the other. In this way one can easily compute the values of $|C \cap H|/|C|$ required in (4.4), as shown in Table 4, where θ, $\phi = 0$ and 2 for the two classes C.

ℓ	q	$H \cong A_4$	$H \cong S_4$	$H \cong A_5$
2		$\frac{3(q-\delta)}{\|G\|}$	$\frac{9(q-\delta)}{\|G\|}$	$\frac{15(q-\delta)}{\|G\|}$
3	$= 3^{2f+1}$	$\frac{8}{q^2-1}$	$-$	$-$
	$= 3^{2f}$	$\frac{8\theta}{q^2-1}$	$\frac{8\theta}{q^2-1}$	$\frac{20\theta}{q^2-1}$
	$\neq 3^e$	$\frac{8}{q(q+\varepsilon)}$	$\frac{8}{q(q+\varepsilon)}$	$\frac{20}{q(q+\varepsilon)}$
4		0	$\frac{6}{q(q+\delta)}$	0
5	$= 5^{2f+1}$	0	$-$	$\frac{24}{q^2-1}$
	$= 5^{2f}$	0	0	$\frac{24\phi}{q^2-1}$
	$\equiv \pm 1(5)$	0	0	$\frac{12}{q(q+\zeta)}$

Table 4 (values of $|C \cap H|/|C|$)

6. Proof of the Theorem, case (i)

Suppose that $G > H \cong A_4$. The only powers $h = g^j$ which can make non-zero contributions to (4.3) have orders $\ell = 1$, 2 or 3. The identity element $h = g^k = 1$ has $\pi(h) = n$. If k is even then by taking $j = \frac{1}{2}k$ we find an involution $h = g^j$, and if k is divisible by 3 there are two mutually inverse elements $h = g^j$ of order $\ell = 3$.

(a) *Assume first that* $p \geq 5$, so $|G| = \frac{1}{2}q(q^2 - 1)$ and $n = q(q^2 - 1)/24$. By (4.4) and Table 4, a power $h = g^j$ of order $\ell = 2$, if it exists, has

$$\frac{\pi(h)}{n} = \frac{6}{q(q+\delta)},$$

and any powers h of order $\ell = 3$ have

$$\frac{\pi(h)}{n} = \frac{8}{q(q+\varepsilon)}.$$

If we introduce the useful convention that for any proposition P

$$[\mathrm{P}] = \begin{cases} 1 & \text{if P is true} \\ 0 & \text{if P is false} \end{cases}$$

we can therefore write (4.3) in the form

$$\frac{\beta(g)}{n} = 1 - \frac{1}{k}\left(1 + [2 \mid k]\frac{6}{q(q+\delta)} + [3 \mid k]\frac{16}{q(q+\varepsilon)}\right). \tag{6.1}$$

This clearly implies that

$$\frac{\beta(g)}{n} \geq 1 - \frac{1}{k}\left(1 + \frac{22}{q(q-1)}\right), \tag{6.2}$$

though for specific values of k we can generally find better bounds. Assuming that $q \geq 7$, we see that

$$\frac{\beta(g)}{n} \geq \frac{2}{5},$$

whereas (4.1′) implies that

$$\sum_{i=1}^{r} \frac{\beta(g_i)}{n} < 2.$$

Thus $r < 5$. Now G is not cyclic, so $r > 2$ and hence $r = 3$ or $r = 4$.

Assume first that $r = 4$. If $q \geq 11$ then

$$\frac{\beta(g)}{n} \geq \frac{26}{55}, \quad \frac{34}{55} \quad \text{for} \quad k = 2, \quad k \geq 3 \text{ respectively},$$

so to satisfy (4.1′) each g_i must have order $k_i = 2$ for $i = 1, \ldots, 4$. However, the group $\Delta(2^{(4)})$ is soluble, so it cannot map onto G, and this case does not arise. If $q = 7$ then

$$\frac{\beta(g)}{n} = \frac{3}{7}, \frac{4}{7}, \frac{5}{7}, \frac{6}{7} \quad \text{for} \quad k = 2,\ 3,\ 4,\ 7$$

(these being the only orders of non-identity elements of G), so the only possibility is that $K = \{2^{(3)}, 3\}$. However, such a 4-tuple $\mathbf{g}$ cannot generate G: if it did, then by applying the Riemann-Hurwitz formula to the representation of G ($\cong \mathrm{PGL}_3(2)$) of degree 7 we would get genus $\gamma = -1$, which is absurd.

Now assume that $r = 3$. First let $K = \{2, 3, 7\}$, so that G is a Hurwitz group, that is, a finite image of $\Delta(2, 3, 7)$. By putting $k = 2$, 3 and 7 in (6.1) we see that

$$\begin{aligned}
\frac{\beta(g_1)}{n} &= \frac{1}{2} - \frac{3}{q(q+\delta)} \geq \frac{1}{2} - \frac{3}{q(q-1)}, \\
\frac{\beta(g_2)}{n} &= \frac{2}{3} - \frac{16}{3q(q+\varepsilon)} \geq \frac{2}{3} - \frac{16}{3q(q-1)}, \\
\frac{\beta(g_3)}{n} &= \frac{6}{7}.
\end{aligned}$$

Now (4.1′) gives
$$\begin{aligned}
2 &> \sum \frac{\beta(g_i)}{n} \\
&\geq \frac{85}{42} - \frac{25}{3q(q-1)},
\end{aligned}$$

so that
$$q(q-1) < 42 \cdot \frac{25}{3} = 350$$

and hence $q \leq 19$. Macbeath [9] has shown that $\mathrm{PSL}_2(q)$ is a Hurwitz group if and only if $q = 7$, or $q = p \equiv \pm 1 \bmod 7$, or $q = p^3$ where $p \equiv \pm 2$ or $\pm 3 \bmod 7$. Thus the only possibilities are $q = 7$ and $q = 13$, and in each case one easily sees that $\mathbf{g}$ satisfies (4.1), so both cases arise.

Now suppose that the triple $K \neq \{2, 3, 7\}$, so

$$\sum \frac{1}{k_i} \leq \frac{23}{24}$$

(attained by $K = \{2,3,8\}$). For $i = 1, 2, 3$ we have

$$\frac{\beta(g_i)}{n} \geq 1 - \frac{1}{k_i}\left(1 + \frac{22}{q(q-1)}\right)$$

by (6.2), so

$$\begin{aligned} 2 &> \sum \frac{\beta(g_i)}{n} \\ &\geq 3 - \left(1 + \frac{22}{q(q-1)}\right)\sum \frac{1}{k_i}. \end{aligned}$$

Thus

$$1 + \frac{22}{q(q-1)} > \left(\sum \frac{1}{k_i}\right)^{-1} \geq \frac{24}{23}$$

and hence $q < 23$, so $q = 19$, 17, 13, 11 or 7.

If $q = 19$ then $n = 285$. The non-identity elements $g \in G$ have orders $k = 2$, 3, 5, 9, 10 and 19, with $\beta(g) = 140$, 186, 228, 252, 256 and 270 respectively. It is straightforward but tedious to check that there is no solution of the 'Knapsack Problem' (4.1)

$$\sum_{i=1}^{3} \beta(g_i) = 2n - 2 = 568,$$

so the case $q = 19$ does not arise. The cases $q = 17$ and $q = 11$ are also eliminated by this method. For $q = 13$ (with $n = 91$) we find only the solution $44 + 58 + 78 = 180$, corresponding to the triple $K = \{2,3,7\}$ already listed. For $q = 7$ (with $n = 14$) the only solutions are $8+8+10 = 26$ and $6 + 8 + 12 = 26$, corresponding to the cases $K = \{2,3,7\}$ (again, listed earlier) and $K = \{3,3,4\}$ where results of Macbeath [9] imply that $PSL_2(7)$ is indeed an image of $\Delta(K)$.

There remains the case $q = 5$. This representation is the same as the natural representation of $PSL_2(4)$, and in [5] it is shown that this has genus 0 for the six multisets K listed in Table 1.

(b) *Now suppose that* $p = 3$. If g has order k then

$$\begin{aligned} \frac{\beta(g)}{n} &= 1 - \frac{1}{k}\left(1 + [2 \mid k]\frac{6}{q(q+\delta)} + [k = 3]\frac{16\theta}{q^2 - 1}\right) \\ &\geq 1 - \frac{1}{k}\left(1 + \frac{32}{q^2 - 1}\right) \end{aligned}$$

where

$$\theta = \begin{cases} 0 \text{ or } 2 & \text{if } e \text{ is even,} \\ 1 & \text{if } e \text{ is odd.} \end{cases}$$

As in (a) we must have $r = 3$ or $r = 4$. If $r = 4$ the only possibility is $K = \{2^{(3)}, 3\}$ with $q = 9$, and as in [5] we can eliminate this case by applying (2.3) to a representation of G ($\cong A_6$) of degree 6, obtaining $\gamma = -1$.

Hence $r = 3$. If $K = \{2, 3, 7\}$ then arguing as in (a) we obtain $q(q-1) < 574$, so $q = 9$, impossible since $\mathrm{PSL}_2(9)$ is not a Hurwitz group [9]. Thus $K \neq \{2, 3, 7\}$, so $\sum k_i^{-1} \leq \frac{23}{24}$, and the argument used in (a) now gives $q \leq 27$. If $q = 27$ we soon find that there are no solutions of (4.1) in G, while for $q = 9$ the only solutions correspond to multisets $K = \{2, \bar{3}, 5\}$, $\{2, 4, 4\}$ and $\{3, \bar{3}, 4\}$ (where 3 and $\bar{3}$ denote generators g_i of order 3 with $\beta(g_i) = 16$ or 20 respectively). Groups generated by triples $\mathbf{g}$ of the first two types must be isomorphic to A_5 or soluble, but the third type can generate G: for instance, we can identify G with A_6 and take $g_1 = (123)$, $g_2 = (145)(236)$, $g_3 = (1543)(26)$.

(c) *Now let* $p = 2$, so $|G| = q(q^2 - 1)$ and $n = q(q^2 - 1)/12$. Since $G > H \cong A_4$, e is even, so $q \equiv 1 \bmod 3$ and $\varepsilon = 1$. If $g \in G$ has order k (= 2 or odd), then

$$\begin{aligned}\frac{\beta(g)}{n} &= 1 - \frac{1}{k}\left(1 + [k = 2]\frac{3}{q^2 - 1} + [3 \mid k]\frac{16}{q(q+1)}\right)\\ &\geq 1 - \frac{1}{k}\left(1 + \frac{16}{q(q-1)}\right).\end{aligned}$$

The case $q = 4$ was dealt with in (a), since $\mathrm{PSL}_2(4) \cong \mathrm{PSL}_2(5)$, so we can assume that $q \geq 16$. Arguing as in (a) and (b) we find that $r = 3$ or 4, and that the latter case leads only to $K = \{2^{(4)}\}$, which is impossible. Thus $r = 3$.

Macbeath [9] has shown that $\mathrm{PSL}_2(2^e)$ is a Hurwitz group only for $e = 3$, so $K \neq \{2, 3, 7\}$. Hence $\sum k_i^{-1} \leq \frac{23}{24}$, and arguing as in (a) and (b) we find that $q(q-1) < 16 \cdot 23$. Thus $q = 16$, and we find that there are no solutions of (4.1) in G.

7. Cases (ii) and (iii)

The proof is similar for $H \cong S_4$ or A_5. The only extra ingredient is that for certain q and K, character theory is used to determine whether or not a particular solution $\mathbf{g} = (g_i)$ of (4.1) generates G. For example, if $H \cong S_4$ and $q = 9$ (so $G \cong A_6$ and $n = 15$) the multiset $K = \{3, 3, 5\}$ satisfies (4.1), where the generators g_1 and g_2 of order 3 are conjugate, with

$\beta(g_1) = \beta(g_2) = 8$ and $\beta(g_3) = 12$; now the character table [1] shows that G has 1440 such triples $\mathbf{g}$ with $g_1 g_2 g_3 = 1$, but since G has 12 subgroups isomorphic to A_5, each generated by 120 such triples, none can generate G.

References

[1] J H CONWAY, R T CURTIS, S P NORTON, R A PARKER AND R A WILSON, *Atlas of Finite Groups* (Oxford University Press, Oxford, 1985).

[2] L E DICKSON, *Linear Groups* (Dover, New York, 1958).

[3] R M GURALNICK AND J G THOMPSON, Finite groups of genus zero, *J. Algebra* **131** (1990) 303-341.

[4] B HUPPERT, *Endliche Gruppen I* (Springer-Verlag, Berlin, Heidelberg, New York, 1967).

[5] G A JONES, The monodromy genus of $\mathrm{PSL}_2(q)$, submitted.

[6] G A JONES AND D SINGERMAN, *Complex Functions: an Algebraic and Geometric Viewpoint* (Cambridge University Press, Cambridge, 1987).

[7] K KUIKEN, On the monodromy groups of Riemann surfaces of genus zero, *J. Algebra* **59** (1979) 481-489.

[8] M W LIEBECK AND J SAXL, Minimal degrees of primitive permutation groups with an application to monodromy groups of covers of Riemann surfaces, *Proc. London Math. Soc.* **63** (1991) 266-314.

[9] A M MACBEATH, Generators of the linear fractional groups, *Proc. Sympos. Pure Math.* **12** (1967) 14-32.

[10] J-P SERRE, Groupes de Galois sur $\mathbb{Q}$, *Astérisque* **161 - 162** (1988) 73-85.

[11] M SUZUKI, *Group Theory I*, (Springer-Verlag, Berlin, Heidelberg, New York, 1982).

Department of Mathematics
University of Southampton
Southampton SO9 5NH

On the rank of NEC groups

Ralf Kaufmann and Heiner Zieschang

To Murray Macbeath on the occasion of his retirement

Abstract. For some NEC groups we show that the geometric rank and the algebraic rank differ seriously, namely their ratio can be $3:2$. Moreover, we give examples of free products of two groups with an amalgamated subgroup where the rank of the product is much smaller than the sum of the ranks of the factors minus the rank of the amalgamated subgroup.

0. Introduction

In 1964 A.M. Macbeath posed the following question to the second author: Does the group $\Gamma = \langle x_1, \ldots, x_5 \mid x_1^2,\ x_2^3,\ x_3^5,\ x_4^7,\ x_5^{11},\ x_1x_2x_3x_4x_5\rangle$ need at least 4 generators? This question was originally considered by J. Nielsen. The group Γ is a Fuchsian group and admits a fundamental polygon with 4 pairs of equivalent sides. Every pair determines a generator, for example, x_2, x_3, x_4, x_5. By simple arguments using the Euler characteristic it can be proved that every system of such "geometric" generators contains at least 4 elements. Call this number the *geometric rank* $\mathrm{gr}(\Gamma)$. The minimal number of generators of a group Γ is called the *rank* of Γ and is denoted by $\mathrm{rg}(\Gamma)$. Clearly, $\mathrm{rg}(\Gamma) \leq \mathrm{gr}(\Gamma)$. Thus the question of Nielsen is whether geometric rank and group-theoretical rank coincide.

This question has been considered for arbitrary Fuchsian groups in [10] where it was claimed that the answer is positive. However slightly later Burns, Karrass, Pietrowski and Solitar showed that the Fuchsian group $\Gamma_0 = \langle c_1, c_2, c_3, c_4 \mid c_1^2,\ c_2^2,\ c_3^2,\ c_4^{2k+1},\ c_1c_2c_3c_4 \rangle$, $k > 0$ is generated by $x = c_1c_2$ and $y = c_1c_3$. Now

$$xy^{-1}x^{-1}y = c_1c_2 \cdot c_3c_1 \cdot c_2c_1 \cdot c_1c_3 = (c_1c_2c_3)^2 = c_4^{-2}.$$

This implies that $c_4 = (c_1c_2c_3)^{-1} \in \langle x, y \rangle$ (the subgroup generated by x, y). A consequence is that $c_3 \in \langle x, y \rangle$ and therefore also $c_1, c_2 \in \langle x, y \rangle$. The geometric rank of Γ_0, however, equals 3. Later it was shown that this is, more or less, the only example of a Fuchsian group where the two concepts of rank differ.

This counterexample has a very interesting consequence. F. Waldhausen asked in [8] whether for a closed orientable 3-manifold M^3 the minimal genus of a Heegaard decomposition of M^3, $h(M^3)$, is equal to the rank of its fundamental group: $h(M^3) = \mathrm{rg}(\pi_1 \mathrm{M}^3)$. (By the Seifert-van Kampen theorem, $\mathrm{rg}(\pi_1 \mathrm{M}^3) \leq \mathrm{h}(\mathrm{M}^3)$.) Using the example of Burns-Karrass-Pietrowski-Solitar it can be shown that the question has a negative answer for some manifolds, namely for Seifert fiber spaces with basis S^2, three exceptional fibres of order 2 and one exceptional fibre of odd order, see [1].

A generalization of Fuchsian groups are discontinuous groups which contain also orientation reversing conformal mappings. These groups are called NEC groups and were considered and classified by Wilkie [9]. The geometric rank of these groups is as one expects, see section 1. If such a group does not contain reflections the rank has also been determined in [7]; it equals the geometric rank. From the Burns-Karrass-Pietrowski-Solitar example it is clear that the situation becomes much more complicated if reflections appear since they define generators of order 2. We will give further examples where the rank is less than the geometric rank. In fact, we describe some NEC groups Γ with $\mathrm{rg}(\Gamma) = \frac{2}{3}\ \mathrm{gr}(\Gamma)$.

Often a Fuchsian or NEC group can be written as free product of two groups with amalgamation: $\Gamma = A *_C B$. An open question is the relationship between the rank of Γ and the ranks of A, B, C. For decompositions of Fuchsian groups, $\mathrm{rg}(\Gamma) \geq \mathrm{rg(A)} + \mathrm{rg(B)} - 2 \cdot \mathrm{rg(C)}$; a similar result is true for the NEC groups considered below. For arbitrary groups Γ, however, there cannot be a general formula of a form $\mathrm{rg}(\Gamma) \geq \mathrm{rg(A)} + \mathrm{rg(B)} - \mathrm{k} \cdot \mathrm{rg(C)}$, $k \in \mathbb{Z}$ as an example of M. Lustig shows.

1. On the geometric rank of NEC groups

A *NEC group* (non-euclidean crystallographic group) is a discontinuous group of motions of the non-euclidean (= Bolyai-Gauß-Lobachevskii or hyperbolic) plane. A model of this plane is the following: Take the upper half-plane of the euclidean plane $\mathsf{H} = \{z \in \mathbb{C} \ : \ \text{Im } z > 0\}$ as point set and half-circles or halflines orthogonal to the real axis of the geometry as lines. Distances are defined using suitable cross ratios. The special group of linear fractional transformations with real coefficients:

$$\text{PSL}(2, \mathbb{R}) = \left\{ w = \frac{az+b}{cz+d} \ : \ a, b, c, d \in \mathbb{R}, \ ad - bc = 1 \right\}$$

is the group of orientation preserving motions. The set of orientation reversing motions is

$$\left\{ w = \frac{a\bar{z}+b}{c\bar{z}+d} \ : \ a, b, c, d \in \mathbb{R}, \ ad - bc = -1 \right\} .$$

A subgroup Γ of the group of motions *operates discontinuously on* H if there exists an open subset $V \subset \mathsf{H}$ such that $V \cap f(V) \neq \emptyset, \ f \in \Gamma$ implies $f = \text{id}_{\mathsf{H}}$.

Every NEC group Γ admits a *fundamental polygon*, that is a polygon $P \subset \mathsf{H}$ with the property that $\mathring{P} \cap f(\mathring{P}) \neq \emptyset, \ f \in \Gamma$ implies $f = 1$. By $\mathring{P}$ we denote the interior of P and by ∂P the boundary of P in H. The sides of P are segments, lines or halflines with respect to the non-euclidean geometry, i.e. segments of euclidean circles or halflines in H which are orthogonal to the real axis. In general, the number of sides can be infinite. There may also appear segments of the closed real axis $\mathbb{R} \cup \{\infty\}$ in the boundary of P if P is considered as subset of $\mathbb{C} \cup \{\infty\}$, but these parts are not interpreted as sides of P. Given a side $\sigma \subset \partial P$, then there is a unique transformation $\gamma \in \Gamma$ such that $P \cap \gamma(P) = \sigma$; hence, every side of P determines an element of Γ. Clearly, $\gamma^{-1}(\sigma)$ is also a side of P and it determines the transformation γ^{-1}. These two sides may be the same, namely if γ has order 2. Then γ is either a rotation of order 2 (if γ preserves orientation) or a reflection along some line (γ reverses the orientation). Let $N(P)$ be the number of such pairs $\{\gamma, \gamma^{-1}\}$. We obtain a system of generators of Γ if we take from each pair one transformation. A system selected this way is called a *geometric system of generators associated to* P. It contains $N(P)$ elements. The minimum of $\{N(P) : P \text{ fundamental polygon of } \Gamma\}$ is called the *geometric rank of* Γ and is denoted by $\text{gr}(\Gamma)$. The NEC group Γ is of *finite type* if the geometric rank is finite, that is if there exists a fundamental polygon for Γ with a finite number of sides. In particular, this is the case when Γ admits a compact fundamental polygon.

NEC groups of finite type were introduced and classified up to topologically equivalent actions by Wilkie [9] for the case of a compact fundamental domain and in [5] and [11] for the general case. The generators described there are geometric; hence, their number is an upper bound for the geometric rank of the NEC group presented.

The classification of NEC groups of finite type is conveniently explained using the orbifold corresponding to the action of Γ on H and the quotient mapping. Let us do it for the case of compact fundamental domain. The orbifold $B = \mathsf{H}/\Gamma$ (= space of orbits with the quotient topology) is a surface of finite type obtained from a closed surface of some genus g by cutting out a number of (open) disks. The result is a compact orientable (+) or non-orientable (−) surface of genus g with a number q of boundary components $C_1, \dots, C_q$. The boundary points of B are the images of points of H which lie on axes of reflections of Γ. The quotient mapping $\pi: \mathsf{H} \to B$ behaves as a covering at all points except at those which are not fixed points of rotations or reflections of Γ. At a rotation center which does not lie on a reflection axis the projection π behaves as a branched covering at a branch point. On B there are only finitely many, say m, points of this type with branching numbers $h_1, \dots, h_m$. If two reflections axes c_1, c_2 intersect in a point $x \in \mathsf{H}$ and the angle between the two lines is α then the product $c_1 c_2$ is a rotation along x of angle 2α. On the boundary component C_i, $1 \le i \le q$ there are the images of some number m_i of inequivalent rotation centers; the case $m_i = 0$ is possible. The images of rotation centers define a cycle $(h_{i,1}, h_{i,2}, \dots, h_{i,m_i})$ of rotation orders. If the surface B is orientable we fix some orientation of B, and this induces an orientation on every boundary component. Now the topological type of Γ is described by the following *signature* (where $\varepsilon = \pm$):

$$(g;\ \varepsilon;\ m : h_1, \dots, h_m;\ q : \{(h_{1,1}, \dots, h_{1,m_1}), \dots, (h_{q,1}, \dots, h_{q,m_q})\}) .$$

Without changing the type of the NEC group we can do the following:

— the enumerations of the h_i and of the boundary components C_i can arbitrarily be changed;

— if the surface B is non-orientable the cycles $(h_{j,1}, \dots, h_{j,m_j})$ can be independently reversed;

— if B is orientable all cycles $(h_{j,1},\ldots,h_{j,m_j})$ can be simultaneously reversed.

Two NEC groups with compact fundamental polygons are topologically equivalent if and only if their signatures are related by the above procedures [9]. This condition is also necessary for NEC groups with compact fundamental polygon to be isomorphic [12, 4.6, 3.4].

By cutting B along suitable arcs we obtain a disk which does not contain images of rotation centers in its interior. Let $\alpha = 2$ if $\varepsilon = +$, otherwise $\alpha = 1$. The minimal number of arcs needed is $\alpha g+m+q-1$ if $m > 0$ or $q > 0$ and αg if $m = q = 0$. Lifting the cut arcs to H we obtain a fundamental polygon P. Each of these arcs is covered by two sides of P, except when the arc ends in the image of a rotation of order 2. The other sides of ∂P are in a 1-1 correspondence with the different segments on the C_i. Thus the polygon P has $2\alpha g + 2m + 2q + \sum_{i=1}^{q} m_i - 2$ or $2\alpha g$ sides, respectively. The number of equivalent pairs of sides is $\alpha g + m + q + \sum_{i=1}^{q} m_i - 1$ or αg, respectively.

1.1 Proposition [6]. *If the NEC group Γ has the signature*

$$(g;\varepsilon;\ m : h_1,\ldots,h_m;\ q : \{(h_{1,1},\ldots,h_{1,m_1}),\ldots,(h_{q,1},\ldots,h_{q,m_q})\}),$$

then

$$\mathrm{gr}(\Gamma) = \begin{cases} \alpha g + m + q + \sum_{i=1}^{q} m_i - 1 & \text{if } m+q>0, \\ \alpha g & \text{otherwise.} \end{cases}$$

Here $\alpha = 2$ if $\varepsilon = +$, or else $\alpha = 1$.

Proof. We obtained above that the expression on the right side is an upper bound for the geometric rank. The other direction has been proved in [6, Theorem 3]. ∎

The (topological) classification of NEC groups of finite type but without compact fundamental polygon is done in a similar fashion; however now the holes may be "open" or contain "open segments"; for details see [5], [12, 4.11].

2. An upper bound for the rank of some NEC groups

As remarked above the rank of Fuchsian groups and of NEC groups without reflections has been determined in [7]. There are many different types of NEC groups with reflections and it seems complicated to deal with them in general. We will consider the simplest interesting new cases, namely the cases with basis S^2 and one boundary component, that is a NEC group with signature $(0;+;0;\ 1:(h_1,\dots,h_n))$. We start with some simple general considerations.

2.1 LEMMA. *Let*

$$H = \langle c_1, c_2, c_3 \mid c_1^2, c_2^2, c_3^2, (c_1c_2)^2, (c_2c_3)^k \rangle \quad \text{with } k = 2\ell + 1.$$

Then $\mathrm{rg(H)} = 2$.

PROOF. Define $x = c_2,\ y = c_1c_3$. It suffices to prove that c_1 is contained in the subgroup $\langle x, y\rangle < H$ generated by x and y. From

$$\begin{aligned} xyx^{-1}y^{-1} &= c_2 \cdot c_1c_3 \cdot c_2^{-1} \cdot c_3^{-1}c_1^{-1} \\ &= c_1c_2c_3c_2c_3c_1 = c_1 \cdot (c_2c_3)^2 \cdot c_1^{-1} \end{aligned}$$

(for the second step the relation $(c_1c_2)^2 = 1$ is crucial) we obtain for $\lambda \in \mathbb{Z}$:

$$\begin{aligned} (xyx^{-1}y^{-1})^\lambda &= c_1(c_2c_3)^{2\lambda}c_1, \\ (xyx^{-1}y^{-1})^\lambda \cdot xy &= c_1(c_2c_3)^{2\lambda}c_1 \cdot c_2c_1c_3 \\ &= c_1(c_2c_3)^{2\lambda+1}; \end{aligned}$$

for $\lambda = \ell$, in particular, it follows that c_1 is a product in x, y. ∎

2.2 COROLLARY. (a) *Let G be a group, $U < G$ a subgroup and $\varphi\colon H \to U$ an epimorphism. Then* $\mathrm{rg(U)} \leq 2$.

(b) *Let $U_1, \dots, U_p$ be subgroups of G each of which is the image of a group H of the form as in 2.1 (possibly with different ℓ's). Then the rank of the subgroup $\langle \cup_{i=1}^p U_i\rangle < G$ is at most $2p$:*

$$\mathrm{rg}(\langle \cup_{i=1}^p \mathrm{U_i}\rangle) \leq 2\mathrm{p}. \qquad ∎$$

Next we apply this result to the NEC groups mentioned above.

2.3 PROPOSITION. *Let*

(1) $G = \langle c_1, \dots c_n \mid c_j^2,\ 1 \leq j \leq n;\ (c_jc_{j+1})^{h_j},\ 1 \leq j \leq n-1\rangle$ *and*
(2) $\Gamma = \langle c_1, \dots c_n \mid c_j^2,\ 1 \leq j \leq n;\ (c_jc_{j+1})^{h_j},\ 1 \leq j \leq n-1,\ (c_nc_1)^{h_n}\rangle$.

Assume that there are p disjoint ('critical') triples $(j, j+1, j+2)$ in which one term equals 2 and one of its neighbours is odd. Then $\mathrm{rg}(\Gamma) \leq \mathrm{rg(G)} \leq \mathrm{n-p}$.

PROOF. This is a direct consequence of 2.2 since G is generated by $c_1, \dots, c_n$ and there are p disjoint triples c_j, c_{j+1}, c_{j+2} where two elements suffice to generate $\langle c_j, c_{j+1}, c_{j+2} \rangle$. ∎

2.4 PROPOSITION.

$$\mathrm{gr}\,(\Gamma) = \mathrm{n}.$$

PROOF. By 1.1, the geometric rank of the group above equals $2 \cdot 0 + 0 + 1 + n - 1 = n$. ∎

2.5 PROPOSITION. *There are NEC groups* Γ *with* $\mathrm{rg}(\Gamma) \leq \frac{2}{3}\,\mathrm{gr}(\Gamma)$. *There are groups with* $\mathrm{rg}(\Gamma) = \frac{2}{3}\,\mathrm{gr}(\Gamma)$.

PROOF. Take $n = 3\nu$, $\nu \in \mathbb{Z}$ and define

$$h_i = \begin{cases} 2\lambda_i + 1 & \text{if } i \equiv 1 \bmod 3 \\ 2 & \text{if } i \equiv 2 \bmod 3 \\ 2\lambda_i & \text{if } i \equiv 0 \bmod 3 \end{cases} \qquad \text{with } \lambda_i > 1, \ 1 \leq i \leq 3\nu.$$

Then $(h_1, \dots, h_n)$ contains ν disjoint critical triples and, therefore,

$$\mathrm{rg}(\Gamma) \leq \mathrm{n} - \nu = \tfrac{2}{3}\,\mathrm{n} = \tfrac{2}{3}\,\mathrm{gr}(\Gamma).$$

We introduce the new relations $c_{3j+1} = c_{3j+2}$, $1 \leq j \leq \nu$ and abelianize the group Γ. This defines an epimorphism $\Gamma \to \mathbb{Z}_2^{2\nu}$; hence $\mathrm{rg}(\Gamma) \geq 2\nu = \frac{2}{3}\mathrm{n}$. ∎

The last argument also proves the following.

2.6 PROPOSITION. *If* Γ *has the signature* $(0; +; 0;\ 1 : (h_1, \dots, h_n))$ *and all* $h_i > 0$ *are even then* Γ *has rank n; hence, by* 2.4, $\mathrm{rg}(\Gamma) = \mathrm{gr}(\Gamma) = \mathrm{n}$. ∎

Even for NEC groups with the special signature $(0; +; 0;\ 1 : (h_1, \dots, h_n))$ the rank is not known. The Nielsen method offers a way to calculate the rank, but this is very cumbersome as we will see when we determine the rank for the simplest cases.

2.7 PROPOSITION. *The NEC groups*

$$G = \langle c_1, c_2, c_3 \mid c_1^2,\ c_2^2,\ c_3^2,\ (c_1 c_2)^h,\ (c_2 c_3)^k \rangle$$

with odd h, $k > 2$ *have rank* 3.

PROOF. Assume that x, y generate G. Write G as a free product with amalgamation:

$$G = \langle c_1, c_2 \mid c_1^2,\ c_2^2,\ (c_1 c_2)^h \rangle *_{\langle c_2 \rangle} \langle c_2, c_3 \mid c_2^2,\ c_3^2,\ (c_2 c_3)^k \rangle = A *_C B.$$

This free product structure defines a length L on G. We apply the theorem on the Nielsen method for free products with amalgamation ([10, Satz 1] or [3, 4.1]) and reduce the generators x, y according to the length as often as possible. Assume that finally we obtain generators, again denoted by x, y, such that every $w \in G$ is written as a word in letters x, y which have length not bigger than $L(w)$. It easily follows that c_1, c_2, c_3 cannot all be expressed as words in x, y.

There are two types of obstructions to getting a situation of this type:
(a) There is an element of $\{x, y\}$, say x, in the amalgamated subgroup $C = \langle c_2 \rangle$ and there is an element $v \in (A-C) \cup (B-C)$ such that $vxv^{-1} \in C$.
(b) One element or both lie in a subgroup of G conjugate to one of the factors A, B and some power of this element or some word in both, respectively, becomes shorter.

Case (a): Since the only non-trivial element of C is c_2, it follows $x = c_2$. Assume that $v \in A$. Then we have to solve the equation $v\ c_2\ v^{-1} = c_2$ in the dihedral group A. The only solutions are $1, c_2 \in C$ since h is odd and > 2. (This is the only place where we use the assumption that h, k are odd and > 2 and this makes the difference to 2.1.) Therefore case (a) cannot appear.

Case (b): Clearly x and y cannot both lie in the same conjugate of A or B. Therefore we may assume that

(1) x lies in a conjugate of A, by [10, (2.4)(a)] and symmetry;

(2) x does not lie in C, by [10, (2.4)(b)];

(3) a power of x is conjugate to a non-trivial element of C, by [10, (2.4)(c)].

After conjugating x and y by the same element, we may assume that

(1′) x lies in A;

(2′) x does not lie in a conjugate of C and in particular $x \notin C$;

(3′) a power of x is conjugate to a non-trivial element of C.

However c_2 is the only non-trivial element of C and is not a proper power of an element in A, contradicting (2′). ∎

By further considerations one can also prove that rg(G) $= 3$ if h, k are allowed to be even ($h > 2$). Similar arguments can also be given for groups with more generators, and for NEC groups of the type 2.3(2), but this becomes rather cumbersome.

3. On the rank of free products with amalgamation

In this section we describe an example of M. Lustig which shows that a result like the Grushko theorem does not hold for free products with amalgamated subgroups.

3.1 PROPOSITION. *The group*

$$X = \langle a_1, \ldots, a_n, b, c_1, \ldots, c_n \mid a_j^2,\ c_j^2,\ b^2,\ (a_j b)^2,\ (bc_j)^{2\kappa_j+1},\ 1 \le j \le n\rangle$$
$$= \mathop{*}_{j=1\,\langle b\rangle}^{n} \langle a_j, b, c_j \mid a_j^2, b_j^2, c_j^2, (a_j b)^2, (bc_j)^{2\kappa_j+1}\rangle$$

is generated by the set $\{x = b,\ y_j = a_j c_j : 1 \le j \le n\}$.

Namely, by 2.1, the elements b and $a_j c_j$ generate

$$\langle a_j, b, c_j \mid a_j^2, b^2, c_j^2, (a_j b)^2, (bc_j)^{2\kappa_j+1}\rangle.$$

3.2 LEMMA. (a) *Let* $U = \langle a_1, \ldots, a_n, b \mid b^2,\ a_j^2,\ (a_j b)^2,\ 1 \le j \le n\rangle$. *Then* $\mathrm{rg(U) = n+1}$.

(b) *Let* $V = \langle b, c_1, \ldots, c_n \mid b^2,\ c_j^2,\ (bc_j)^{h_j},\ h_j \ge 2,\ 1 \le j \le n\rangle$. *Then* $\mathrm{rg(V)} \ge \frac{1}{2} \cdot (\mathrm{n}+1)$.

(c) *If there is a number* $p > 1$ *such that* $p|h_j$ *for* $1 \le j \le n$ *then* $\mathrm{rg(V) = n+1}$.

PROOF. (a) The group U has the group $\mathbb{Z}_2^{n+1}$ as a factor group and is generated by $n+1$ elements.

(b) Consider the homomorphism

$$\omega\colon V \to \mathbb{Z}_2 = \{1, -1\}, \quad b, c_1, \ldots, c_n \mapsto -1.$$

By the Reidemeister-Schreier method, $\ker(\omega)$ is generated by

$$x = b^2,\ y_j = c_j b^{-1},\ z_j = bc_j \qquad \text{with } 1 \le j \le n,$$

and the defining relations are

$$x,\ y_j z_j,\ y_j^{h_j},\ z_j^{h_j},\ 1 \le j \le n;$$

hence

$$\ker\, \omega = \langle y_1, \ldots, y_n \mid y_1^{h_1}, \ldots, y_n^{h_n}\rangle \cong \mathbb{Z}_{h_1} * \ldots * \mathbb{Z}_{h_n}.$$

By the Grushko theorem, $\mathrm{rg}(\ker\omega) = \mathrm{n}$. If V has rank r then every subgroup of index 2 has a rank of at most $2\cdot(r-1)+1$ as follows from the formula for subgroups of free groups (see [12, 1.6.1]). Therefore

$$2(r-1)+1 \geq n \quad \Longrightarrow \mathrm{rg(V)} \geq \frac{1}{2}(\mathrm{n}+1).$$

(c) Consider the homomorphism

$$\rho\colon \mathbb{Z}V \to \mathbb{Z}/p\mathbb{Z}, \qquad b, c_1, \ldots, c_n \mapsto -1.$$

We consider this as a mapping to the ring of 1×1- matrices with coefficients in the ring $\mathbb{Z}/p\mathbb{Z}$. Then the conditions of [4, Corollary 1] are fulfilled; in particular, the images of the Fox derivatives [2, p. 123-125] of the defining relations vanish:

$$\left(\frac{\partial c_i^2}{\partial c_j}\right)^{\rho} = \begin{cases} \rho(1+c_i) \equiv 0 \bmod \mathrm{p} & \text{if } i=j, \\ \rho(0)=0 & \text{otherwise;} \end{cases}$$

$$\left(\frac{\partial (c_i b)^{h_i}}{\partial c_j}\right)^{\rho} = \begin{cases} \rho\left(1+\sum_{k=1}^{h_i-1}(c_i b)^k\right) = h_i \equiv 0 \bmod \mathrm{p} & \text{if } i=j, \\ 0 & \text{otherwise;} \end{cases}$$

$$\left(\frac{\partial b^2}{\partial c_i}\right)^{\rho} = 0,$$

and similarly for the derivative $\partial/\partial b$. By [4, Corollary 1] it follows that $\mathrm{rg(V)} = \mathrm{n}+1$. ∎

3.3 PROPOSITION. *For every function $d\colon \mathbb{N} \to \mathbb{N}$ there exist groups U, V, W such that*

$$\mathrm{rg}\,(\mathrm{U} *_{\mathrm{W}} \mathrm{V}) \leq \mathrm{rg(U)} + \mathrm{rg(V)} - \mathrm{d(rg(W))}.$$

PROOF. Take the group X from 3.1 and the groups U and V from 3.2. Moreover let $W = \langle b\rangle$. Then $X = U *_W V$, $\mathrm{rg(X)} = \mathrm{n}+1$ and

$$\mathrm{rg(U)} + \mathrm{rg(V)} - \mathrm{d}(1) \geq (\mathrm{n}+1) + \frac{1}{2}(\mathrm{n}+1) - \mathrm{d}(1).$$

For $n > 2d(1)$ we have the required inequality. ∎

For a free product $\mathrm{rg(U * V)} = \mathrm{rg(U)} + \mathrm{rg(V)}$ by the Grushko theorem. A similar statement with subtraction of a multiple of the rank of the amalgamated subgroup does not hold for free products with amalgamation as pointed out in Proposition 3.3.

References

[1] Boileau, M.; Zieschang, H.: *Heegaard genus of orientable Seifert fibre spaces.* Invent. math. **76** (1984), 455-468

[2] Burde, G.; Zieschang, H.: *Knots.* Berlin-New York: de Gruyter 1985

[3] Collins, D.J.; Zieschang, H.: *On the Nielsen method in free products with amalgamated subgroups.* Math. Z. **197**, (1987), 97-118

[4] Lustig, M.: *On the Rank, the Deficiency and the Homological Dimension of Groups: Computation of a Lower Bound via Fox Ideals.* In: Topology and Combinatorial Group Theory, Lecture Notes in Mathematics **1440**, 164-174. Berlin- Heidelberg-New York: Springer Verlag 1990

[5] Macbeath, A. M.; Hoare, A..M.: *Groups of hyperbolic crystallography.* Math. Proc. Cambridge Phil. Soc. **79** (1976), 235-249

[6] Martinez, E.: *Convex fundamental regions for NEC groups.* Arch. Math. **47** (1986), 457-464

[7] Peczynski, N.; Rosenberger, G.; Zieschang, H: *Über Erzeugende ebener diskontinuierlicher Gruppen.* Invent. math. **29** (1975), 161-180

[8] Waldhausen, F.: *Some problems on 3-manifolds.* Proc. Symposia in Pure Math. **32** (part 2) (1978), 313-322

[9] Wilkie, H. C.: *On noneuclidean crystallographic groups.* Math. Z. **91** (1966), 87-102

[10] Zieschang, H.: *Über die Nielsensche Kürzungsmethode in freien Produkten mit Amalgam.* Invent. math. **10** (1970), 4-37

[11] Zieschang, H.: *On decompositions of discontinuous groups of the plane.* Math. Z. **151** (1976), 165-188

[12] Zieschang, H.; Vogt, E.; Coldewey, H.-D.: *Surfaces and planar discontinuous groups.* Lecture Notes Math. **835**. Berlin-Heidelberg-New York: Springer-Verlag 1980

Ralf Kaufmann
Wilhelmstr. 13a
4300 Essen-Kettwig

Heiner Zieschang
Fakultät für Mathematik
Ruhr-Universität Bochum
Postfach 102148
4630 Bochum 1

The geometry of bending quasi-Fuchsian groups

Christos Kourouniotis

1. Introduction

In this article we begin the study of properties of the deformation of quasi-Fuchsian structures defined in [K2]. We calculate the first and second variation of the length and the rotation angle associated with a simple closed geodesic when the structure undergoes the deformation determined by bending along another simple closed geodesic, and prove certain basic relations between these variations.

Let S be a closed surface of negative Euler characteristic. We consider homomorphisms $\rho : \pi_1(S) \to \mathrm{PSL}(2,\mathbf{C})$ such that $\Gamma = \mathrm{im}\rho$ is a quasi-Fuchsian group and $\mathcal{H}^3/\Gamma \cong S \times (0,1)$. Two homomorphisms ρ_1, ρ_2 define the same *quasi-Fuchsian structure* on S if there is an inner automorphism $\mathrm{ad}A$ of $\mathrm{PSL}(2,\mathbf{C})$ such that $\rho_1 = \mathrm{ad}A \circ \rho_2$. We shall denote the space of quasi-Fuchsian structures on S by $Q(S)$. Inside $Q(S)$ lies the subset of Fuchsian points, which form the Teichmüller space $T(S)$ of the surface.

The limit set of the group Γ is a Jordan curve. In [K2] the limit set of Γ was used to define a deformation on the space $Q(S)$. For every measured geodesic lamination on S and complex number with sufficiently small imaginary part this defines a mapping $Q(S) \to Q(S)$. This deformation generalises to $Q(S)$ the Fenchel-Nielsen deformation studied by Wolpert [W1], Thurston's earthquakes, and the deformation of bending defined for hyperbolic manifolds of arbitrary dimension by Kourouniotis [K1] and Johnson and Millson [JM], and studied in greater generality in the case of hyperbolic surfaces by Epstein and Marden [EM].

Wolpert [W2] and Kerckhoff [Ke] have used the Fenchel-Nielsen deformation and the geodesic length functions on $T(S)$ to study the geometry of the Teichmüller space, in particular to give an intrinsic geometric interpretation to the Weil-Petterson metric on $T(S)$. In this article, we define complex length functions on $Q(S)$ determined by simple closed geodesics and calculate their variations under bending. The formulae obtained reduce in the special case of the Fenchel-Nielsen deformation to the well-known formulae of Wolpert. Goldman [G] has obtained similar results for the first variation in a more general setting, without direct reference to the hyperbolic geometry of the quasi-Fuchsian structure.

In Section 2 we define the notion of complex distance between two geodesics in hyperbolic space: a complex number whose real part refers to the distance between the geodesics and whose imaginary part refers to the angle between them. This complex distance satisfies relations analogous to those of hyperbolic trigonometry in the plane, [B], [K3]. In Section 3 we consider the case of the derivatives of the complex displacement of a loxodromic isometry under elementary bending deformations. In Section 4 this calculation is extended to the case of the derivatives of the complex length of a simple closed geodesic, under bending along a simple closed geodesic. In Section 5 the complex length function associated to a simple closed geodesic is defined on $Q(S)$. Its first and second variations with respect to bending are calculated and are shown to satisfy certain reciprocity relations, analogous to those of Wolpert [W2] for real length functions.

2. The geometry of geodesics in $\mathcal{H}^3$.

We shall use the upper half plane $\mathcal{H}^2 = \{z \in \mathbf{C},\ \mathrm{Im}(z) > 0\}$ and the upper half space $\mathcal{H}^3 = \{(x,y,z) \in \mathbf{R}^3,\ z > 0\}$, with their respective Poincaré metrics. $\overline{\mathcal{H}^2}$ and $\overline{\mathcal{H}^3}$ will denote the closure of $\mathcal{H}^2$ and $\mathcal{H}^3$ inside $\hat{\mathbf{C}} = \mathbf{C} \cup \{\infty\}$ and $\mathbf{R}^3 \cup \{\infty\}$ respectively. The boundary of hyperbolic space will be denoted by $\partial\mathcal{H}^2 = \overline{\mathcal{H}^2} - \mathcal{H}^2$ and $\partial\mathcal{H}^3 = \overline{\mathcal{H}^3} - \mathcal{H}^3$. The mapping $\pi : \mathrm{SL}(2,\mathbf{C}) \to \mathrm{PSL}(2,\mathbf{C})$ will be the projection. Elements of $\mathrm{PSL}(2,\mathbf{C})$ will be identified with the corresponding isometries of $\mathcal{H}^3$. All isometries of $\mathcal{H}^3$ will be orientation preserving and non-parabolic.

For u and v distinct points of $\mathbf{C}$ and $\varphi \in \mathbf{C}$, we denote by $A(u,v,\varphi)$ the matrix

$$\begin{pmatrix} \cosh\frac{1}{2}\varphi + \frac{v+u}{v-u}\sinh\frac{1}{2}\varphi & \frac{-2uv}{v-u}\sinh\frac{1}{2}\varphi \\ \frac{2}{v-u}\sinh\frac{1}{2}\varphi & \cosh\frac{1}{2}\varphi - \frac{v+u}{v-u}\sinh\frac{1}{2}\varphi \end{pmatrix},$$

and by $A(\varphi) = A(0, \infty, \varphi)$ the matrix

$$A(\varphi) = \begin{pmatrix} e^{\frac{1}{2}\varphi} & 0 \\ 0 & e^{-\frac{1}{2}\varphi} \end{pmatrix}.$$

Note that $A(u, v, \varphi) = A^{-1}(u, v, -\varphi) = A^{-1}(v, u, \varphi)$.

We shall use the following subset of $\mathbf{C}$.

$$\begin{aligned} \Sigma = \{w \in \mathbf{C}, \pi > \operatorname{Im} w > 0\} \\ \cup \{w \in \mathbf{C}, \operatorname{Im} w = 0, \operatorname{Re} w \geq 0\} \\ \cup \{w \in \mathbf{C}, \operatorname{Im} w = \pi, \operatorname{Re} w \leq 0)\}. \end{aligned}$$

Let u, v, p, q, be four pairwise distinct points of $\mathbf{C}$. We define the *cross ratio*

$$[u, p, q, v] = \frac{(u-q)(p-v)}{(u-p)(q-v)}$$

and extend it in the obvious way when one of the points is ∞. We identify $\hat{\mathbf{C}}$ with $\partial\mathcal{H}^3$. If u, v are in $\partial\mathcal{H}^3$, we denote by (u, v) the oriented geodesic from u to v in $\mathcal{H}^3$. Let $\alpha = (u, v)$ and $\beta = (p, q)$ be oriented geodesics in $\mathcal{H}^3$ with no common endpoints. There is a unique complex number $\sigma(\alpha, \beta)$ in Σ such that

$$\cosh \sigma(\alpha, \beta) = \frac{[u, p, q, v] + 1}{[u, p, q, v] - 1}$$

We call $\sigma(\alpha, \beta)$ the *complex distance* between the oriented geodesics α and β.We have

$$\sigma(\alpha, \beta) = \sigma(\beta, \alpha),$$

and if $-\alpha$ denotes the geodesic (v, u),

$$\sigma(-\alpha, \beta) = -\sigma(\alpha, \beta) + i\pi.$$

We also define $\sigma(\alpha, \alpha) = 0$ and $\sigma(\alpha, -\alpha) = i\pi$.

LEMMA 2.1. *(i) Let α and β be geodesics in $\mathcal{H}^3$ without common endpoints. Then the hyperbolic distance between α and β is $|\operatorname{Re}(\sigma(\alpha, \beta))|$. If α and β intersect, the angle between the positive rays of α and β is $\operatorname{Im}(\sigma(\alpha, \beta))$. If they do not intersect let α' be the geodesic which intersects β and is obtained by parallel translation of α along the common perpendicular of α and β. Then the angle between the positive rays of α' and β is $\operatorname{Im}(\sigma(\alpha, \beta))$.*

(ii) If $f \in$ PSL(2,C) then $\sigma(f(\alpha), f(\beta)) = \sigma(\alpha, \beta)$. ∎

If α and β are two geodesics in $\mathcal{H}^3$ without common endpoints, they have a unique common perpendicular. We define an orientation on the common perpendicular $\kappa_{\alpha\beta}$ in the following way. If $\mathrm{Re}(\sigma(\alpha,\beta)) > 0$ the orientation of $\kappa_{\alpha\beta}$ is from α to β. If $\mathrm{Re}(\sigma(\alpha,\beta)) < 0$ the orientation of $\kappa_{\alpha\beta}$ is from β to α. If $\mathrm{Re}(\sigma(\alpha,\beta)) = 0$, then α and β have a common point x. We define the orientation of $\kappa_{\alpha\beta}$ so that the positive directions of α, β and $\kappa_{\alpha\beta}$ form a right handed system at x. Note that $\kappa_{\alpha\beta} = -\kappa_{\beta\alpha} = -\kappa_{-\alpha\beta}$.

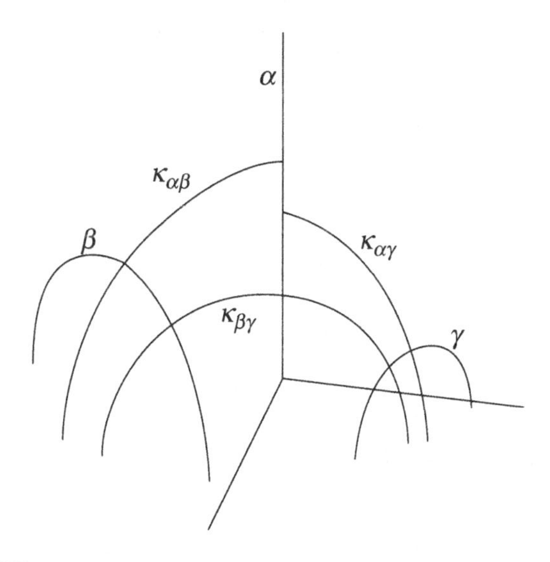

Fig. 1. The geodesics α, β, γ and their common perpendiculars

Let α, β, γ be oriented geodesics with pairwise distinct endpoints. We shall define a sign $\varepsilon_{\alpha\beta\gamma} \in \{+1, -1\}$, which describes the relative orientation of the geodesics. If α, β, γ have a common perpendicular, then $\kappa_{\alpha\beta}, \kappa_{\beta\gamma}$ coincide as unoriented geodesics. We define $\varepsilon_{\alpha\beta\gamma}$ so that $\kappa_{\alpha\beta} = \varepsilon_{\alpha\beta\gamma}\kappa_{\beta\gamma}$. Otherwise $\kappa_{\alpha\beta}$ and $\kappa_{\beta\gamma}$ have no common endpoints and we define $\varepsilon_{\alpha\beta\gamma}$ so that their common perpendicular is $\varepsilon_{\alpha\beta\gamma}\beta$. Finally we define

$$\delta(\alpha, \beta, \gamma) = \varepsilon_{\alpha\beta\gamma}\sigma(\kappa_{\alpha\beta}, \kappa_{\beta\gamma}).$$

LEMMA 2.2. *If α, β, γ are oriented geodesics with pairwise distinct end points then the sign $\varepsilon_{\alpha\beta\gamma}$ satisfies*

i. $\varepsilon_{(-\alpha)\beta\gamma} = -\varepsilon_{\alpha\beta\gamma} = \varepsilon_{\alpha\beta(-\gamma)}$.

ii. $\varepsilon_{\alpha\beta\gamma} = \varepsilon_{\alpha(-\beta)\gamma}$.

iii. If α, β, γ *have no common perpendicular,* $\varepsilon_{\alpha\beta\gamma} = -\varepsilon_{\gamma\beta\alpha}$. ∎

The proofs of the following propositions are given in [K3].

PROPOSITION 2.3. (HYPERBOLIC COSINE RULE) *Let* α, β, γ *be geodesics with pairwise distinct end points. Then*

$$\sinh\sigma(\alpha,\beta)\sinh\sigma(\beta,\gamma)\cosh\delta(\alpha,\beta,\gamma) = \cosh\sigma(\alpha,\gamma) - \cosh\sigma(\alpha,\beta)\cosh\sigma(\beta,\gamma).$$ ∎

PROPOSITION 2.4. (HYPERBOLIC SINE RULE) *Let* α, β, γ *be geodesics with pairwise distinct end points, and let* ρ *be a permutation of* $\{\alpha, \beta, \gamma\}$. *Then*

$$\sinh\sigma(\alpha,\gamma)\sinh\delta(\rho(\alpha),\rho(\beta),\rho(\gamma)) = \mathrm{sign}\rho\,\sinh\delta(\alpha,\beta,\gamma)\sinh\sigma(\rho(\alpha),\rho(\gamma)).$$ ∎

COROLLARY 2.5. *If* α, β, γ *have a common perpendicular, then*

$$\cosh\sigma(\alpha,\gamma) = \cosh(\sigma(\alpha,\beta) + \varepsilon_{\alpha\beta\gamma}\,\sigma(\beta,\gamma)).$$

PROOF. When α, β, γ have a common perpendicular, then $\cosh\delta(\alpha,\beta,\gamma) = \varepsilon_{\alpha\beta\gamma}$. ∎

Let f be a non-parabolic isometry of $\mathcal{H}^3$ different from the identity. Then f has an *axis* α_f. Let β be a geodesic perpendicular to α_f. The *oriented axis* of f is the geodesic α_f with the orientation of $\kappa_{\beta\, f(\beta)}$. We define the *complex displacement* of f to be

$$\omega_f = \sigma(\beta, f(\beta)).$$

LEMMA 2.6. *If* f *is a non-parabolic isometry of* $\mathcal{H}^3$ *different from the identity then*

(i) ω_f *is independent of the choice of* β.

(ii) $\omega_f = \omega_{gfg^{-1}}$ *if* g *is an orientation preserving isometry of* $\mathcal{H}^3$. *In particular* $\omega_f = \omega_{f^{-1}}$.

(iii) f *is conjugate to the isometry* $\pi(A(\omega_f))$.

PROOF. (i) Let γ be an oriented geodesic perpendicular to α_f. Then there is an orientation preserving isometry g of $\mathcal{H}^3$ preserving α_f and mapping β to γ. Moreover g commutes with f and

$$\sigma(\gamma, f(\gamma)) = \sigma(g(\beta), gf(\beta)) = \sigma(\beta, f(\beta)) = \omega_f.$$

(ii) The axis of gfg^{-1} is $g(\alpha_f)$ and $g(\beta)$ is perpendicular to it.

$$\omega_{gfg^{-1}} = \sigma(g(\beta), gf(\beta)) = \sigma(\beta, f(\beta)) = \omega_f.$$

(iii) A non-parabolic isometry f is conjugate to $\pi(A(w))$ for some $w \in \mathbf{C}$. Note that $\pi(A(w))$ is conjugate to $\pi(A(z))$ if and only if $z = \pm w + 2k\pi i$ for some $k \in \mathbf{Z}$. So there is a unique $w \in \Sigma$ for which f is conjugate to $\pi(A(w))$. We must show that if $w \in \Sigma$ then $\omega_{\pi(A(w))} = w$. For this it is enough to check that

$$\cosh \sigma((-1, 1), (e^{-w}, e^w)) = \cosh w. \qquad \blacksquare$$

The complex displacement and the oriented axis of a non parabolic isometry of $\mathcal{H}^3$ determine it completely.

LEMMA 2.6. *The isometry* f *with oriented axis* $\alpha_f = (u, v)$ *and complex displacement* $\omega_f = \varphi$ *is represented by the matrix* $A_f = A(u, v, \varphi)$.

$\blacksquare$

3. The derivatives of complex displacement

In this Section we shall use loxodromic isometries f, g, h with complex displacements φ, χ, ψ and axes $\alpha, \beta = (u, v), \gamma = (p, q)$ respectively.

A *deformation* of f is an analytic mapping $F : U \to \mathrm{PSL}(2,\mathbf{C})$ where U is a neighbourhood of 0 in $\mathbf{R}^n$ or in $\mathbf{C}^n$, and $F(0) = f$. In this Section we calculate the first and second derivatives of the complex displacement $\omega_{F(t)}$ at 0 for certain classes of deformations associated to bending.

Let $a \in \Sigma$ and let U be a neighbourhood of 0 in $\mathbf{C}$ such that $aU \subset \Sigma \cup (-\Sigma)$. For $t \in U$ define $G(t) = A(u, v, ta)$. If $ta \in \Sigma$, then $\pi(G(t))$ has complex displacement ta and axis α_g, while if $ta \in -\Sigma$ then $\pi(G(t))$ has complex displacement $-ta$ and axis $-\alpha_g$. Define the deformation $F(t) = \pi(G(t))f$ and the functions

$$\tau(t) = \mathrm{tr}(G(t)A_f)$$

$$w(t) = 2\,\mathrm{arccosh}\,\tfrac{1}{2}\tau(t)$$

where we choose the sheet of arccosh that gives $w(0) = \varphi$. Then

$$\sinh \tfrac{1}{2}w\,\frac{dw}{dt} = \frac{d\tau}{dt}\,.$$

If ω_f is in the interior of Σ, then $F(t)$ has complex displacement $w(t)$ for t near 0.

LEMMA 3.1. *With the above notation w is analytic at 0 and*

$$\frac{dw(0)}{dt} = a\cosh\sigma(\alpha_f, \alpha_g).$$

PROOF. Assume that $\alpha = (0, \infty)$. Then

$$\begin{aligned}\frac{d\tau}{dt} &= \mathrm{tr}\left(\frac{dG(t)}{dt}A_f\right)\\ &= a\,\frac{v+u}{v-u}\sinh\tfrac{1}{2}\varphi\,. \qquad (1)\end{aligned}$$

Hence $dw(0)/dt = a(v+u)/(v-u) = a\cosh\sigma((u,v),(0,\infty))$. The lemma follows by invariance. ∎

Now we consider a two parameter deformation $F(s,t)$ of f. F is an analytic mapping $F : U \to \mathrm{PSL(2,C)}$ where U is a neighbourhood of 0 in $\mathbf{R}^2$ or in $\mathbf{C}^2$.

Let $a, b \in \Sigma$ and let U be a neighbourhood of 0 in $\mathbf{C}^2$ such that if $(s,t) \in U$ then sb and ta lie in $\Sigma \cup (-\Sigma)$. Let $H(s) = A(p,q,sb)$ and define $U(s)$ and $V(s)$ to be the fixed points of $\pi(H(s))g$ on $\hat{\mathbf{C}}$, with $U(0) = u, V(0) = v$. Let $K(s,t) = A(U(s), V(s), ta)$. Define the deformation $L(s,t) = \pi(K(s,t)H(s))f$ and the functions

$$\tau(s,t) = \mathrm{tr}(K(s,t)H(s)A_f),$$

$$w(s,t) = 2\mathrm{arccosh}\tfrac{1}{2}\tau(s,t),$$

where we choose the sheet of arccosh that gives $w(0,0)=\varphi$. At the point (s,t)

$$\begin{aligned}\frac{\partial^2\tau}{\partial s\partial t} &= \frac{\partial}{\partial s}\left(\sinh\tfrac{1}{2}w\frac{\partial}{\partial t}w\right)\\ &= \tfrac{1}{2}\cosh\tfrac{1}{2}w\frac{\partial w}{\partial t}\frac{\partial w}{\partial s}+\sinh\tfrac{1}{2}w\frac{\partial^2 w}{\partial s\partial t}\end{aligned}$$

and

$$\sinh\tfrac{1}{2}w\frac{\partial^2 w}{\partial s\partial t} = \frac{\partial^2\tau}{\partial s\partial t}-\tfrac{1}{2}\frac{\cosh\frac{1}{2}w}{\sinh^2\frac{1}{2}w}\frac{\partial\tau}{\partial t}\frac{\partial\tau}{\partial s}. \tag{2}$$

The rest of this section is devoted to the proof of the following lemma.

LEMMA 3.2. *With the above notation, w is analytic at $(0,0)$ and*

$$\begin{aligned}&\frac{\partial^2 w(0,0)}{\partial s\partial t}\\ &\quad= \frac{ab}{2\sinh\frac{1}{2}\varphi}\sinh\sigma(\alpha_g,\alpha_f)\sinh\sigma(\alpha_f,\alpha_h)\cosh(\tfrac{1}{2}\varphi-\delta(\alpha_g,\alpha_f,\alpha_h))\\ &\qquad+\frac{ab}{2\sinh\frac{1}{2}\chi}\sinh\sigma(\alpha_f,\alpha_g)\sinh\sigma(\alpha_g,\alpha_h)\cosh(\tfrac{1}{2}\chi-\delta(\alpha_f,\alpha_g,\alpha_h)).\end{aligned}$$

PROOF. Assume that $\alpha=(0,\infty)$. Note that $H(0)=id$, $K(s,0)=id$ and $K(0,t)=G(t)$. Hence $L(s,0)=\pi(H(s))f$ and $L(0,t)=\pi(G(t))f=F(t)$. From Lemma 3.1 we have

$$\frac{\partial\tau(0,0)}{\partial s} = \left.\frac{d\tau_{H(s)f}}{ds}\right|_0 = b\sinh\tfrac{1}{2}\varphi\cosh\sigma(\alpha,\gamma) \tag{3}$$

$$\frac{\partial\tau(0,0)}{\partial t} = \left.\frac{d\tau_{G(t)f}}{dt}\right|_0 = a\sinh\tfrac{1}{2}\varphi\cosh\sigma(\alpha,\beta). \tag{4}$$

It remains to calculate $\partial^2\tau/\partial s\partial t$. A rather long but elementary calculation gives

$$\frac{\partial^2 w(0,0)}{\partial s\partial t} = R-S \tag{5}$$

where

$$R=\frac{ab}{2}\left[\frac{2(pq-uv)}{(v-u)(q-p)}-\coth\tfrac{1}{2}\varphi\frac{2(pq+uv)}{(v-u)(q-p)}\right]$$

and

$$S=\frac{ab}{2}\left[\frac{2(pq-uv)}{(v-u)(q-p)}-B\coth\tfrac{1}{2}\chi\right]$$

where

$$B = \frac{q+p}{q-p} - \frac{(v+u)^2(q+p)}{(v-u)^2(q-p)} + \frac{2(v+u)(pq+uv)}{(v-u)^2(q-p)}.$$

Now we want to replace the terms involving u, v, p, q by expressions in the complex distance between the geodesics α, β, γ and their perpendiculars. We have

$$\frac{v+u}{v-u} = \cosh\sigma(\alpha,\beta), \qquad \frac{q+p}{q-p} = \cosh\sigma(\alpha,\gamma) \tag{6}$$

and from the Hyperbolic Cosine Rule

$$\begin{aligned}\frac{-2(pq+uv)}{(v-u)(q-p)} &= \cosh\sigma(\beta,\gamma) - \cosh\sigma(\beta,\alpha)\cosh\sigma(\alpha,\gamma)\\ &= \sinh\sigma(\beta,\alpha)\sinh\sigma(\alpha,\gamma)\cosh\delta(\beta,\alpha,\gamma)\end{aligned} \tag{7}$$

and

$$\begin{aligned}\frac{2(pq-uv)}{(v-u)(q-p)} &= \frac{-2((ip)(iq)+uv)}{-i(iq-ip)(v-u)}\\ &= i\sinh\sigma(\beta,\alpha)\sinh\sigma(\alpha,i\gamma)\cosh\delta(\beta,\alpha,i\gamma)\end{aligned}$$

where $i\gamma = (ip, iq)$. Multiplication by i leaves α invariant, so $\sigma(\alpha, i\gamma) = \sigma(\alpha,\gamma)$. Note that $\kappa_{\alpha(i\gamma)} = i\kappa_{\alpha\gamma}$ and hence $\sigma(\kappa_{\alpha,\gamma}, \kappa_{\alpha(i\gamma)}) = i\pi/2$ and $\kappa_{\kappa_{\alpha\gamma}\,\kappa_{\alpha(i\gamma)}} = \alpha$. So in this case we have $\varepsilon_{\kappa_{\beta\alpha}\,\kappa_{\alpha\gamma}\,\kappa_{\alpha(i\gamma)}} = \varepsilon_{\beta\alpha\gamma}$ and by Corollary 2.5

$$\begin{aligned}\cosh\delta(\beta,\alpha,i\gamma) &= \cosh\left[\delta(\beta,\alpha,\gamma) + \varepsilon_{\beta\alpha\gamma}\, i\pi/2\right]\\ &= i\sinh\delta(\beta,\alpha,\gamma).\end{aligned}$$

Hence

$$\frac{2(pq-uv)}{(v-u)(q-p)} = -\sinh\sigma(\alpha,\beta)\sinh\sigma(\alpha,\gamma)\sinh\delta(\beta,\alpha,\gamma). \tag{8}$$

We substitute (7) and (8) in R to obtain

$$R = \frac{ab}{2\sinh\frac{1}{2}\varphi}\sinh\sigma(\beta,\alpha)\sinh\sigma(\alpha,\gamma)\cosh\left[\tfrac{1}{2}\varphi - \delta(\beta,\alpha,\gamma)\right]$$

Using (6) and (7) we have

$$\begin{aligned}B &= \cosh\sigma(\alpha,\gamma) - \cosh\sigma(\alpha,\beta)\cosh\sigma(\beta,\gamma)\\ &= \sinh\sigma(\alpha,\beta)\sinh\sigma(\beta,\gamma)\cosh\delta(\alpha,\beta,\gamma)\end{aligned}$$

and hence

$$S = \frac{-ab}{2\sinh\frac{1}{2}\chi}\left[\sinh\tfrac{1}{2}\chi\sinh\sigma(\beta,\alpha)\sinh\sigma(\alpha,\gamma)\sinh\delta(\beta,\alpha,\gamma)\right.$$
$$\left. + \cosh\tfrac{1}{2}\chi\sinh\sigma(\alpha,\beta)\sinh\sigma(\beta,\gamma)\cosh\delta(\alpha,\beta,\gamma)\right].$$

By the Hyperbolic Sine Rule

$$\sinh\sigma(\beta,\gamma)\sinh\delta(\gamma,\beta,\alpha) = \sinh\sigma(\gamma,\alpha)\sinh\delta(\beta,\alpha,\gamma)$$

and hence

$$\begin{aligned} S &= \frac{-ab}{2\sinh\frac{1}{2}\chi}\sinh\sigma(\alpha,\beta)\sinh\sigma(\beta,\gamma) \\ &\quad \left[\cosh\tfrac{1}{2}\chi\cosh\delta(\alpha,\beta,\gamma) + \sinh\tfrac{1}{2}\chi\sinh\delta(\gamma,\beta,\alpha)\right] \\ &= \frac{-ab}{2\sinh\frac{1}{2}\chi}\sinh\sigma(\alpha,\beta)\sinh\sigma(\beta,\gamma)\cosh\left[\tfrac{1}{2}\chi + \delta(\gamma,\beta,\alpha)\right]. \end{aligned}$$

Hence

$$\begin{aligned} &\frac{\partial^2 w(0,0)}{\partial s\partial t} \\ &\quad = \frac{ab}{2\sinh\frac{1}{2}\varphi}\sinh\sigma(\beta,\alpha)\sinh\sigma(\alpha,\gamma)\cosh\left[\tfrac{1}{2}\varphi - \delta(\beta,\alpha,\gamma)\right] \\ &\qquad + \frac{ab}{2\sinh\frac{1}{2}\chi}\sinh\sigma(\alpha,\beta)\sinh\sigma(\beta,\gamma)\cosh\left[\tfrac{1}{2}\chi - \delta(\alpha,\beta,\gamma)\right]. \qquad \blacksquare \end{aligned}$$

4. Bending along simple closed geodesics

Let S be a closed surface of negative Euler characteristic. We choose a Fuchsian structure on S, $\rho : \pi(S) \to \Gamma \subset \mathrm{PSL}(2,\mathbf{R})$. We shall consider S with the hyperbolic structure defined by ρ, and we shall use ρ to identify $\pi(S)$ with Γ. Let α, β, γ be distinct simple closed geodesics on S, and Λ_α, Λ_β, Λ_γ the corresponding discrete geodesic laminations on $\mathcal{H}^2$.

Let ρ_0 be a *quasi-Fuchsian structure* on S, that is $\rho_0 : \Gamma \to \Gamma_0 \subset \mathrm{PSL}(2,\mathbf{C})$ is an isomorphism which is induced by a mapping $\xi : \overline{\mathcal{H}^2} \to \overline{\mathcal{H}^3}$, which is a homeomorphism onto its image, and $\rho_0(f)|_{\xi(\mathcal{H}^2)} = \xi \circ f \circ \xi^{-1}|_{\xi(\mathcal{H}^2)}$, [K2]. Then $\mathcal{H}^3/\Gamma_0$ is a hyperbolic manifold of infinite volume, homeomorphic to $S \times (0,1)$. The set $\xi(\partial\mathcal{H}^2)$ is a Jordan curve and ξ induces a natural mapping on the geodesics in $\mathcal{H}^2$: if $\lambda = (u,v)$ is a geodesic in $\mathcal{H}^2$, let $\xi_*(\lambda)$ denote the geodesic $(\xi(u), \xi(v))$ in $\mathcal{H}^3$. Define $\xi_*(\Lambda_\alpha)$ in the obvious way.

Consider the orientation on $\partial\mathcal{H}^2$ given by $(0, 1, \infty)$. If $p, q \in \partial\mathcal{H}^2$, then $(p, q)_\partial$ denotes the open arc on $\partial\mathcal{H}^2$ from p to q.

Without loss of generality we may assume that the geodesic $(0, \infty)$ is contained in $\xi_*(\Lambda_\alpha)$, that $\xi(0) = 0$, $\xi(\infty) = \infty$, and that $i \in \mathcal{H}^2$ does not belong to $\Lambda_\alpha \cap (\Lambda_\beta \cup \Lambda_\gamma)$.

Let f be the element of Γ_0 corresponding to the geodesic α, with axis $\alpha_f = (0, \infty)$ and let $f^0 = \rho_0^{-1}(f)$. Number the leaves of Λ_β intersecting the geodesic segment $[i, f^0(i)] \subset \mathcal{H}^2$ in order $\tilde{\beta}_1, \ldots, \tilde{\beta}_k$, so that $\tilde{\beta}_1$ is closest to i. Let $\beta_j = \xi_*(\tilde{\beta}_j)$.

For $j = 1, \ldots, k$ define u_j, v_j so that $\beta_j = (u_j, v_j)$ and $u_j \in \xi(0, \infty)_\partial$, $v_j \in \xi(\infty, 0)_\partial$. For a and U as in Section 3, define $G_j(t) = A(u_j, v_j, ta)$. Define the deformation of f, $F(t) = \pi(G_1(t) \ldots G_k(t))f$ and the functions

$$\tau(t) = \operatorname{tr}(G_1(t) \ldots G_k(t) A_f)$$

$$w(t) = 2 \operatorname{arccosh} \tfrac{1}{2}\tau(t).$$

Let $\alpha\#\beta = \{1, \ldots, k\}$ be the set of intersection points of the geodesics α and β on S. Since i belongs to the axis of f^0, $\alpha\#\beta$ is in bijective correspondence with $\{\beta_j, j = 1, \ldots, k\}$. Define

$$\sigma_j = \sigma(\alpha_f, \beta_j).$$

THEOREM 4.1. *With the above notation w is analytic at 0 and*

$$\frac{dw(0)}{dt} = \sum_{j \in \alpha\#\beta} a \cosh \sigma_j.$$

PROOF. We have

$$\frac{d\tau(t)}{dt} = \sum_{j=1}^{k} \operatorname{tr}\left(G_1 \ldots \frac{dG_j}{dt} \ldots G_k A_f\right).$$

At $t = 0$, $G_j(0) = I$ and

$$\frac{d\tau(0)}{dt} = \sum_{j=1}^{k} \operatorname{tr}\left(\frac{dG_j}{dt} A_f\right) = \sum_{j=1}^{k} \frac{a(v_j + u_j)}{v_j - u_j} \sinh \tfrac{1}{2}\varphi.$$

Hence

$$\frac{dw(0)}{dt} = \sum_{j=1}^{k} \frac{a(v_j + u_j)}{v_j - u_j} = \sum_{j \in \alpha \# \beta} a \cosh \sigma_j. \qquad \blacksquare$$

Consider the leaves of Λ_γ which intersect $[i, f^0(i)]$ and number them in order, $\tilde{\gamma}_{01}, \ldots, \tilde{\gamma}_{0\ell_0}$, starting from the leaf nearest i. Let $\gamma_{0n} = \xi_*(\tilde{\gamma}_{0n})$. For $n = 1, \ldots, \ell_0$, define p_{0n}, q_{0n} so that $\gamma_{0n} = (p_{0n}, q_{0n})$ and $p_{0n} \in \xi(0, \infty)_\partial$, $q_{0n} \in \xi(\infty, 0)_\partial$.

Let g_j be the element of Γ_0 which corresponds to the geodesic β and has oriented axis β_j. Let $g_j^0 = \rho_0^{-1}(g_j)$. Consider the leaves of Λ_γ which intersect $[i, g_j^0(i)]$ and number them in order, $\tilde{\gamma}_{j1}, \ldots, \tilde{\gamma}_{j\ell_j}$, starting from the leaf nearest i. Let $\gamma_{jm} = \xi_*(\tilde{\gamma}_{jm})$. For $j = 1, \ldots, k$ and $m = 1, \ldots, \ell_j$ define p_{jm}, q_{jm} so that $\gamma_{jm} = (p_{jm}, q_{jm})$, $p_{jm} \in \xi(\xi^{-1}(u_j), \xi^{-1}(v_j))_\partial$ and $q_{jm} \in \xi(\xi^{-1}(v_j), \xi^{-1}(u_j))_\partial$.

Let $a, b \in \Sigma$ and U a neighbourhood of 0 in $\mathbf{C}^2$ such that if $(s,t) \in U$ then sb and ta lie in $\Sigma \cup (-\Sigma)$. Define $H_{jm}(s) = A(p_{jm}, q_{jm}, sb)$, $j = 0, \ldots, k$; $m = 1, \ldots, \ell_j$. Define $U_j(s)$ and $V_j(s)$ to be the fixed points of $\pi(H_{j1}(s) \ldots H_{j\ell_j}(s))g_j$ on $\hat{\mathbf{C}}$, so that $U_j(0) = u_j$, $V_j(0) = v_j$. Let $K_j(s,t) = A(U_j(s), V_j(s), ta)$. Define the deformation

$$L(s,t) = \pi(K_1(s,t) \ldots K_k(s,t) H_{01}(s) \ldots H_{0\ell_0}(s)) f$$

and

$$\tau(s,t) = \operatorname{tr}(K_1(s,t) \ldots K_k(s,t) H_{01}(s) \ldots H_{0\ell_0}(s) A_f)$$

$$w(s,t) = 2 \operatorname{arccosh} \tfrac{1}{2}\tau(s,t).$$

Since $i \in \alpha_f$, the set $\alpha \# \gamma$ is in bijective correspondence with $\{\gamma_{0n}, n = 1, \ldots, \ell_0\}$. Define

$$\zeta_n = \sigma(\alpha_f, \gamma_{0n}).$$

Now consider $\beta \# \gamma = \{1, \ldots, \ell\}$. For each $j \in \{1, \ldots, k\}$, there is a number n_j such that $\beta \# \gamma$ is in bijective correspondence with the set $\{\gamma_{jm}, n_j < m \leq \ell_j - n_j\}$. For $m \in \beta \# \gamma$, define

$$\theta_{jm} = \sigma(\beta_j, \gamma_{j(n_j+m)}).$$

The geodesics $\xi_*^{-1}(\gamma_{jn})$ for $n \leq n_j$ and $n > \ell_j - n_j$ do not intersect the axis of g_j^0. They come in pairs in the sense that for $n = 1, \ldots, n_j$, $\gamma_{j(\ell_j - n + 1)} = -g_j(\gamma_{jn})$. Therefore $\sigma(\beta_j, \gamma_{j(\ell_j - n + 1)}) = -\sigma(\beta_j, \gamma_{jn}) + \pi i$.

Finally, for $j \in \alpha\#\beta$, $n \in \alpha\#\gamma$, define

$$\mu_{jn} = \delta(\beta_j, \alpha_f, \gamma_{0n})$$

and for $j \in \alpha\#\beta$, $m \in \beta\#\gamma$, define

$$\nu_{jm} = \delta(\alpha_f, \beta_j, \gamma_{j(n_j+m)}).$$

THEOREM 4.2.

$$\frac{\partial^2 w(0,0)}{\partial s \partial t} = \frac{ab}{2\sinh\frac{1}{2}\varphi} \sum_{j\in\alpha\#\beta} \sum_{n\in\alpha\#\gamma} \sinh\zeta_n \sinh\sigma_j \cosh(\tfrac{1}{2}\varphi - \mu_{jn})$$
$$+ \frac{ab}{2\sinh\frac{1}{2}\chi} \sum_{j\in\alpha\#\beta} \sum_{m\in\beta\#\gamma} \sinh\theta_{jm} \sinh\sigma_j \cosh(\tfrac{1}{2}\chi - \nu_{jm})$$

PROOF. At $(s,t) = (0,0)$, $\partial K_j(0,0)/\partial s = 0$ and hence

$$\frac{\partial^2 \tau(0,0)}{\partial s \partial t} = \operatorname{tr}\left[\sum_{j=1}^{k}\left(\frac{\partial^2 K_j(0,0)}{\partial s\partial t} + \sum_{n=1}^{\ell_0} \frac{dG_j(0)}{dt}\frac{dH_{0n}(0)}{ds}\right) A_f\right].$$

A computation similar to the one in Section 3 gives

$$\operatorname{tr}\left(\frac{\partial^2 K_j(0,0)}{\partial s\partial t} A_f\right)$$
$$= \frac{ab}{2}\sinh\tfrac{1}{2}\varphi \sum_{m=1}^{\ell_j}\left[\frac{-2(p_{jm}q_{jm} - u_j v_j)}{(v_j - u_j)(q_{jm} - p_{jm})} + B_{jm}\coth\tfrac{1}{2}\chi\right] \quad (9)$$

where

$$B_{jm} = \frac{q_{jm}+p_{jm}}{q_{jm}-p_{jm}} - \frac{(v_j-u_j)^2(q_{jm}+p_{jm})}{(v_j-u_j)^2(q_{jm}-p_{jm})} + \frac{2(v_j+u_j)(p_{jm}q_{jm}+u_j v_j)}{(v_j-u_j)^2(q_{jm}-p_{jm})}$$

and

$$\operatorname{tr}\left(\frac{dG_j(0)}{dt}\frac{dH_{0n}(0)}{ds} A_f\right)$$
$$= \frac{ab}{2}\left[\left(\frac{(v_j+u_j)(q_{0n}+p_{0n})}{(v_j-u_j)(q_{0n}-p_{0n})} - \frac{2(u_j v_j + p_{0n}q_{0n})}{(v_j-u_j)(q_{0n}-p_{0n})}\right)\cosh\tfrac{1}{2}\varphi - \frac{2(u_j v_j - p_{0n}q_{0n})}{(v_j-u_j)(q_{0n}-p_{0n})}\sinh\tfrac{1}{2}\varphi\right]. \quad (10)$$

From (2) we have

$$\sinh\tfrac{1}{2}\varphi\frac{\partial^2 w(0,0)}{\partial s\partial t} = \sum_{j=1}^{k}\mathrm{tr}\left(\frac{\partial^2 K_j}{\partial s\partial t}A_f\right) + \sum_{j=1}^{k}\sum_{n=1}^{\ell_0}\mathrm{tr}\left(\frac{dG_j}{dt}\frac{dH_{0n}}{ds}A_f\right)$$
$$- \frac{ab}{2}\cosh\tfrac{1}{2}\varphi\sum_{j=1}^{k}\sum_{n=1}^{\ell_0}\frac{(q_{0n}+p_{0n})(v_j+u_j)}{(q_{0n}-p_{0n})(v_j-u_j)}. \quad (11)$$

We substitute (9) and (10) in (11) and obtain

$$\frac{\partial^2 w(0,0)}{\partial s\partial t}$$
$$= \frac{ab}{2}\sum_{j=1}^{k}\sum_{n=1}^{\ell_0}\left[\frac{2(p_{0n}q_{0n}-u_jv_j)}{(v_j-u_j)(q_{0n}-p_{0n})} - \frac{2(p_{0n}q_{0n}+u_jv_j)}{(v_j-u_j)(q_{0n}-p_{0n})}\coth\tfrac{1}{2}\varphi\right]$$
$$- \frac{ab}{2}\sum_{j=1}^{k}\sum_{m=1}^{\ell_j}\left[\frac{2(p_{jm}q_{jm}-u_jv_j)}{(v_j-u_j)(q_{jm}-p_{jm})} - B_{jm}\coth\tfrac{1}{2}\chi\right].$$

A calculation similar to the one in Section 3 gives

$$\frac{\partial^2 w(0,0)}{\partial s\partial t} = \frac{ab}{2\sinh\frac{1}{2}\varphi}\sum_{j=1}^{k}\sum_{n=1}^{\ell_0}\sinh\sigma(\beta_j,\alpha_f)\sinh\sigma(\alpha_f,\gamma_{0n})$$
$$\cosh\left(\tfrac{1}{2}\varphi - \delta(\beta_j,\alpha_f,\gamma_{0n})\right)$$
$$+ \frac{ab}{2\sinh\frac{1}{2}\chi}\sum_{j=1}^{k}\sum_{m=1}^{\ell_j}\sinh\sigma(\alpha_f,\beta_j)\sinh\sigma(\beta_j,\gamma_{jm})$$
$$\cosh\left(\tfrac{1}{2}\chi - \delta(\alpha_f,\beta_j,,\gamma_{jm})\right).$$

For $1 \leq m \leq n_j$, let $\gamma'_{jm} = \gamma_{j(\ell_j-m+1)}$. Then $\gamma'_{jm} = -g(\gamma_{jm})$ and hence $\sigma(\beta_j,\gamma'_{jm}) = i\pi - \sigma(\beta_j,\gamma_{jm})$ and

$$\delta(\alpha_f,\beta_j,\gamma'_{jm}) = \delta(\alpha_f,\beta_j,\gamma_{jm}) \pm i\pi.$$

The terms corresponding to m and $\ell_j - m + 1$ for $1 \leq m \leq n_j$, cancel out and we are left with the terms for $n_j < m \leq \ell_j - n_j$, which are in bijective correspondence with $\beta\#\gamma$. ∎

5. The derivatives of complex length functions

Let α be a simple closed geodesic on S corresponding to an element $f^0 \in \Gamma$. We define the *complex length* of α in the quasi-Fuchsian structure ρ_0,

$$\lambda_\alpha(\rho_0) = \omega_{\rho_0(f^0)}.$$

In this way every simple closed geodesic α on S defines a mapping

$$\lambda_\alpha : Q(S) \to \Sigma.$$

If μ is a measured lamination on S, there is a neighbourhood W of $0 \in \mathbf{C}$, on which we can define a quasiconformal deformation of ρ_0, called *bending along* μ, [K2].

$$B_\mu(\cdot, \rho_0) : W \to Q(S) : t \mapsto B_\mu(t, \rho_0).$$

We define the *first variation of the length of* α *under bending along* μ by

$$T_\mu \lambda_\alpha = \frac{d}{dt} \left[\lambda_\alpha(B_\mu(t, \rho_0))\right]_{t=0}.$$

If ν is a measured lamination on S, we define the *second variation of the length of* α *under bending along* μ and ν by

$$T_\mu T_\nu \lambda_\alpha = \frac{\partial^2}{\partial s \partial t} \left[\lambda_\alpha(B_\mu(t, B_\nu(s, \rho_0)))\right]_{(0,0)}.$$

THEOREM 5.1. *Let α and β be simple closed geodesics on S. Then*

$$T_\beta \lambda_\alpha = \sum_{j \in \alpha \# \beta} \cosh \sigma_j.$$

PROOF. B_β is given on a set $\{f_j, j = 1, \dots, k\}$ of generators of Γ, by analytic mappings $F_{j,\beta} : W \to \mathrm{PSL}(2,\mathbf{C})$. The mapping $F_{j,\beta}$ is the one parameter deformation defined in Section 4 if α is the geodesic corresponding to f_j and $a = 1$. Then

$$T_\beta \lambda_\alpha = \frac{dw(0)}{dt} = \sum_{j \in \alpha \# \beta} \cosh \sigma_j. \qquad \blacksquare$$

THEOREM 5.2. *Let α, β, γ be simple closed geodesics on S. Then*

$$T_\beta T_\gamma \lambda_\alpha = \frac{1}{2 \sinh \frac{1}{2}\lambda_\alpha} \sum_{j \in \alpha \# \beta} \sum_{n \in \alpha \# \gamma} \sinh \sigma_j \sinh \zeta_n \cosh\left(\tfrac{1}{2}\lambda_\alpha - \mu_{jn}\right)$$
$$+ \frac{1}{2 \sinh \frac{1}{2}\lambda_\beta} \sum_{j \in \alpha \# \beta} \sum_{m \in \beta \# \gamma} \sinh \sigma_j \sinh \theta_{jm} \cosh\left(\tfrac{1}{2}\lambda_\beta - \nu_{jm}\right).$$

PROOF. The two parameter deformation defined in Section 4 gives the bending deformation $B_\beta(t, B_\gamma(s,\rho_0))$. It follows that

$$T_\beta T_\gamma \lambda_\alpha = \frac{\partial^2 w(0,0)}{\partial s \partial t}.$$

The variations of the complex length of a geodesic satisfy the following relations, which generalise Wolpert's reciprocity relations, [W2].

THEOREM 5.3. *Let α, β, γ be simple closed geodesics on S. Then*

(i) $T_\alpha \lambda_\beta + T_\beta \lambda_\alpha = 0$

(ii) $T_\beta T_\gamma \lambda_\alpha + T_\gamma T_\alpha \lambda_\beta + T_\alpha T_\beta \lambda_\gamma = 0.$

PROOF. (i) Choose an orientation for α. For each $j \in \alpha\#\beta$, denote by $\bar{\alpha}_j$ the geodesic α with the chosen orientation and by $\bar{\beta}_j$ the geodesic β with the orientation induced by the definition of bending α along β, that is so that it crosses α from right to left.

Now consider bending β along α. If β has the orientation $\bar{\beta}_j$, then the induced orientation on α is $-\bar{\alpha}_j$. Then

$$T_\alpha \lambda_\beta = \sum_{j \in \alpha\#\beta} \cosh \sigma(\bar{\beta}_j, -\bar{\alpha}_j) = -T_\beta \lambda_\alpha.$$

(ii) For each $(j,n) \in (\alpha\#\beta) \times (\alpha\#\gamma)$ denote by $\bar{\alpha}_{jn}, \bar{\beta}_{jn}, \bar{\gamma}_{jn}$ the orientations on α, β, γ involved in the calculation of $T_\beta T_\gamma \lambda_\alpha$, that is $\bar{\beta}_{jn}$ and $\bar{\gamma}_{jn}$ cross $\bar{\alpha}_{jn}$ from right to left at j and n respectively. The corresponding triplets in the calculation of $T_\gamma T_\alpha \lambda_\beta$ and $T_\alpha T_\beta \lambda_\gamma$ are $\bar{\beta}_{jn}, \bar{\gamma}_{jn}, -\bar{\alpha}_{jn}$ and $\bar{\gamma}_{jn}, -\bar{\alpha}_{jn}, -\bar{\beta}_{jn}$ respectively. We use the properties of $\varepsilon_{\alpha\beta\gamma}$ under change of sign (Lemma 2.2) and the result of Theorem 5.2 to complete the proof. ∎

References

[B] A. F. Beardon. *The geometry of discrete groups.*(Springer-Verlag, New York, 1983)

[EM] D. B. A. Epstein, A. Marden. Convex hulls in hyperbolic space, a theorem of Sullivan, and measured pleated surfaces, in : D. B. A. Epstein (ed.) *Analytical and geometric aspects of hyperbolic space.* London Math. Soc. Lecture Note Series **111** (Cambridge University Press, 1987), 113–253.

[G] W. M. Goldman. Invariant functions on Lie groups and Hamiltonian flows of surface group representations. *Invent. Math.* **85** (1986) 263–302.

[JM] D. Johnson, J. J. Millson. Deformation spaces associated to compact hyperbolic manifolds, in : R. Howe (ed.) *Discrete groups in geometry and analysis,* Papers in honour of G. D. Mostow on his sixtieth birthday. Progress in Math. **67** (Birkhäuser, Boston, 1987) 48–106.

[Ke] S. P. Kerckhoff. The Nielsen realization problem. *Ann. of Math.* **117** (1983) 235–265.

[K1] C. Kourouniotis. Deformations of hyperbolic structures. *Math. Proc. Camb. Phil. Soc.* **98** (1985) 247–261.

[K2] C. Kourouniotis. Bending in the space of quasi-Fuchsian structures. *Glasgow Math. J.* **33** (1991) 41–49.

[K3] C. Kourouniotis. Complex distance and skew hexagons in hyperbolic 3-space. University of Crete preprint.

[W1] S. Wolpert. The Fenchel-Nielsen deformation. *Ann. of Math.* **115** (1982) 501–528.

[W2] S. Wolpert. On the symplectic geometry of deformations of a hyperbolic surface. *Ann. of Math.* **117** (1983) 207–234.

Department of Mathematics
University of Crete
GR 714 09 Iraklio Crete
Greece
Email: chrisk@talos.cc.uch.gr

Farey series and sums of continued fractions

Joseph Lehner

To Murray Macbeath, friend and colleague, on his retirement

In 1947 Marshall Hall [H] proved that every real number has a fractional part that is representable as the sum of two regular continued fractions $CF(0, a_1, a_2, \ldots)$ whose partial quotients a_i do not exceed 4. In a sort of complementary result published in 1971, T. Cusick [C], using Hall's method, considered continued fractions with partial quotients a_i not less than 2. He showed that every real number has a fractional part representable as the sum of two such continued fractions. These results were obtained by constructing a type of Cantor set of a certain interval and then proving that the sum of two such sets exhausts the real numbers modulo 1.

In the present note we observe that such a Cantor set may be obtained by partitioning the interval by a type of Farey series. This leads to a new proof of Cusick's result.

1. The basic theorem of [H] may be stated as follows.

THEOREM H. Suppose from a closed interval A a middle open interval M_{12} is removed, leaving an interval M_1 on the left and an interval M_2 on the right. Let the process be continued indefinitely, removing a middle interval from each left-hand and from each right-hand interval. In a typical step of the process let the following condition hold:

$$m_{12} \leq m_1 , \qquad m_{12} \leq m_2 \tag{1}$$

where m_1 is the length of M_1, *etc.* Denote the Cantor set that remains after all middle intervals have been removed by $C(A)$. Then

$$C(A) + C(A) = A + A . \tag{2}$$

Here the sum of two sets A and B is defined by

$$A + B = \{a + b : a \in A , b \in B\}.$$

2. We now make use of a type of Farey series to construct a set $C(A)$. Two reduced rational fractions h/k, h'/k', are said to be *neighbors* if $hk' - kh' = \pm 1$. The *mediant* of two neighbors is defined to be

$$\mu\left(\frac{h}{k}, \frac{h'}{k'}\right) = \frac{h+h'}{k+k'} ; \tag{3}$$

it is a reduced rational fraction lying between them and is a neighbor of both of them.

We shall use mediants to effect the desired subdivision of A. Consider the following lines. (Note 0=0/1 .)

F_1 $$0, \frac{1}{2}$$

F_2 $$0, \frac{1}{3}, \frac{2}{5}, \frac{1}{2}$$

F_3 $$0, \frac{1}{4}, \frac{2}{7}, \frac{1}{3}, \frac{2}{5}, \frac{5}{12}, \frac{3}{7}, \frac{1}{2}$$

F_4 $$0, \frac{1}{5}, \frac{2}{9}, \frac{1}{4}, \frac{2}{7}, \frac{5}{17}, \frac{3}{10}, \frac{1}{3}, \frac{2}{5}, \frac{7}{17}, \frac{12}{29}, \frac{5}{12}, \frac{3}{7}, \frac{7}{16}, \frac{4}{9}, \frac{1}{2}$$

Removing the middle interval $(1/3 = \mu(0, 1/2), 2/5 = \mu(1/3, 1/2))$ from F_1 leaves a left-hand interval $(0, 1/3)$ and a right-hand interval

$(2/5,\ 1/2)$. From the left-hand interval $(0,\ 1/3)$ we remove a middle interval $(1/4 = \mu(0,\ 1/3),\ 2/7 = \mu(1/4,\ 1/3))$ – that is, we take mediants from left to right. This leaves in F_3 a left-hand interval $(0,\ 1/4)$ and a right-hand interval $(2/7,\ 1/3)$. From the right-hand interval $(2/5,\ 1/2)$ we remove a middle interval $(3/7 = \mu(2/5,\ 1/2),\ 5/12 = \mu(3/7,\ 2/5))$ – that is, we take mediants from right to left. This leaves in F_3 a left-hand interval $(2/5,\ 5/12)$ and a right-hand interval $(3/7,\ 1/2)$.

In general each left-hand interval, say $(h_1/k_1,\ h_2/k_2)$, is partitioned by the points

$$\frac{h_1}{k_1}; \qquad \frac{h_3}{k_3} = \mu\left(\frac{h_1}{k_1}, \frac{h_2}{k_2}\right); \qquad \frac{h_4}{k_4} = \mu\left(\frac{h_3}{k_3}, \frac{h_2}{k_2}\right); \qquad \frac{h_2}{k_2} \tag{4}$$

into a left-hand, middle, and right-hand interval, and the middle interval $(h_3/k_3,\ h_4/k_4)$ is removed. Of course $h_3 = h_1 + h_2,\ h_4 = h_2 + h_3 = h_1 + 2h_2$, *etc.* Each right-hand interval, say $(h_5/k_5,\ h_6/k_6)$, is similarly partitioned by the points

$$\frac{h_5}{k_5}; \qquad \frac{h_8}{k_8} = \mu\left(\frac{h_7}{k_7}, \frac{h_5}{k_5}\right); \qquad \frac{h_7}{k_7} = \mu\left(\frac{h_6}{k_6}, \frac{h_5}{k_5}\right); \qquad \frac{h_6}{k_6} \tag{5}$$

Condition (1) is easily seen to be fulfilled. For a left hand interval we have, since adjacent fractions are neighbors:

$$m_1 = \frac{h_1 + h_2}{k_1 + k_2} - \frac{h_1}{k_1} = \frac{1}{k_1(k_1 + k_2)}, \tag{6}$$

$$m_2 = \frac{h_2}{k_2} - \frac{h_3 + h_2}{k_3 + k_2} = \frac{1}{k_2(k_3 + k_2)} = \frac{1}{k_2(k_1 + 2k_2)},$$

$$m_{12} = \frac{h_3 + h_2}{k_3 + k_2} - \frac{h_3}{k_3} = \frac{1}{k_3(k_3 + k_2)} = \frac{1}{(k_1 + k_2)(k_1 + 2k_2)}$$

For a right-hand interval,

$$m_1 = \frac{h_7 + h_5}{k_7 + k_5} - \frac{h_5}{k_5} = \frac{h_6 + 2h_5}{k_6 + 2k_5} - \frac{h_5}{k_5} = \frac{1}{k_5(k_6 + 2k_5)}, \tag{7}$$

$$m_2 = \frac{h_6}{k_6} - \frac{h_6 + h_5}{k_6 + k_5} = \frac{1}{k_6(k_6 + k_5)},$$

$$m_{12} = \frac{h_6 + h_5}{k_6 + k_5} - \frac{h_6 + 2h_5}{k_6 + 2k_5} = \frac{1}{(k_6 + k_5)(k_6 + 2k_5)}$$

The points of division form a Cantor set $C(0,\ 1/2)$. Since condition (1) is satisfied, we have from Theorem H:

THEOREM 1. $C(0,\ 1/2) + C(0,\ 1/2) = [0,1]$.

3. Finally we wish to show that the division points obtained in [C] and those developed above by Farey series are the same.

Let

$$CF(0,\ a_1,\ \ldots,\ a_n,\ \ldots) = \frac{1}{a_1 + \frac{1}{a_2 + \ldots}}$$

be the regular continued fraction with partial quotients a_i. We assume n even and $a_i \geq 2$. Cusick starts with the interval A $= [0,\ 1/2]$ and in subdividing A introduces two types of intervals. Intervals of the first type – these correspond to our left-hand intervals – have the form

$$[CF(0,\ a_1, \ldots,\ a_n),\ CF(0,\ a_1, \ldots,\ a_{n+1})] \tag{8}$$

Intervals of the second type – these correspond to our right-hand intervals – have the form

$$[CF(0,\ a_1, \ldots,\ a_n),\ CF(0,\ a_1, \ldots,\ a_{n-1})]\,. \tag{9}$$

From each interval (8) we remove the interval

$$(CF(0,\ a_1, \ldots,\ a_n,\ a_{n+1} + 1),\ CF(0,\ a_1, \ldots,\ a_{n+1},\ 2)) \tag{10}$$

and from each interval (9) we remove the interval

$$(CF(0,\ a_1, \ldots,\ a_n,\ 2),\ CF(0,\ a_1, \ldots,\ a_{n-1},\ a_n + 1))\,. \tag{11}$$

In both cases the removal of the middle interval leaves behind an interval of the first type on the left side and an interval of the second type on the right side. The indefinite continuation of this process leads to a Cantor set.

Let $a_1 = 2$. The first steps of Cusick's scheme are

B_1 $$A = \left[CF(0) = 0,\ CF(0,2) = \frac{1}{2}\right]$$

B_2 $$CF(0) = 0,\ CF(0,3) = \frac{1}{3},\ CF(0,2,2) = \frac{2}{5},\ CF(0,2) = \frac{1}{2}$$

B_3 $CF(0) = 0,\ CF(0,4) = \frac{1}{4},\ CF(0,3,2) = \frac{2}{7},\ CF(0,3) = \frac{1}{3},$

$$CF(0,2,2) = \frac{2}{5},\ CF(0,2,2,2) = \frac{5}{12},\ CF(0,2,3) = \frac{3}{7},\ CF(0,2) = \frac{1}{2}.$$

These agree with F_1, F_2, F_3 – see the lines following (3). Notice that in each line intervals of the first type alternate with those of the second type.

Now suppose C_i and F_i agree for $i \leq n$. Denote the jth convergent of $CF(0, a_1, a_2, \ldots)$ by p_j/q_j. We subdivide a typical interval of the first type in C_n to get

$$CF(0, a_1, \ldots, a_j) = \frac{p_j}{q_j}, \tag{12}$$

$$CF(0, a_1, \ldots, a_{j+1} + 1) = \frac{(a_{j+1} + 1)p_j + p_{j-1}}{(a_{j+1} + 1)q_j + q_{j-1}},$$

$$CF(a_1, \ldots, a_{j+1}, 2) = \frac{2p_{j+1} + p_j}{2q_{j+1} + q_j},$$

$$CF(0, \ldots, a_{j+1}) = \frac{p_{j+1}}{q_{j+1}}.$$

Using the recurrence relations

$$p_{j+1} = a_{j+1}p_j + p_{j-1}, \qquad q_{j+1} = a_{j+1}q_j + q_{j-1}$$

we find

$$\frac{(a_{j+1} + 1)p_j + p_{j-1}}{(a_{j+1} + 1)q_j + q_{j-1}} = \frac{p_{j+1} + p_j}{q_{j+1} + q_j}.$$

By the inductive assumption the extreme terms of (12) agree with those of (4), *i. e.*

$$\frac{p_j}{q_j} = \frac{h_1}{k_1}, \qquad \frac{p_{j+1}}{q_{j+1}} = \frac{h_2}{k_2}.$$

Hence the middle terms also agree, namely,

$$\frac{p_{j+1} + p_j}{q_{j+1} + q_j} = \mu\left(\frac{p_j}{q_j}, \frac{p_{j+1}}{q_{j+1}}\right) = \mu\left(\frac{h_1}{k_1}, \frac{h_2}{k_2}\right) = \frac{h_3}{k_3},$$

$$\frac{2p_{j+1} + p_j}{2q_{j+1} + q_j} = \mu\left(\frac{p_{j+1} + p_j}{q_{j+1} + q_j}, \frac{p_{j+1}}{q_{j+1}}\right) = \mu\left(\frac{h_2 + h_1}{k_2 + k_1}, \frac{h_2}{k_2}\right) = \frac{h_4}{k_4}.$$

Similar calculations apply to intervals of the second type (our right-hand intervals). This completes the induction and the proof of

THEOREM 2. *The Cantor sets $C(0, 1/2)$ as given in [C] and in Section 2 above are identical.*

References

[C] T. W. CUSICK. Sums and products of continued fractions. *Proc. Amer. Math. Soc.* **27** (1971) 35-38.

[H] M. HALL. On the sum and product of continued fractions. *Ann. of Math. (2)* **48** (1947) 966-993.

314-N Sharon Way
Jamesburg, New Jersey, 08831
USA

Commensurability classes of two-generator Fuchsian groups

C. Maclachlan and G. Rosenberger

To Murray Macbeath on the occasion of his retirement

0. Introduction

A Fuchsian group is a discrete subgroup of $PSL_2(\mathbb{R})$ and two such groups Γ_1, Γ_2 are commensurable if and only if the intersection $\Gamma_1 \cap \Gamma_2$ is of finite index in both Γ_1 and Γ_2. In this paper we determine when two two-generator Fuchsian groups of finite covolume are commensurable and in addition, the relationship between two such groups, by obtaining part of the lattice of the commensurability class which contains one representative from each conjugacy class, in $PGL_2(\mathbb{R})$, of two-generator groups. By the result of Margulis (see e.g. [Z]) that a non-arithmetic commensurability class contains a unique maximal member, the non-arithmetic cases reduce to a compilation of earlier results on determining which two-generator Fuchsian groups occur as subgroups of finite index in other two-generator Fuchsian groups [Sc], [S2], [R]. For the arithmetic cases, all two-generator arithmetic Fuchsian groups have been determined [T2], [T4], [MR] and one can immediately read off when two such groups are commensurable from the structure of the corresponding quaternion algebra (see e.g. [T3]). The relationship between such groups is more difficult to determine and we utilise structure theorems for arithmetic Fuchsian groups [B], [V]. The relationship between arithmetic triangle groups was determined in [T3].

1. Two-generator Fuchsian groups

A non-elementary Fuchsian group of finite covolume has a presentation of the form

$$\text{Generators}: \quad a_1, b_1, \cdots a_g, b_g, x_1, x_2, \cdots, x_r, p_1, p_2, \cdots, p_s \tag{1}$$

$$\text{Relations}: \quad x_i^{m_i} = 1 (i = 1, 2, \cdots, r) \prod_{i=1}^{r} x_i \prod_{j=1}^{s} p_j \prod_{k=1}^{g} [a_k, b_k] = 1$$

This presentation determines the signature of the group $(g; m_1, m_2, \cdots m_r; s)$. The group will be cocompact if and only if $s = 0$. If Γ is a two-generator Fuchsian group, then Γ has signature of one of the following forms

(i) $(0; m_1, m_2, \cdots m_r; s)$ where $r + s = 3$

(ii) $(1; q; 0)$, $(1; -; 1)$

(iii) $(0; 2, 2, 2, e; 0)$ where e is an odd integer.

Now if Γ_1 is a subgroup of Γ_2 of finite index, the inclusion map defines an embedding $T(\Gamma_2) \longrightarrow T(\Gamma_1)$, where $T(\Gamma)$ denotes the Teichmuller space of Γ. Furthermore, this embedding will be surjective whenever the dimensions of the two Teichmuller spaces are equal. All situations when this arises are given in [S2] and in these cases, every Fuchsian group with the same signature as Γ_1 occurs as a subgroup of a group with the same signature as Γ_2. In particular, whenever Γ_1 has signature $(1; q; 0)$ (resp. $(1; -; 1)$) then Γ_1 is a subgroup of index 2 in a group with signature $(0; 2, 2, 2, 2q; 0)$ (resp. $(0; 2, 2, 2; 1)$).

We denote the signature $(0; 2, 2, 2, e; 0)$ (resp. $(0; 2, 2, 2; 1)$) where e is even or odd by σ_e (resp. σ_∞). Thus if a commensurability class contains a two-generator group, then it contains a triangle group or a group of signature σ_e where $(3 \leq e \leq \infty)$.

Using the theorem below, the problem divides into two cases depending on whether the Fuchsian group is arithmetic or not. The definition and relevant results on arithmetic groups will be given in section 3. For the moment it suffices to note that arithmeticity is a commensurability and conjugacy invariant notion.

$$Comm(\Gamma) = \{t \in PGL_2(\mathbf{R}) \mid \Gamma, t\Gamma t^{-1} \text{ are commensurable}\}$$

THEOREM (Margulis) *Let Γ be a finite covolume Fuchsian group. Then Γ is a subgroup of finite index in $Comm(\Gamma)$ if and only if Γ is non-arithmetic.*

2. The non-arithmetic cases

Let Λ be a non-arithmetic commensurability class of Fuchsian groups which contains a two-generator Fuchsian group Γ of finite covolume. By the Margulis Theorem Λ has a unique maximal member $Comm^{+}(\Gamma) = \Gamma_0$. Recall that if Γ_1, Γ_2 are two Fuchsian groups with $\Gamma_2 \subset \Gamma_1$, $[\Gamma_1 : \Gamma_2] < \infty$ and Γ_2 is a triangle group, then Γ_1 is also a triangle group. Thus if Λ contains a triangle group then Γ_0 is a triangle group. If Γ_0 is not a triangle group then $\Gamma_0 \supset \Gamma$ where Γ has signature σ_e $(3 \leq e \leq \infty)$ and is of finite index in Γ_0. Thus Γ_0 must have an element of order e (resp. be non-cocompact if $e = \infty$). But then a simple volume calculation shows that Γ_0 must have signature σ_e and so $\Gamma = \Gamma_0$. Again a simple volume calculation shows that in this case Γ_0 cannot contain a subgroup of signature $\sigma_{e'}$ with $e' \neq e$.

PROPOSITION 2.1 *Let Γ_0 be the unique maximal group in a commensurability class which contains a two-generator non-arithmetic Fuchsian group. Then either Γ_0 is a triangle group or Γ_0 has signature σ_e and the commensurability class contains exactly one two-generator group*

In the cases where Γ_0 is a triangle group, we must determine those triangle groups which contain some other two-generator Fuchsian groups as subgroups of finite index. This problem has been resolved by Schulenberg[Sc], Singerman[S2] and Rosenberger[R]. By compiling their results we obtain a finite number of families of triangle groups and the lattice of their subgroups which contain two-generator groups. One family is doubly-infinite and the other families depend on a single parameter t. For each of these families we obtain a finite group G, independent of t, such that for all t,

$$G \cong \frac{\Gamma}{N}$$

where Γ is the maximal triangle group and N is the maximal normal subgroup which is contained in all two-generator subgroups of Γ, and so each two-generator subgroup of Γ corresponds to a subgroup of G.

For each family, we give the signature of the maximal triangle group Γ, the signature of the maximal normal subgroup N and the finite group G. The excluded values of the parameter correspond to those families which are arithmetic, this being immediately deducible from the results in [T3].

Note, however, that when the value of the parameter does correspond to an arithmetic group, the lattice given below will be a sublattice of the full lattice of two-generator groups in these cases. Finally we actually only give three lattice diagrams as the others can be regarded as subdiagrams of these, suitably interpreted. On the full diagrams, the normal subgroups corresponding to the subdiagrams are labelled and the subdiagram consists of all the groups between that and the maximal triangle group.

Notation

(i) All groups here are cocompact and so signatures are reduced to the form

$$(g; m_1, m_2, \cdots, m_r)$$

Additionally, if a period n is repeated r times, this is indicated $n^{(r)}$.

(ii) Only one representative in the conjugacy class in $PGL_2(\mathbf{R})$ of each of the groups involved is indicated on the lattice diagram.

	Maximal triangle group	Maximal normal subgroup	factor group	excluded values
A	$\Gamma = (0; 2, t_1, 2t_2)$, $t_1 \neq 3, 4$	$N = (0; t_1, t_1, t_2)$	$\mathbf{Z}_2$	$(5, n)$ $n = 3, 4, 5, 10, 15$
B	$\Gamma = (0; 2, 3, 2t)$, $(t, 6) = 1$	$N_1 = (0; t, t, t)$	S_3	$5, 7$
	$\Gamma = (0; 2, 3, 3t)$, $(t, 2) = 1$	$N_2 = (0; t, t, t, t)$	A_4	3
	$\Gamma = (0; 2, 3, 4t)$, $(t, 3) = 1$	$N_3 = (0; t^{(6)})$	S_4	$2, 4$
	$\Gamma = (0; 2, 3, 6t)$, $(t, 2) = 1$	$N_4 = (1; t^{(12)})$	G_4	$3, 5$
	$\Gamma = (0; 2, 3, 12t)$	$N_5 = (13; t^{(24)})$	G_5	$1, 2$
C	$\Gamma = (0; 2, 4, 2t)$, $(t, 2) = 1$	$N_1 = (0; t, t, t, t)$	D_4	$3, 5, 9$
	$\Gamma = (0; 2, 4, 4t)$	$N_2 = (1; t, t, t, t)$	H_2	$2, 3$

Table 1

In this table the finite groups are $\mathbf{Z}_2$, the cyclic group of order 2, S_3, S_4, the symmetric groups on 3 and 4 letters, A_4 the alternating group on 4 letters, D_4 the dihedral group of 8 elements and the remaining groups are described below.

$$G_4 = \{x, y \mid x^2 = y^3 = (xy)^6 = [x, y]^6 = [x, y]^2[x, y^{-1}]^2 = 1\}$$

$$G_5 = \{x, y \mid x^2 = y^3 = (xy)^{12} = ((xy^{-1})^2(xy)^2)^2 = (xy^{-1})^{12} = 1$$
$$(yx)^2(xy)^6 = (xy)^6(yx)^2, \quad (xy)^2(yx)^6 = (yx)^6(xy)^2,$$
$$(y^{-1}x)^6 = (yx)^4x(yx)^6xyx\}$$
$$H_2 = \{x, y \mid x^2 = y^4 = (xy)^4 = 1, \quad xy^2 = y^2x\}$$

Case A

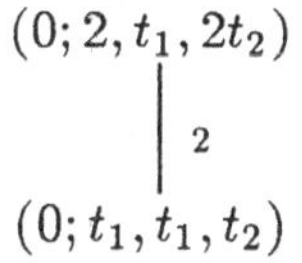

$(0; 2, t_1, 2t_2)$

$\quad | \ 2$

$(0; t_1, t_1, t_2)$

Figure 1

Case B

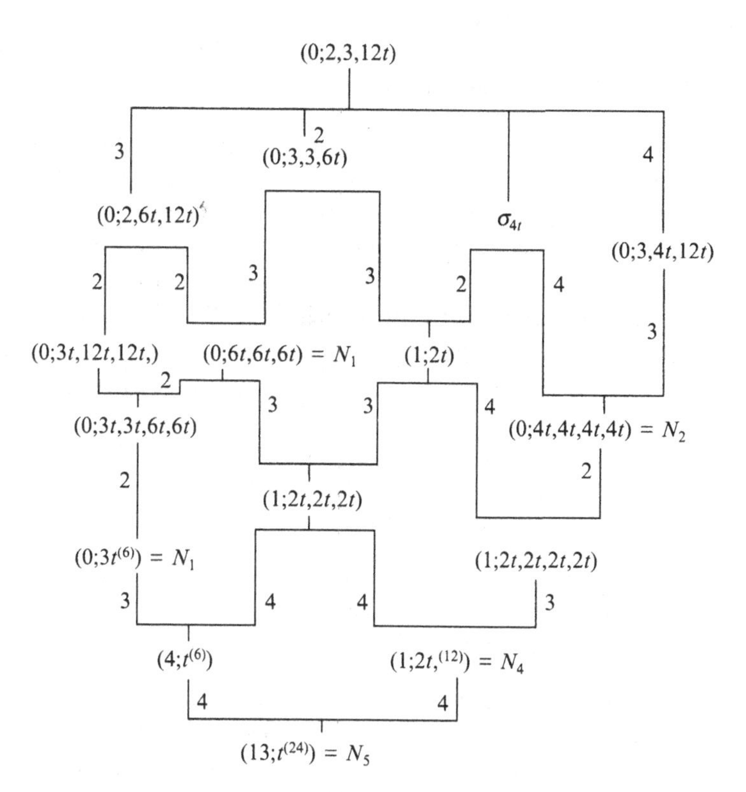

Figure 2

Case C

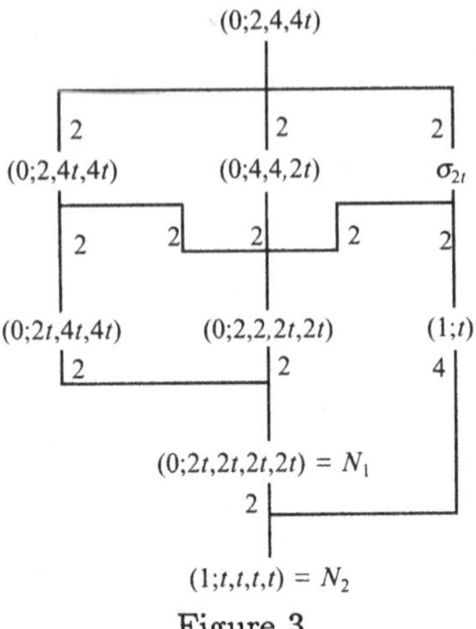

Figure 3

3. The arithmetic cases

In this section we determine the commensurability classes of the arithmetic two-generator Fuchsian groups. Recall, that arithmetic Fuchsian groups are defined as follows (e.g. [V], [B]).

Let k be a totally real number field and A a quaternion algebra defined over k which is ramified at all real places except one. Let ρ be a representation $\rho : A \to M_2(\mathbb{R})$ and $\mathcal{O}$ an order in A. Then if $\mathcal{O}^1$ denotes the elements of norm 1 in $\mathcal{O}$, the group $P\rho(\mathcal{O}^1)$ is a Fuchsian group of finite covolume and the class of *arithmetic Fuchsian groups* is the class of all Fuchsian groups commensurable with some such $P(\rho(\mathcal{O}^1))$. If Γ is an arithmetic Fuchsian group, let $\Gamma^{(2)}$ be the subgroup generated by $\gamma^2, \gamma \in \Gamma$. Then the isomorphism classes of the quaternion algebra can be recovered from Γ as

$$A(\Gamma^{(2)}) = \{\sum a_i \gamma_i \mid a_i \in \mathbb{Q}(\operatorname{tr}\gamma, \gamma \in \Gamma^{(2)})\}$$

and is uniquely determined by the arithmetic group [T1], [T3].

THEOREM (Takeuchi) *Let Γ_1, Γ_2 be two arithmetic Fuchsian groups. Then Γ_1 is commensurable with a conjugate of Γ_2 if and only if the associated quaternion algebras are isomorphic.*

Now two quaternion algebras over the same field are isomorphic if and only if they have the same set of ramified places. All two-generator arithmetic Fuchsian groups together with the defining field and ramification set of the corresponding quaternion algebra have been given by Takeuchi [T2]

[T4] and Maclachlan - Rosenberger [MR]. Thus one can read off from these lists exactly when two arithmetic two-generator groups are commensurable. However, we also determine that part of the commensurability lattice which contains one representative from each $PGL_2(\mathbb{R})$ conjugacy class of two-generator groups.

A. Non-compact case. From the lists mentioned above, this commensurability class contains groups of the following signatures : $(0;2,3;1)$, $(0;2,4;1)$, $(0;3,3;1)$, $(0;2,6;1)$, $(0;4,4;1)$, $(0;2;2)$, $(0;6,6;1)$, $(0;-;3)$ and four classes of groups of signature $(1;-;1)$. Recall that each group of signature $(1;-;1)$ is contained as a subgroup of index 2 in a group of signature $(0;2,2,2;1)$. The relationship between the triangle groups has been given in [T3] and by the analogues of Figures 2 and 3 above we see that $(0;2,3;1)$ contains a subgroup of signature $(0;2,2,2;1)$ as does $(0;2,4;1)$. Since the second of these is normal in $(0;2,4;1)$ which is maximal in $PSL_2(\mathbb{R})$, these two groups of signature $(0;2,2,2;1)$ cannot be conjugate.

Now groups commensurable with the classical modular group have been studied in [H], [M] in terms of the groups $\Gamma_0(n)$ and their normalizers $\Gamma_N(n)$. From [M] the groups $\Gamma_N(5)$, $\Gamma_N(6)$ have signatures $(0;2,2,2;1)$ and cannot be conjugate to each other or to the groups of the same signature already described. Now $\Gamma_N(5) \cap \Gamma_0(1) = \Gamma_0(5)$ which has signature $(0;2,2;2)$ while $\Gamma_0(6)$ has signature $(0;-;4)$. We thus obtain the following commensurability diagram.

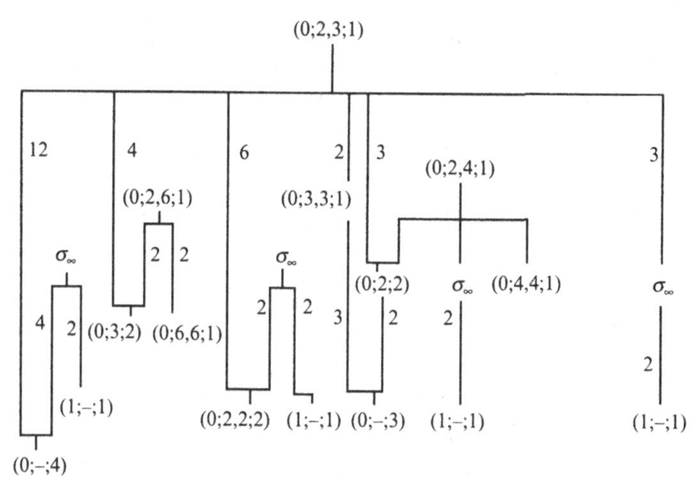

Figure 4

B. Cocompact classes involving triangle groups. The arguments in these cases all follow the same lines, so we give details in only one case - corresponding to the triangle group $(0;2,3,8)$ - and only appropriate comments in the other cases. Similar arguments were employed in [MR].

From the lists in [T2], [T4], [MR] the quaternion algebra is defined over $\mathbf{Q}(\sqrt{2})$ and its finite ramification is at $\mathcal{P}_2$. Furthermore, groups of the following signatures are in the commensurability class : $(0;2,3,8)$, $(0;3,3,4)$, $(0;2,4,8)$, $(0;2,6,8)$, $(0;2,8,8)$, $(0;4,4,4)$, $(0;3,8,8)$, $(0;4,6,6)$, $(0;4,8,8)$, three classes of signature $(0;2,2,2,3)$, one class of signature $(1;2)$ and three of signature $(1;4)$.

The relationship between the triangle groups is known [T3] (Note, however that in the appropriate diagram $(0;4,8,8)$ should be shown as a subgroup of $(0;2,8,8)$) and from the $(0;2,4,4t)$ lattice in figure 3, the group of signature $(1;2)$ is a subgroup of index 4 in $(0;2,4,8)$.

Now let us consider the three classes of groups of signature $(0;2,2,2,3)$. If such a group is to occur as a subgroup of Γ_0 which has signature $(0;2,3,8)$ then it will correspond to the stabiliser of 1 in a suitable permutation representation $\phi : \Gamma_0 \to S_4$ since the index would have to 4.(see e.g.[S1]) The obvious homomorphism onto S_4 does in fact yield such a subgroup and one can easily show that there is only one class of such homomorphisms and so one conjugacy class of subgroups of Γ_0 of signature $(0;2,2,2,3)$ (c.f. [MR])

Let Γ have signature $(0;2,2,2,3)$ and be such that it is not conjugate in $PGL_2(\mathbf{R})$ to a subgroup of Γ_0. Since Γ cannot be a subgroup of another triangle group commensurable with Γ_0 by a simple volume calculation, it follows that Γ is maximal in $PSL_2(\mathbf{R})$. We now make extensive use of results of Borel [B], which necessitate working in $PGL_2(\mathbf{R})$. In the tree of maximal orders at each prime ν unramified in the quaternion algebra A, choose a vertex $\mathcal{P}_\nu$. Choose also an adjoining vertex $\mathcal{P}'_\nu$ and edge e_ν. Let S and S' be two finite disjoint subsets of the set of primes which are unramified in A. Define

$$\Gamma_{S,S'} = \{\alpha \in A^* \mid \alpha \text{ fixes } \mathcal{P}_\nu, \nu \notin S \cup S', \alpha \text{ fixes } e_\nu, \nu \in S, \alpha \text{ fixes } \mathcal{P}'_\nu, \nu \in S'\}$$

In these cases where the maximal group is a triangle group, the group of smallest covolume in the commensurability class can be taken to be $\Gamma_{\phi,\phi}$ and $[\Gamma_{\phi,\phi} : \Gamma^+_{\phi,\phi}] = 2$ where $\Gamma^+_{\phi,\phi} = \Gamma_{\phi,\phi} \cap PSL_2(\mathbf{R})$. Now $\Gamma \subset \Gamma_{max}$ of index $l = 1$ *or* 2 where Γ_{max} is a maximal group in the commensurability class in $PGL_2(\mathbb{R})$ of $\Gamma_{\phi,\phi}$. Now Γ_{max} will be conjugate to some group $\Gamma_{S,S'}$ with generalized index given by

$$[\Gamma_{\phi,\phi} : \Gamma_{S,S'}] = 2^{-m} \prod_{\nu \in S} (N\nu + 1)$$

where $0 \leq m \leq \mid S \mid$and if $\mid S \mid= 1$ then $m = 1$. In our case by a volume calculation

$$[\Gamma_{\phi,\phi} : \Gamma_{S,S'}] = \frac{8}{l}$$

By examining the primes of small norm in $\mathbb{Q}(\sqrt{2})$, we find that we must have $S = \{\mathcal{P}_7\}$ or $\{\mathcal{P}'_7\}$ the two primes of norm 7, and $l = 2$. We can assume that $\Gamma_{max} = \Gamma_{S,\phi}$. Since the tree of maximal orders in the case of $\mathcal{P}_7$ or $\mathcal{P}'_7$ has valency 8, it follows that $[\Gamma_{\phi,\phi} : \Gamma_{\phi,\phi} \cap \Gamma_{S,\phi}] = 8$ and so $[\Gamma_{S,\phi} : \Gamma_{\phi,\phi} \cap \Gamma_{S,\phi}] = 2$. Since $l = 2$, $[\Gamma_{S,\phi} : \Gamma^+_{S,\phi}] = 2$ and so $[\Gamma^+_{S,\phi} : \Gamma_{\phi,\phi} \cap \Gamma^+_{S,\phi}] = 2$. Since every subgroup of index 2 in Γ has signature $(0; 2, 2, 3, 3)$, it follows that $\Gamma_{\phi,\phi} \cap \Gamma^+_{S,\phi}$ has signature $(0; 2, 2, 3, 3)$. Furthermore, for $S_1 = \{\mathcal{P}_7\}$, $\Gamma_{S_1,\phi}$ will contain an element odd at $\mathcal{P}_7$ (for the definition of 'odd' and its role with respect to the groups $\Gamma_{S,S'}$ see [B]. See also [MR section 6]). It follows that $\Gamma_{S_1,\phi}$ cannot be conjugate to $\Gamma_{S_2,\phi}$ where $S_2 = \{\mathcal{P}'_7\}$.

Now consider, in this case, the groups of signature $(1; 4)$, each of which is contained in a group of signature $(0; 2, 2, 2, 8)$. As above, we find that there is just one class of such groups which are subgroups of $(0; 2, 3, 8)$ of index 9. In the other two cases, we argue as above and find that the groups $\Gamma_{S,\phi}$, where $S = \{\mathcal{P}_{17}\}$ *or* $\{\mathcal{P}'_{17}\}$ are such that $[\Gamma_{S,\phi} : \Gamma^+_{S,\phi}] = 2$ and $\Gamma^+_{S,\phi}$ has signature $(0; 2, 2, 2, 8)$. Also $[\Gamma^+_{S,\phi} : \Gamma_{\phi,\phi} \cap \Gamma^+_{S,\phi}] = 2$. Thus $\Gamma_{\phi,\phi} \cap \Gamma^+_{S,\phi}$ has one of the three signatures $(1; 4)$, $(0; 2, 2, 8, 8)$ or $(0; 2, 2, 2, 2, 4)$. Note that $\Gamma_{\phi,\phi} \cap \Gamma^+_{S,\phi}$ is normal in $\Gamma^+_{S,\phi}$, which is maximal in $PSL(2, \mathbb{R})$. Thus as a subgroup of $\Gamma^+_{\phi,\phi}$, the group $\Gamma_{\phi,\phi} \cap \Gamma^+_{S,\phi}$ must be self-normalizing. By employing CAYLEY† we obtained that the group of signature $(0; 2, 3, 8)$ contained 2 conjugacy classes of self-normalizing subgroups of signature $(0; 2, 2, 8, 8)$ and none of signature $(1; 4)$. But each group of signature $(0; 2, 2, 8, 8)$ is a subgroup of index 2 in a group of signature $(0; 2, 2, 2, 8)$. It thus follows that these groups of signature $(0; 2, 2, 2, 8)$ must be the groups $\Gamma^+_{S,\phi}$ where $S = \{\mathcal{P}_{17}\}$ *or* $\{\mathcal{P}'_{17}\}$ and as above cannot be conjugate in $PSL(2, \mathbb{R})$.

C. Cocompact arithmetic classes. The lattice diagrams for all cocompact arithemtic classes are given on the pages which follow.

---

† The authors are grateful to Alan Reid for help with the implementation of this.

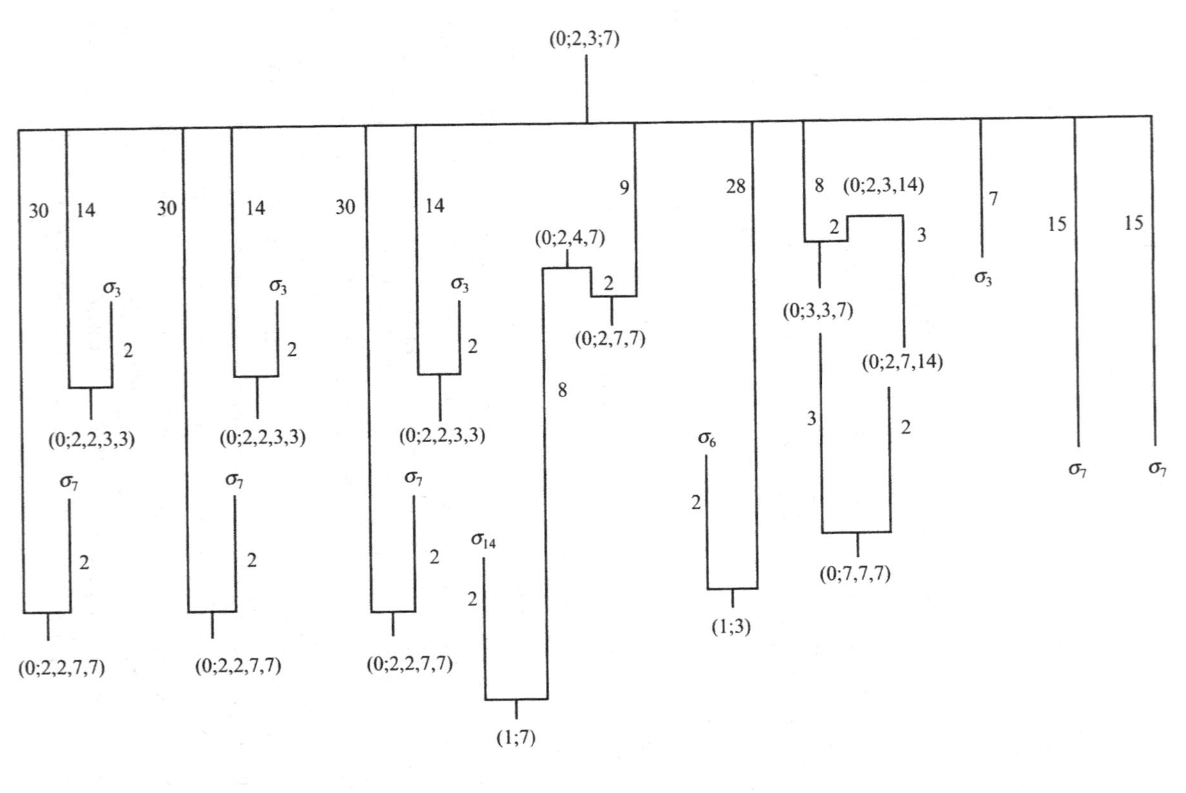

(0;2,3;7)
30
14
σ3
2
(0;2,2,3,3)
σ7
2
(0;2,2,7,7)
30
14
σ3
2
(0;2,2,3,3)
σ7
2
(0;2,2,7,7)
30
14
σ3
2
(0;2,2,3,3)
σ7
2
(0;2,2,7,7)
9
(0;2,4,7)
2
(0;2,7,7)
8
σ14
2
(1;7)
28
σ6
2
(1;3)
8
(0;2,3,14)
2
3
(0;3,3,7)
(0;2,7,14)
3
2
(0;7,7,7)
7
σ3
15
σ7
15
σ7

Figure 5

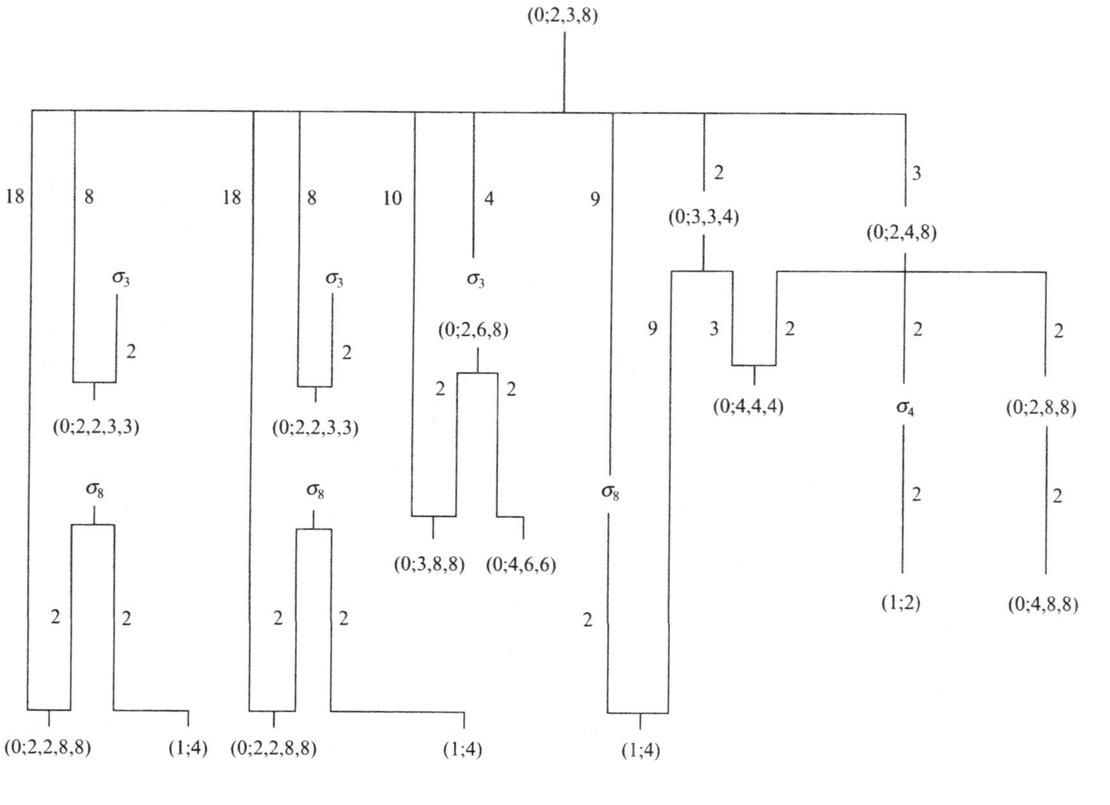
(0;2,3,8)
18
8
18
8
10
4
9
2
3
(0;3,3,4)
(0;2,4,8)
σ3
σ3
σ3
(0;2,6,8)
9
3
2
2
2
2
2
2
2
(0;2,2,3,3)
(0;2,2,3,3)
(0;4,4,4)
σ4
(0;2,8,8)
σ8
σ8
σ8
2
2
(0;3,8,8)
(0;4,6,6)
2
2
2
2
2
2
2
(1;2)
(0;4,8,8)
(0;2,2,8,8)
(1;4)
(0;2,2,8,8)
(1;4)
(1;4)

Figure 6

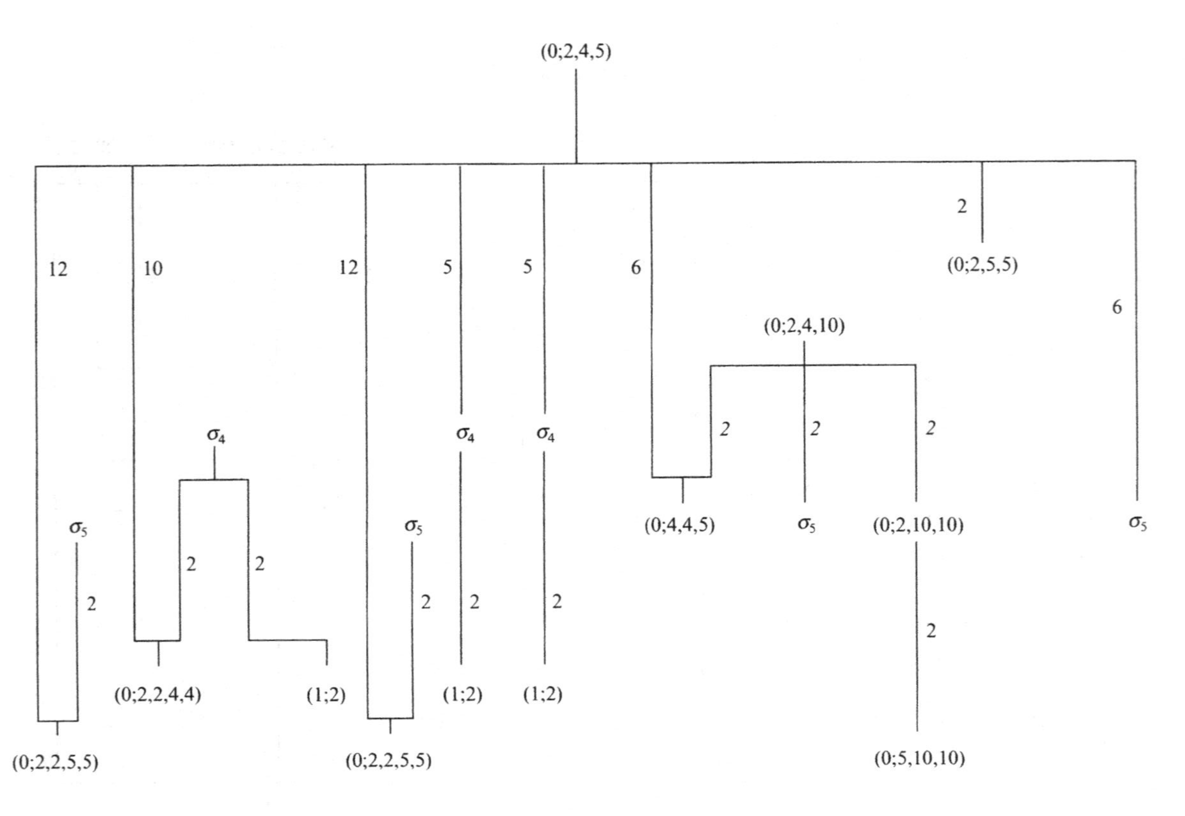
(0;2,4,5)
12
10
12
5
5
6
2
(0;2,5,5)
6
(0;2,4,10)
σ4
σ4
σ4
2
2
2
σ5
σ5
(0;4,4,5)
σ5
(0;2,10,10)
σ5
2
2
2
2
2
2
2
(0;2,2,4,4)
(1;2)
(1;2)
(1;2)
(0;2,2,5,5)
(0;2,2,5,5)
(0;5,10,10)

Figure 7

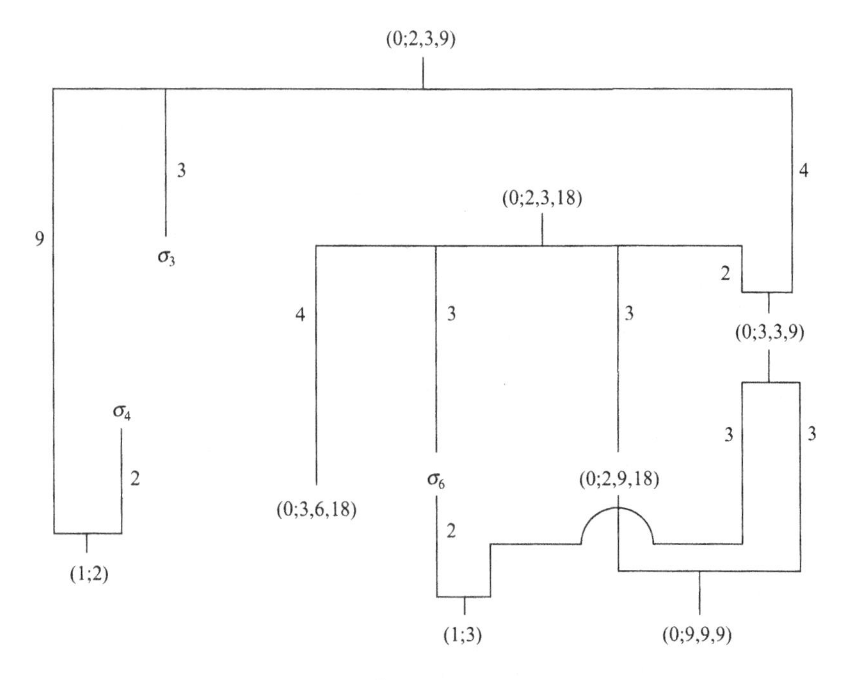

Figure 8

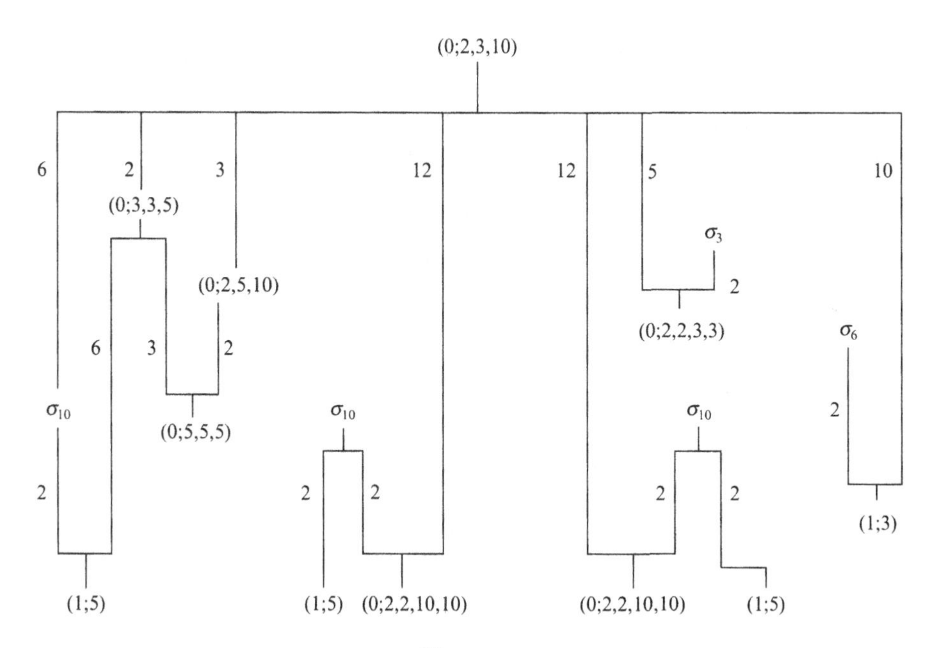

Figure 9

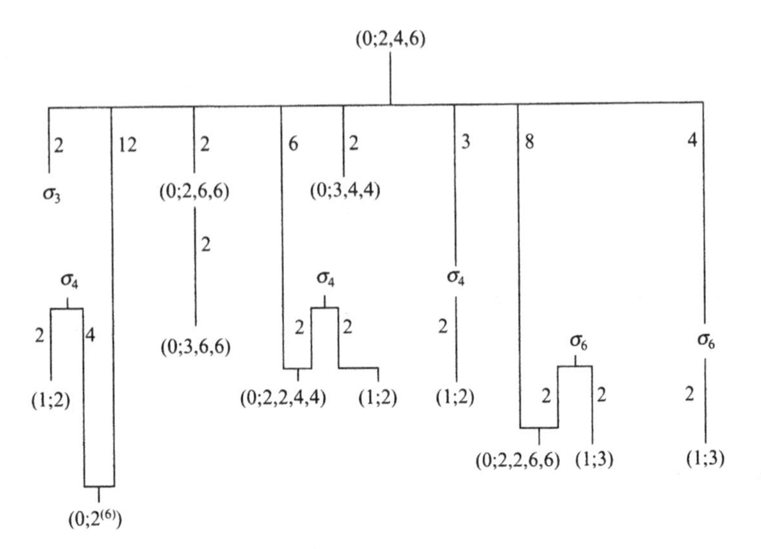

Figure 10

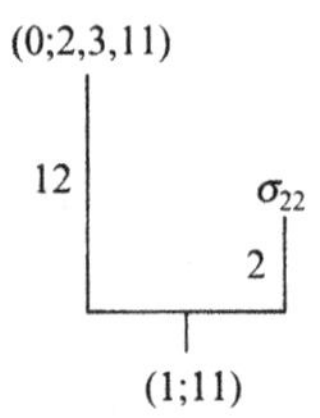

Figure 11

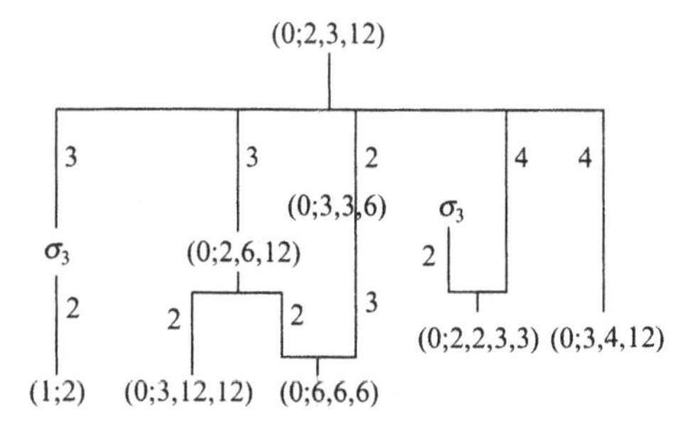

Figure 12

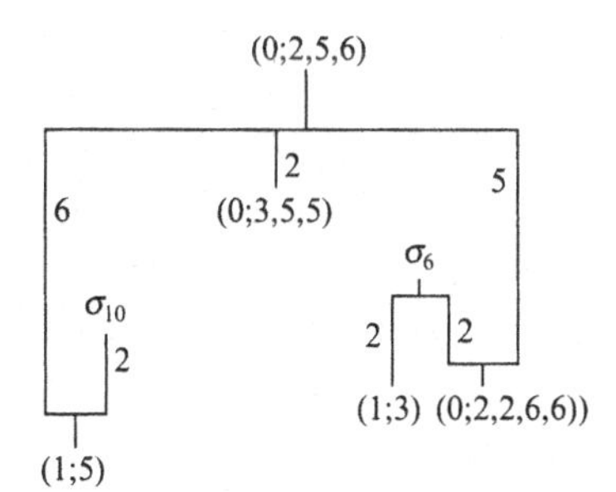

Figure 13

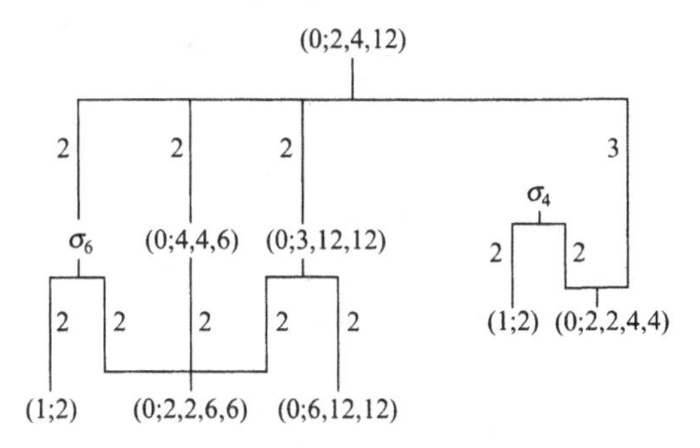

Figure 14

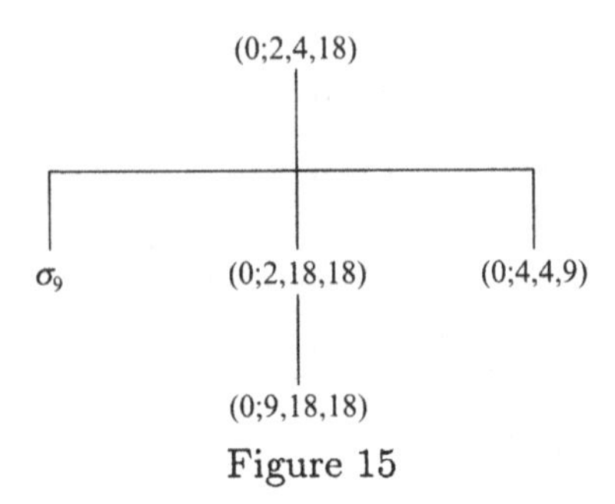

Figure 15

Remarks on these diagrams

1. It will be noted from the arguments involved in obtaining the commensurability lattice for the group $(0;2,3,8)$, that if a group Γ of signature $(0;2,2,2,e)$ where e is odd, is known to be commensurable with a maximal triangle group (of smaller covolume) and the intersection with the maximal group has index 2 in Γ, then one immediately deduces that the intersection has signature $(0;2,2,e,e)$. However, if e is even and the same conditions hold, there are three possible signatures which the intersection may have. It is then necessary to investigate the self-normalizing subgroups of the appropriate index as was done in the $(0;2,3,8)$ case. Note that this tactic employing CAYLEY was also used in the cases $(0;2,3,10)$, $(0;2,4,5)$, $(0;2,4,6)$.

2. In many cases the results of Schulenberg [Sc] and Rosenberger [R] were employed to determine when groups of signature $(1;q)$ or $(0;2,2,2,e)$ with e odd were subgroups of triangle groups.

3. Our detailed investigations revealed some small errors in [T4]. We note only one of significance – the group of signature $(1;5)$ with defining field $\mathbb{Q}(\sqrt{5})$ and ramification set $\{\mathcal{P}_2\}$ should have ramification set $\{\mathcal{P}_3\}$.

4. The diagrams above are ordered such that the covolumes of the maximal groups increase but we have omitted 6 maximal arithmetic triangle groups which we now comment upon. The two-generator commensurability diagram corresponding to the groups of signatures $(0;2,3,16)$, $(0;2,5,20)$, $(0;2,5,30)$ and $(0;2,5,8)$ involve only triangle groups and are already given in [T3]. In the cases corresponding to the groups $(0;2,3,24)$ and $(0;2,3,30)$ the diagram is exactly as was obtained in the non-arithmetic cases $(0;2,3,12t)$ and $(0;2,3,6t)$ in section 1.

D. Commensurability classes not involving triangle groups. Most classes of such groups just contain one conjugacy class of two-generator groups. There are eleven classes which contain more than one conjugacy class of two- generator groups and seven of these contain pairs of groups of different signatures.

Of these seven, we treat one example briefly, the others following a similar pattern. Let A be defined over $\mathbb{Q}(\sqrt{13})$ and ramified at $\mathcal{P}_3$ (or $\mathcal{P}_3'$). This gives rise to one class of groups of each of the signatures $(0;2,2,2,3)$ and $(1;3)$. Again Borel's results [B] are employed to give

$$\frac{\Gamma_{\mathcal{O}}}{\Gamma_{\mathcal{O}}^1} \cong \frac{R_{f,\infty}^*}{R_f^{*2}} \qquad \frac{\Gamma_{\mathcal{O}}^+}{\Gamma_{\mathcal{O}}^1} \cong \frac{R_{f,+}^*}{R_f^{*2}}$$

in his notation. A fundamental unit is $\frac{3+\sqrt{13}}{2}$ and a $\mathcal{P}_3$ unit is $\frac{5+\sqrt{13}}{2}$ so that the first group above has order 4 while the second has order 2. Now using the notation adopted in C. above, this yields that

$$[\Gamma_{\phi,\phi} : \Gamma_{S,\phi}] = \frac{4}{l} = 2^{-m} \prod_{\nu \in S} (N\nu + 1)$$

where $\Gamma_{S,\phi}$ contains the maximal (in $PSL(2,\mathbb{R})$) group of signature $(0;2,2,2,6)$. The only posssible solution is $S = \{\mathcal{P}_3'\}$ (resp. $\mathcal{P}_3$) and $l = 2$. As before we obtain that $\Gamma_{\phi,\phi} \cap \Gamma_{S,\phi}^+$ has index 2 in $\Gamma_{S,\phi}^+$ and index 4 in $\Gamma_{\phi,\phi}^+$. One calculates directly that the intersection cannot have signature $(0;2,2,6,6)$ (clearly) or $(1;3)$. Thus we obtain the commensurability diagram

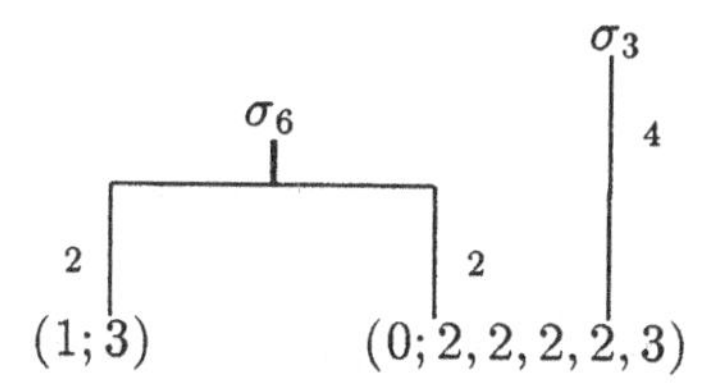

Figure 16

The same diagram arises for A defined over $\mathbb{Q}$ and ramified at $(2),(5)$.

If A is defined over $\mathbb{Q}(\sqrt{17})$ and ramified at $\mathcal{P}_2$ or $\mathcal{P}_2'$ we obtain

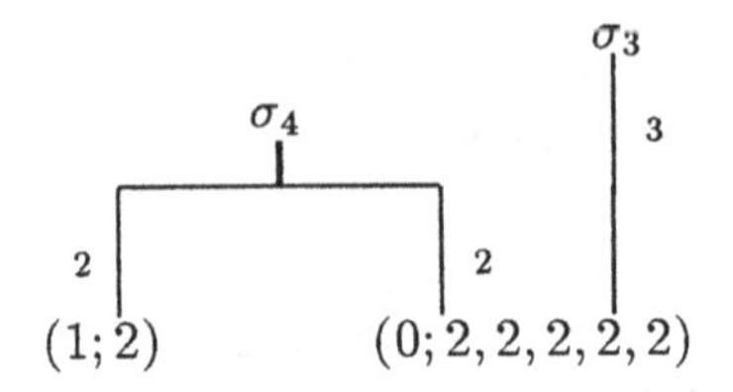

Figure 17

If A is defined over $\mathbb{Q}(\sqrt{2})$ and ramified at $\mathcal{P}_7$ *or* $\mathcal{P}_7'$ this gives

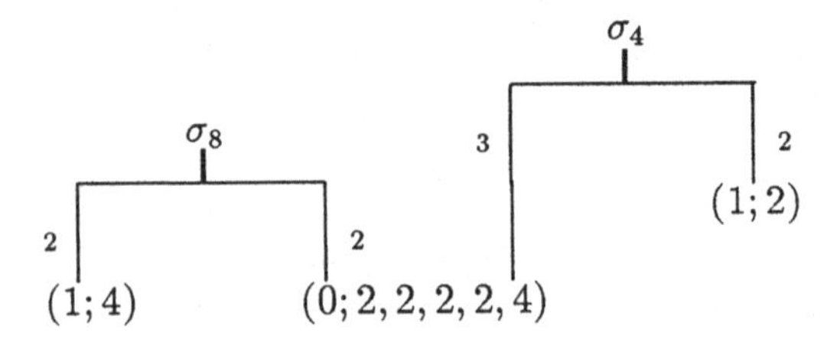

Figure 18

The remaining four cases arise from quaternion algebras whose type number is 2 and in all cases this gives rise to commensurable pairs of groups of signature $(0; 2, 2, 2, 3)$. They arise in the following cases (for more details see [MR]).

field discriminant	finite ramification	real ramification
36497	ϕ	x_2, x_3, x_4, x_5
38569	ϕ	x_1, x_2, x_3, x_4
229	ϕ	x_2, x_3
257	ϕ	x_1, x_2

(The roots of the polynomial generating an integer basis as given in [PZW] are ordered such that $x_1 < x_2 < \cdots < x_n$)

In these cases there is no group which is the obvious intersection of the two maximal groups since we could conjugate one maximal order while leaving the other fixed.

References

[B] A.Borel. Commensurability classes and volumes of hyperbolic three-manifolds *Ann. Sc. Norm. Sup. Pisa* **8** (1981) 1-33

[H] H.Helling. On the commensurability class of the rational modular group. *J. London Math. Soc.* **2** (1970) 67-72

[M] C.Maclachlan. Groups of units of zero ternary quadratic forms. *Proc. Royal Soc. Edinburgh* **88A** (1981) 141-157

[MR] C.Maclachlan and G.Rosenberger. Two-generator arithmetic Fuchsian groups *to appear*

[PZW] M.Pohst, H.Zassenhaus and P.Weiler. On effective computation of fundamental units II. *Math. Comp.* **38** (1982) 293-329

[R] G.Rosenberger. Von Untergruppen der Triangel-Gruppen. *Illinois J. Maths.* **22** (1978) 404-413

[Sc] E.Schulenberg. Die Erweiterungen der Grenzkreisgruppen mit zwei Erzeugenden. *Abh. Math. Sem. Univ. Hamburg* **13** (1939) 144-199

[S1] D.Singerman. Subgroups of Fuchsian groups and finite permutation groups. *Bull. London Math. Soc.* **2** (1970) 319-323

[S2] D.Singerman. Finitely maximal Fuchsian groups. *J. London Math. Soc.* **6** (1972) 29-38

[T1] K.Takeuchi. A characterization of arithmetic Fuchsian groups. *J. Math. Soc. Japan* **27** (1975) 600-612

[T2] K.Takeuchi. Arithmetic triangle groups. *J. Math. Soc. Japan* **29** (1977) 91-106

[T3] K.Takeuchi. Commensurability classes of arithmetic triangle groups. *J. Fac. Sc. Univ. Tokyo* Sect. 1A **24** (1977) 201-222

[T4] K.Takeuchi. Arithmetic Fuchsian groups of signature $(1; e)$. *J. Math. Soc. Japan* **35** (1983) 381-407

[V] M-F.Vigneras. Arithmetique des Algebres de Quaternions. *Lecture Notes In Maths.* Vol. **800** Springer Berlin (1980)

[Z] R.Zimmer. *Ergodic theory and semi-simple groups.* Birkhauser Boston (1984)

Department of Mathematical Sciences
Edward Wright Building
University of Aberdeen
Aberdeen AB9 2TY
Scotland

Fachbereich Mathematik
der Universitat Dortmund
Postfach 500500
4600 Dortmund 50
Germany

Limit points via Schottky pairings

P. J. Nicholls and P. L. Waterman

Dedicated to A.M. Macbeath

1. Introduction

Let Γ be a discrete group of Möbius transformations preserving the upper half of Euclidean n-space:

$$H^n = \{ x = (x_1, x_2, \ldots, x_n) : x_n > 0 \}.$$

We denote by j the *canonical point* $(0, 0, \ldots, 0, 1)$, and define the *limit set* of Γ to be the set of accumulation points of $\{\gamma(j) : \gamma \in \Gamma\}$. A well known consequence of discreteness is that the limit set is a subset of the hyperplane $\{ x \in \mathbf{R}^n : x_n = 0 \}$. In fact, the limit set is closed, Γ-invariant, and, unless the group is elementary, is a perfect set.

Much work has been done on the classification of various types of limit points, usually in connection with the rate of orbital approach to the point in question; see [Nicholls [4], Chapter 2], for example. However, the construction of concrete examples of limit points exhibiting certain important characteristics is by no means easy. The purpose of this note is to describe some fairly general constructive techniques that the authors have found particularly useful.

We now define those particular classes of limit point that we propose to consider. For this purpose, we assume that the point under consideration is the point at infinity; conjugation by a Möbius transformation shows that this involves no loss of generality. Indeed, for the remainder of this paper we will assume that our limit point is at infinity.

If infinity is a limit point for the group Γ, then clearly the orbit of j, and indeed of any a in H^n, must be unbounded. However, we are interested in the n-th coordinate so let $\gamma(a) = (\gamma_1(a),\ \gamma_2(a),\ \ldots,\ \gamma_n(a))$. We say that infinity is a ***horospherical limit point*** for Γ if $\sup\{\ \gamma_n(a) : \gamma \in \Gamma\ \} = \infty$ for some, and hence every, a in H^n . Recall that a horosphere is a Euclidean sphere in H^n that is tangent to the base plane $\{\ x \in \mathbf{R}^n : x_n = 0\ \}$ and so we see that for infinity to be a horospherical limit point we are requiring that the orbit of j enter every horosphere tangent to the base plane at infinity. Such horospheres are just hyperplanes $x_n = \mathit{constant}$.

If infinity is not a horospherical limit point, then $\sup\{\ \gamma_n(a) : \gamma \in \Gamma\ \}$ is bounded. If this bound is attained for all a in H^n , we say that infinity is a *Dirichlet point* for Γ. If for some a in H^n the bound is not attained we say that infinity is a *Garnett limit point* with respect to the orbit of a. Such points are shown to exist in [3]. The reason for the term "Dirichlet point" is that when infinity is of this type, then it must be represented on the boundary of every classical Dirichlet region for Γ . These conditions are explained in much greater detail in [Nicholls [4], Chapter 2]. In particular, it is shown that the classical Ford construction yields a fundamental region for Γ if and only if infinity is a Dirichlet point. In fact, the Ford construction amounts to the selection of the "highest" point from each orbit and this is clearly only going to yield a fundamental region if each orbit possesses a point of maximum height. Indeed, the Ford domain is void if infinity is a horospherical limit point. When considering the nature of a limit point at infinity a useful construction is to form a Schottky group by pairwise identifying disjoint spheres, and to analyse the orbit of some canonical point. We give simple criteria on a set of disjoint spheres which guarantee a Schottky pairing generating a discrete group for which infinity is of the desired type. This result is then used to show that in hyperbolic four-space a parabolic fixed point can also be a Garnett limit point.

2. Construction criteria for limit points

Given (Euclidean) spheres S and Σ a *Schottky pairing* is a Möbius transformation T mapping the outside of S onto the inside of Σ . An infinite family of disjoint spheres may thus be paired, in many ways, by Möbius transformations generating a *Schottky group*. It follows from elementary combination theorems that this group is discrete and that the exterior of the spheres (non-void when the spheres are regarded as hyperbolic planes) is a packing (no pair of distinct points are group equivalent) [Maskit [2]]. Further, if the hyperbolic distance between any pair of spheres is uniformly bounded below then, as in [Maskit [2], p172] the exterior is actually a fundamental region.

Crucial to our results are the following observations which generalise familiar results concerning Möbius transformations of the plane. Their proofs are minor modifications of the standard complex arguments.

LEMMA 1. *If, utilising Ahlfors' description of a Möbius transformation via Clifford algebras* [1], [5],

$$T = \begin{pmatrix} a & b \\ c & d \end{pmatrix}$$

then T *has isometric sphere* $|cw + d| = 1$ *and*

$$\mathcal{I}m(Tj) = \frac{1}{|c|^2 + |d|^2}$$

where $\mathcal{I}m$ *denotes the height above the boundary and* j *the canonical point.*

LEMMA 2. *If* T *is a Schottky pairing of spheres* $\mathcal{S}$, *centre* α *and radius* r, *and* Σ, *centre* β *and radius* ρ, *then its isometric sphere is* $|w - \alpha| = \sqrt{r\rho}$. *Indeed, for some* λ *with* $|\lambda| = 1$,

$$T = \begin{pmatrix} \dfrac{\beta\lambda}{\sqrt{r\rho}} & -\left[\dfrac{\beta\lambda\alpha^*}{\sqrt{r\rho}} + \sqrt{r\rho}\lambda'\right] \\ \dfrac{\lambda}{\sqrt{r\rho}} & \dfrac{-\lambda\alpha}{\sqrt{r\rho}} \end{pmatrix}.$$

The isometric spheres of T and its inverse thus have as radii the geometric mean of the radii of $\mathcal{S}$ and Σ. In particular, $\mathcal{S}$ and Σ are the isometric spheres of T and T^{-1} if and only if $r = \rho$.

Our method consists of adeptly utilising the geometric averaging of radii, possible by Lemma 2, to obtain information about the isometric spheres of the generating transformations. We are thus able to prescribe the various kinds of limiting behaviour at infinity.

THEOREM 1. *Given disjoint spheres* $\mathcal{S}_n$, *centre* α_n *and radius* r_n *with* $sup\{r_n\} = \infty$:

(I) *There exists a Schottky pairing* $\{T_j\}$ *of the* $\mathcal{S}_n$ *so that infinity is a horospherical limit point for* $\Gamma = < T_1, T_2, \ldots >$.

(II) *There exists a Schottky pairing* $\{T_j\}$ *of spheres* $\hat{\mathcal{S}}_n$*, centre* α_n *and radius* $\rho_n \leq r_n$*, so that infinity is a Dirichlet limit point for* $\Gamma = < T_1, T_2, \ldots >$.

(III) *There exists a Schottky pairing* $\{T_j\}$ *of spheres* $\hat{\mathcal{S}}_{n_m}$*, centre* α_{n_m} *and radius* $\rho_{n_m} \leq r_{n_m}$*, so that infinity is a Garnett limit point for* $\Gamma =< T_1, T_2, \ldots >$.

PROOF.

(I) Let ${}_mT_n$ be a Schottky pairing of $\mathcal{S}_m$ and $\mathcal{S}_n$ then by Lemma 2:

$$\sup\{\ \mathcal{I}m[{}_mT_n(j)]\ \} = \sup\{\ \frac{r_m r_n}{1 + |\alpha_m|^2}\ \} = \infty.$$

Hence a Schottky pairing exists for which infinity is a horospherical limit point.

(II) Any pairing of spheres for which the isometric spheres are uniformly bounded apart in the hyperbolic metric will generate a discrete Möbius group for which the Ford domain is a fundamental region obtained by a Schottky pairing of isometric spheres [2]. Infinity is thus a Dirichlet point. Hence, there exists a Schottky pairing $\{T_j\}$ of spheres $\hat{\mathcal{S}}_n$ centre α_n and radius $\rho_n \leq r_n$ for which infinity is a Dirichlet limit point.

(III) Re-normalising, we consider a subset of the $\mathcal{S}_n$ whose centres α_n satisfy: $|\alpha_n| \to \infty$ monotonically in a cone $\mathcal{C}$ of opening less than π and with vertex at the origin. See figure 1.

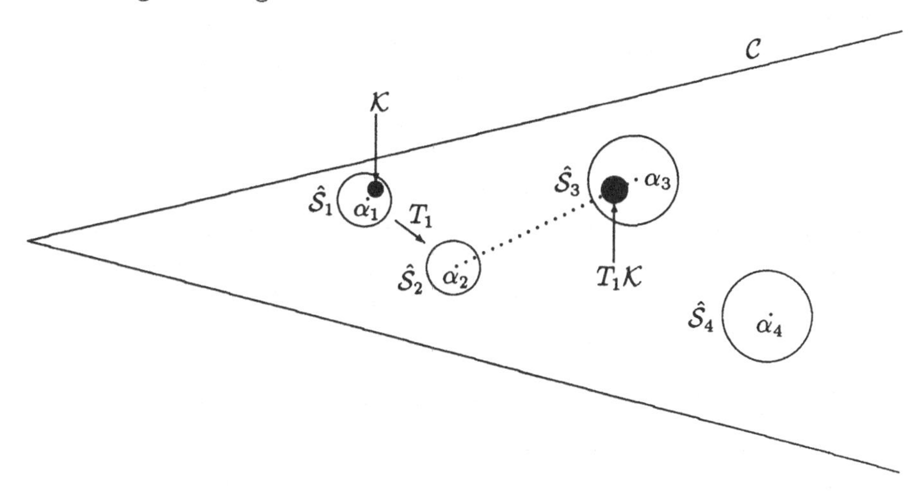

Figure 1. Schottky pairing of spheres.

Now replace $\mathcal{S}_1$ and $\mathcal{S}_2$ with spheres $\hat{\mathcal{S}}_1$ and $\hat{\mathcal{S}}_2$, still centred at α_1 and α_2 respectively, and each having radius $\rho_1 \leq \sqrt{r_1 r_2}$. Further insist that every point of $\hat{\mathcal{S}}_1$ has absolute value less than any point of $\hat{\mathcal{S}}_2$.

Let $\mathcal{K}$ be a closed ball inside $\hat{\mathcal{S}}_1$ not containing α_1 . Choose, re-labeling if necessary, α_3 so large that it has absolute value larger than any point of $\hat{\mathcal{S}}_2$ and of $T\mathcal{K}$, for any Schottky pairing T of $\hat{\mathcal{S}}_1$ and $\hat{\mathcal{S}}_2$.

Let T_1 be a Schottky pairing of $\hat{\mathcal{S}}_1$ and $\hat{\mathcal{S}}_2$ for which $T_1\mathcal{K}$ has centre on the line joining α_2 and α_3. Observe that $\hat{\mathcal{S}}_1$ and $\hat{\mathcal{S}}_2$ are the isometric spheres of T_1, T_1^{-1} respectively.

Choose α_4 so large that there exist spheres $\hat{\mathcal{S}}_3$ and $\hat{\mathcal{S}}_4$ of radius $\rho_2 \leq \sqrt{r_3 r_4}$, centred at α_3 and α_4 and with $T_1\mathcal{K} \subset \hat{\mathcal{S}}_3$; $\hat{\mathcal{S}}_1$, $\hat{\mathcal{S}}_2$, $\hat{\mathcal{S}}_3$, $\hat{\mathcal{S}}_4$ disjoint and having every point of $\hat{\mathcal{S}}_3$ of absolute value less than any point of $\hat{\mathcal{S}}_4$. Repeat this process to obtain a Schottky group $\Gamma = < T_1, T_2, \ldots >$ with Ford domain $\mathcal{F}$: the exterior of the $\hat{\mathcal{S}}_i$. Observe that $\mathcal{F}$, though non-void, is not a covering: no point of $\mathcal{F}$ is Γ-equivalent to a point of $\mathcal{K}$. It follows that infinity is a Garnett limit point.

As an application of our technique we see that in H^n, $n \geq 4$, a parabolic fixed point can, unlike the situation for $n \leq 3$, be either a horospherical limit point [5] or a Garnett limit point.

THEOREM 2 *There exists a discrete group of Möbius transformations of hyperbolic four-space for which infinity is both a parabolic fixed point and a Garnett limit point.*

PROOF. In [5] a region $\mathcal{P}$ is constructed with the following properties: (I) $\mathcal{P}$ is a packing for $< T >$, where T is a specified screw translation. (II) There is a family of disjoint spheres, of unbounded radii, contained in $\mathcal{P}$ and centred on a line orthogonal to the axis of T. Thus, if $\mathcal{R}$ is the subset of $\mathcal{P}$ exterior to the spheres and Γ_1 any Schottky group pairing the spheres, then $\mathcal{R}$ is a packing for $\Gamma = < T, \Gamma_1 >$ which is therefore discrete. After shrinking and deleting some of the spheres, as required by Theorem 1, one may construct a new Schottky group Γ_1^*, pairing the new spheres and giving a packing $\mathcal{R}^*$, say, for the discrete group $\Gamma^* = < T, \Gamma_1^* >$. Note that the Ford domain for Γ_1^* is, by the construction of Theorem 1, obtained by a Schottky pairing of spheres centred on a line orthogonal to the axis of T. Since infinity is a Garnett limit point for Γ_1^* it is either a Garnett limit point or a horospherical limit point for Γ^*. That infinity is indeed a Garnett limit point follows by observing that if $\mathcal{F}$ is the Ford domain for Γ_1^* then $\mathcal{F} \cap \mathcal{P} \neq \emptyset$.

Acknowledgement

The second author acknowledges with gratitude the hospitality of Fred Gehring and the University of Michigan where part of this research was done.

References

[1] L. V. AHLFORS. Möbius Transformations and Clifford Numbers. In: *Differential Geometry and Complex Analysis*, editors I. Chavel and H. M. Farkas, (Springer-Verlag, Berlin, 1985) 65-73.

[2] B. MASKIT. *Kleinian Groups.* (Springer- Verlag, Berlin, 1988).

[3] P. J. NICHOLLS. Garnett points for Fuchsian groups. *Bull. London Math. Soc.* **12** no. 3 (1980) 216-218.

[4] P. J. NICHOLLS. *The Ergodic Theory of Discrete Groups*, (Cambridge University Press, Cambridge, 1989).

[5] P. L. WATERMAN. Möbius Transformations in Several Dimensions, to appear, *Adv. in Math.*

College of Arts and Sciences
Kansas State University
Manhattan, Kansas 66502
USA

Institut des Hautes Etudes Scientifiques
35, Routes de Chartres
91440 Bures-sur-Yvette
FRANCE

Diagonalizing Eisenstein Series III

R. A. Rankin

To Murray Macbeath on the occasion of his retirement

13. Introduction and Recapitulation.

This paper is a continuation of [3] and [4]. The first of these two papers was concerned with the structure of the space $\mathcal{E}(\chi)$ of Eisenstein series of integral weight $k > 2$ belonging to the group $\Gamma_0(N)$ and having multiplier system the character χ modulo N, where $\chi(-1) = (-1)^k$. In that paper the action of the Hecke operators T_n ($n \in \mathbf{N}$) on the members of the space was studied in detail. In [4] an inner product on $\mathcal{E}(\chi)$ was defined, and it was shown that, in the case when N is squarefree, a basis, whose members were eigenforms for all the operators T_n, could be constructed.

The purpose of the present paper is to consider the problem for general level N and character χ. It will be shown that diagonalization is not possible in every case, but can only occur for certain specified N and χ, so that the problem is completely solved.

The same notation as in the earlier papers will be used, and section and formulae numbers will be continued. However, it will be convenient for the reader if certain results and definitions are recapitulated.

$\mathcal{E}(\chi)$ is a vector space of mutually orthogonal subspaces $\mathcal{E}(\chi, t_1)$, where

t_1 is a divisor of N, and we put

$$(13.1) \qquad t_1 t_2 = N\,, \quad h = (t_1, t_2).$$

The subspace $\mathcal{E}(\chi, t_1)$ contains only the zero form, unless t_1 is unbranched for χ, and this occurs if and only if $N(\chi)$, the conductor of χ, divides N/h. When t_1 is unbranched the space $\mathcal{E}(\chi, t_1)$ is invariant under the operators T_n with $(n, N) = 1$, and has dimension $\phi(h)$, where ϕ is Euler's function. It is spanned by the $\phi(h)$ basic functions $f_\epsilon(z, t_1)$, defined by

$$f_\epsilon(z, t_1) = \sum_u \epsilon(u) G(W^{ut_1}, \chi; z)\,,$$

where ϵ is any character modulo h. Here u runs through a set of $\phi(h)$ different integers prime to t_2 and such that no two different values are congruent modulo h; see (5.14) of [3]. To define the Eisenstein series $G(W^{ut_1}, \chi; z)$, we put

$$L = W^{ut_1} = \begin{bmatrix} 1 & 0 \\ ut_1 & 1 \end{bmatrix}\,,$$

and note that, since t_1 is unbranched, the character χ can be expressed as $\chi = \chi_1 \chi_2$, where χ_1 and χ_2 are characters modulo t_1 and modulo t_2, respectively. Then, we have

$$G(W^{ut_1}, \chi; z) = \chi_2(u) \sum_{T \in \mathcal{R}_L} \overline{\chi}(T)(LT : z)^{-k}\,,$$

where $\mathcal{R}_L$ is an arbitrary right transversal of the group $\langle -I, L^{-1}U^n L\rangle$ in $\Gamma_0(N)$, and where $n = n_L = t_2/(t_1, t_2)$ is the width of the cusp $L^{-1}\infty$. Also $LT : z$ is, as usual, the denominator of the bilinear map $LT(z)$.

The basic functions $f_\epsilon(z, t_1)$ are mutually orthogonal eigenforms for the operators T_n when $(n, N) = 1$. As we shall be dealing with a variety of different characters ϵ bearing different suffixes, it is convenient to modify the notation, drop the independent variable z and write

$$(13.2) \qquad f(\epsilon; t_1) \quad \text{in place of} \quad f_\epsilon(z, t_1).$$

We then have

$$(13.3) \qquad f(\epsilon; t_1) | T_n = \lambda(n; t_1, \epsilon, \chi) f(\epsilon; t_1) \quad \text{for} \quad (n, N) = 1,$$

where

$$(13.4) \qquad \lambda(n; t_1, \epsilon, \chi) = \sum_{n = ad} a^{k-1} \chi_1(a) \chi_2(d) \overline{\epsilon}(a) \epsilon(d).$$

As stated above, t_i is the modulus of the character χ_i for $i = 1, 2$. When we need to emphasize this we shall write, for any q,

$$\chi_i(q, t_i) \quad \text{in place of} \quad \chi_i(q) \quad (i = 1,\ 2). \tag{13.5}$$

When $h = 1$, or when ϵ is the principal character modulo h, we may write 1 in place of ϵ.

In Lemma 3.2 the characters χ_1 and χ_2 were defined originally (as $\psi \circ \psi_1$ and ψ_2) for moduli N_0N_1 and N_2, respectively, where

$$(N_0N_1, N_2) = 1, \quad N/h = N_0N_1N_2, \quad N_0N_1|t_1, \quad N_2|t_2.$$

These characters were then extended to the moduli t_1 and t_2. We shall write

$$\chi_1 = \chi_1^*\phi_1\,, \quad \chi_2 = \chi_2^*\phi_2, \tag{13.6}$$

where ϕ_1 and ϕ_2 are principal characters to the moduli t_1 and t_2, respectively and χ_1^* and χ_2^* are primitive characters. We have

$$N(\chi_1^*)|N_0N_1\,, \quad N(\chi_2^*)|N_2. \tag{13.7}$$

As in (3.10) we write, for any prime q dividing N,

$$\chi = \epsilon_q\epsilon_0, \tag{13.8}$$

where the modulus of ϵ_q is a power of q, and the modulus of ϵ_0 is the greatest factor of N prime to q.

14. The action of the operator T_q

Throughout this section it is assumed that q is a prime divisor of N, and that t_1 is an unbranched divisor of N. We define α, β, δ, and ν by

$$q^\alpha \| t_1, \quad q^\beta \| t_2, \quad q^\delta \| N(\epsilon_q), \qquad \nu = \alpha + \beta. \tag{14.1}$$

As in [3], we put

$$q_\alpha = \begin{cases} 1 - q^{-1} & (\alpha = 1), \\ 1 & (\alpha > 1). \end{cases} \tag{14.2}$$

The following two lemmas are easily proved by considering the cases $\beta \le \alpha - 2$, $\beta = \alpha - 1$ and $\beta \ge \alpha$ separately.

LEMMA 14.1. *Let $\alpha \geq 1$. Then t_1/q is branched if and only if*

$$\beta \leq \alpha - 2 \quad \text{and} \quad \delta = \alpha. \tag{14.3}$$

LEMMA 14.2. *Let ϵ be a character modulo h and assume that $\alpha \geq 1$. Put*

$$h' = (t_1/q, t_2q), \tag{14.4}$$

and define ρ by

$$q^\rho \| N(\epsilon). \tag{14.5}$$

Then ϵ is not a character modulo h' if and only if

$$\beta \geq \alpha \quad \text{and} \quad \rho = \alpha. \tag{14.6}$$

We note that when $h' = qh$ (so that $\beta \leq \alpha - 2$) ϵ can be extended to a character modulo h' by multiplying it by the principal character modulo q.

We now assume that $q|t_1$ and that ϵ is a character modulo h, as in Lemma 14.2, and define a character η as follows:

$$\eta = \begin{cases} \epsilon\bar{\epsilon}_q & \text{if } \; 0 < \delta \leq \beta = \alpha, \\ \epsilon\bar{\epsilon}_q & \text{if } \; 0 < \delta \leq \beta = \alpha - 1, \\ \epsilon & \text{otherwise.} \end{cases} \tag{14.7}$$

We note that in the first two cases t_1/q is certainly unbranched, but that in general η may not be a character modulo h'. We now define $\xi_q(\eta)$ to be 1 if and only if

$$t_1 \text{ and } t_1/q \text{ are unbranched and } \eta \text{ is a character modulo } h'. \tag{14.8}$$

Otherwise we put $\xi_q(\eta) = 0$. It is easily verified that this definition of $\xi_q(\eta)$ is equivalent to the definition of $\xi_q(\eta)$ in equation (8.3) of [3].

As shown in [3], the subspaces $\mathcal{E}(\chi, t_1)$ are not in general invariant under the operators T_q, and we now state Theorem 8.1 in simpler form:

THEOREM 14.3. *Let t_1 be unbranched and ϵ a character modulo h. Then, for any prime q dividing N we have*

$$f(\epsilon; t_1)|T_q = q^{k-1}\chi_1(q)\bar{\epsilon}(q)f(\epsilon; t_1) \qquad (\alpha = 0), \tag{14.9}$$

$$f(\epsilon; t_1)|T_q = q_\alpha \xi_q(\eta) f(\eta; t_1/q) \qquad (0 < \alpha < \nu), \tag{14.10}$$

$$f(\epsilon; t_1)|T_q = \chi_2(q)\epsilon(q)f(\epsilon; t_1) + q_\nu \xi_q(\eta) f(\eta; t_1/q) \qquad (\alpha = \nu). \tag{14.11}$$

Note that, in (14.11), $\eta = \epsilon$, and that $\xi_q(\eta) = 0$ when $\delta = \nu \geq 2$.

If we put

$$\mathcal{E}_q(\chi, t) = \bigoplus_{\alpha=0}^{\nu} \mathcal{E}(\chi, q^\alpha t) \qquad (t|q^{-\nu}N), \tag{14.12}$$

then by the Theorem,

$$\mathcal{E}_q(\chi, t)|T_q \subseteq \mathcal{E}_q(\chi, t). \tag{14.13}$$

Note that the subspaces $\mathcal{E}_q(\chi, t)$ are mutually orthogonal for different values of t.

We now consider the possible eigenvalues that an eigenform for T_q may possess.

THEOREM 14.4. *Let t divide $Nq^{-\nu}$ and let g be an eigenform on $\mathcal{E}_q(\chi, t)$ for the operator T_q, and let its associated eigenvalue be λ. Then (i) $|\lambda| = q^{k-1}$ if $g \in \mathcal{E}(\chi, t)$, (ii) $|\lambda| = 1$ if g is not orthogonal to $\mathcal{E}(\chi, tq^\nu)$, and otherwise (iii) $\lambda = 0$.*

PROOF. We suppose that g is not the zero form.
(*i*) If $g \in \mathcal{E}(\chi, t)$, then

$$g = \sum_\epsilon a(\epsilon) f(\epsilon; t),$$

the sum being taken over all appropriate characters ϵ. Accordingly,

$$\lambda g = g|T_q = \sum_\epsilon a(\epsilon) q^{k-1} \chi_1(q, t)\bar{\epsilon}(q) f(\epsilon; t),$$

so that

$$\lambda a(\epsilon) = q^{k-1}\chi_1(q, t)\bar{\epsilon}(q) a(\epsilon).$$

This proves (*i*).

(*ii*) We can write, similarly,

$$g = \sum_\epsilon b(\epsilon) f(\epsilon; q^\nu t) + G$$

where G is orthogonal to $\mathcal{E}(\chi, tq^\nu)$. Then, if $q^\nu tt' = N$,

$$\lambda g = g|T_q = \sum_\epsilon b(\epsilon)\{\chi_2(q, t')\epsilon(q) f(\epsilon; q^\nu t) + q_\nu \xi_q(\eta) f(\eta; q^{\nu-1}t)\} + G|T_q.$$

Since $f(\eta, q^{\nu-1}t)$ and $G|T_q$ are orthogonal to $\mathcal{E}(\chi, tq^\nu)$, we have, for some ϵ, $\lambda = \chi_2(q, t')\epsilon(q)$, which has unit modulus.

(*iii*) Otherwise we have

$$g = \sum_{i=0}^{m} g_i \,,$$

where $g_i \in \mathcal{E}(\chi, tq^i)$ and $0 \le m \le \nu$. Then, by (14.10),

$$\lambda g = \sum_{i=0}^{m} g_i|T_q \in \bigoplus_{i<m} \mathcal{E}(\chi, tq^i).$$

This is a contradiction unless $\lambda = 0$.

15. The first main theorem.

We shall say that $\mathcal{E}(\chi)$ is diagonalizable if it has a basis consisting of eigenforms for all the Hecke operators T_n ($n \in \mathbf{N}$).

THEOREM 15.1. *$\mathcal{E}(\chi)$ is not diagonalizable if, for some prime divisor q of N, we have, in the notation of* (14.1), *either*

$$\delta \le 2 \quad \text{when} \quad \nu = 3\,, \tag{15.1}$$

or

$$\delta \le \nu - 2 \quad \text{when} \quad \nu > 3\,. \tag{15.2}$$

PROOF. Let a prime divisor q of N satisfy either (15.1) or (15.2). Take $t_1 = q^2$, and observe that both q^2 and q are unbranched, and that $q \| h'$ and $q|h$. Our object is to choose a character ϵ modulo h such that

$$f(\epsilon; q^2)|T_q = f(1, q). \tag{15.3}$$

For $\nu = 3$ we have $h = q$ and we take $\epsilon = 1$, if $\delta = 0$ or 2, and $\epsilon = \epsilon_q$ if $\delta = 1$. In either case $\eta = 1$, by (14.7). For $\nu = 4$, we have $h = q^2$ and we take $\epsilon = \epsilon_q$. Finally, when $\nu \ge 5$, we have $h = q^2$ and take $\epsilon = 1$, which gives $\eta = 1$. Then (15.3) follows in all cases, by (14.10).

Note that $f(\epsilon; q^2)$ and $f(1; q)$ are in $\mathcal{E}_q(\chi, 1)$. If $\mathcal{E}(\chi)$ is diagonalizable, then $f(\epsilon; q^2)$ is a linear combination of eigenforms for the operator T_q. Projecting from $\mathcal{E}(\chi)$ to $\mathcal{E}_q(\chi, 1)$ we see that $f(\epsilon; q^2)$ is a linear combination of eigenforms in $\mathcal{E}_q(\chi, 1)$ for T_q. It follows from Theorem 14.4 that $f(1; q)$

is a linear combination of eigenforms, which are either in $\mathcal{E}(\chi, 1)$, or not orthogonal to $\mathcal{E}(\chi, q^\nu)$. But any eigenform of the latter type is clearly unique (to within a scalar factor) and is a nonzero multiple of a function of the form

$$f(1; q^\nu) + F,$$

where F is orthogonal to $\mathcal{E}(\chi, q^\nu)$. This is a contradiction, and the theorem is proved.

16. The second theorem.

Theorem 5.1 gives sufficient conditions for nondiagonalizability. We now show that these conditions are also necessary. This involves a rather more detailed examination of the structure of the level N.

THEOREM 16.1. *$\mathcal{E}(\chi)$ is diagonalizable when, for every prime factor q of N, either*

$$(16.1) \qquad (a)\ \delta = \nu \geq 1, \quad (b)\ \delta = \nu - 1 \geq 3, \quad \text{or} \quad (c)\ \delta \leq \nu - 1 \leq 1.$$

PROOF. We may suppose that $N > 1$, since the theorem is trivial in this case, as $\mathcal{E}(\chi)$ then consists of scalar multiples of a single eigenform $E_k(z)$.

Note first that the conditions (16.1) are the negation of (15.1) and (15.2). When (16.1) holds, every prime divisor q of N must belong to one of the following four subsets:

$$(16.2) \qquad A = \{q|N : \delta = \nu\},$$

$$(16.3) \qquad B = \{q|N : \delta = \nu - 1 \geq 3\},$$

$$(16.4) \qquad C = \{q|N : \delta \leq 1,\ \nu = 2\},$$

$$(16.5) \qquad D = \{q|N : \delta = 0,\ \nu = 1\}.$$

We take a fixed unbranched divisor t_1 of N, and write

$$(16.6) \qquad A_1 = \{q \in A : q|t_1\}, \qquad D_1 = \{q \in D : q|t_1\},$$

$$(16.7) \qquad B_i = \{q \in B : q^i \| t_1\} \quad (i = 1, \nu - 1, \nu),$$

$$(16.8) \qquad C_{10} = \{q \in C : q\|t_1,\ \delta = 0\}, \qquad C_{11} = \{q \in C : q\|t_1,\ \delta = 1\},$$

$$(16.9) \qquad C_{20} = \{q \in C : q^2|t_1,\ \delta = 0\}, \qquad C_{21} = \{q \in C : q^2|t_1,\ \delta = 1\}$$

Define $D_1(t_1)$ to be the product of all primes in the sets B_1, B_ν, C_{10}, C_{11}, C_{21}, and D_1, and $D_2(t_1)$ to be the product of all primes in C_{20}. Write

$$(16.10) \qquad D(t_1) = D_1(t_1) D_2^2(t_1).$$

LEMMA 16.1. *If $d|D(t_1)$ then t_1/d is unbranched.*

This is clear from the previous definitions. We divide t_1 successively by the prime factors (or their squares) of d and note that in each case the inequality $\delta \leq \max(\alpha, \beta)$ is satisfied.

Now take any character ϵ modulo h. The following table shows the values of the associated character η defined by (14.7) in the different cases.

Column	1	2	3	4	5	6	7
Set	B_1	B_ν	C_{10}	C_{11}	C_{20}	C_{21}	D_1
h'	h/q	hq	h/q	h/q	hq	hq	h
δ	$\nu-1$	$\nu-1$	0	1	0	1	0
η	ϵ	ϵ	ϵ	$\epsilon\bar{\epsilon}_q$	ϵ	ϵ	ϵ

The character η is a character modulo h' (see (14.4)) in columns 2, 5, 6 and 7; in columns 1 and 3 provided that $q \nmid N(\epsilon)$ and in column 4 provided that $q \nmid N(\epsilon\bar{\epsilon}_q)$.

For any divisor d of t_1 we write

$$h(d) = (t_1/d,\ dt_2). \tag{16.11}$$

We now define, for $d|D(t_1)$, the character η_d inductively as follows:

$$\eta_1 = \epsilon, \quad \eta_{qd} = \eta_d \quad \text{if} \quad q \notin C_{11} \text{ and } q \nmid d, \tag{16.12}$$

$$\eta_{qd} = \eta_d\bar{\epsilon}_q \quad \text{if} \quad q \in C_{11} \text{ and } q \nmid d, \tag{16.13}$$

$$\eta_{dq^2} = \eta_{dq}\bar{\epsilon}_q = \eta_d\bar{\epsilon}_q \quad \text{if} \quad q \in C_{20} \cup C_{21} \text{ and } q \nmid d. \tag{16.14}$$

However, η_{dq^2} is a character modulo $h(dq^2)$ only if $q \in C_{20}$.

LEMMA 16.2. *Let $d|D(t_1)$. Then*

$$\chi_1(n, t_1)\bar{\epsilon}(n) = \chi_1(n, t_1/d)\bar{\eta}_d(n) =: \lambda_1(n; t_1, \epsilon, \chi) \tag{16.15}$$

for $(n, t_1) = 1$, and

$$\chi_2(n, t_2)\epsilon(n) = \chi_2(n, dt_2)\eta_d(n) =: \lambda_2(n; t_1, \epsilon, \chi) \tag{16.16}$$

for $(n, dt_2) = 1$.

PROOF. It is sufficient to prove this when d is a prime q, and then, by replacing t_1 by t_1/q, etc. the general case follows. In this particular case the lemma follows from the last six lines at the bottom of p. 433 of [3], since, for example, in the third and fourth lines we have

$$\chi_1\bar{\epsilon} = \epsilon_q \eta_1 \phi_1 \bar{\epsilon}, \qquad \chi_1' \bar{\eta} = \epsilon_q \eta_1 \phi_1' \bar{\epsilon}.$$

We now define, for $d|D(t_1)$ the multiplicative function $C(d)$ as follows. Put $C(1) = 1$, and for any prime $q|D(t_1)$ take

$$C(q) = q_\alpha \{\chi_2(q)\epsilon(q) - q^{k-1}\chi_1(q, t_1/q)\bar{\eta}_q(q)\}^{-1}. \tag{16.17}$$

When $q^2|D(t_1)$ we put

$$C(q^2) = q_1 C(q)\{\chi_2(q)\epsilon(q) - q^{k-1}\chi_1(q, t_1/q^2)\bar{\epsilon}(q)\}^{-1}. \tag{16.18}$$

We take

$$C(mn) = C(m)C(n) \quad \text{for} \quad (m, n) = 1. \tag{16.19}$$

In this way $C(d)$ is defined for all $d|D(t_1)$. We note that $C(d)$ depends on t_1, η_d, k, χ and ϵ as well as on d.

When $q \in D_1$ both terms in curly brackets on the right of (16.17) are nonzero, but, for the other prime factors q of $D(t_1)$ exactly one vanishes. Also, $\eta_q = \epsilon$ except when $q \in C_{11}$ and then $\eta_q = \epsilon\bar{\epsilon}_q$. We can bring the notation of (16.17) into line with that used in [4], for the simpler case when N is squarefree, by noting that, in the notation of (13.6),

$$\chi_1(q, t_1/q) = \chi_1^*(q), \qquad \chi_2(q) = \chi_2^*(q), \qquad \text{when} \quad q|D(t_1).$$

Now define

$$g_\epsilon(t_1) = \sum_{d|D(t_1)} C(d) f(\eta_d; t_1/d). \tag{16.20}$$

It is clear that the set of functions $g_\epsilon(t_1)$, for unbranched $t_1|N$ and characters ϵ modulo h, span the space $\mathcal{E}(\chi)$. We shall show that they are eigenforms for all the operators T_n ($n \in \mathbf{N}$). In what follows we shall omit the condition $d|D(t_1)$ under the summation sign, and replace it, where necessary, by subsidiary conditions such as $q|d$ or $q \nmid d$. For certain values of d, η_d may not be a character modulo $h(d)$, and then we take $f(\eta_d; t_1/d)$ to be the zero function.

First of all we consider the action of T_q on $g_\epsilon(t_1)$, where q is a prime divisor of N. There are several cases to consider.

(i) $q \nmid t_1$. By (14.9),

$$g_\epsilon(t_1)|T_q = \sum q^{k-1} C(d)\chi_1(q, t_1/d)\overline{\eta}_d(q) f(\eta_d; t_1/q)$$
$$= q^{k-1}\lambda_1(q; t_1, \epsilon, \chi) g_\epsilon(t_1),$$

by Lemma 16.2.

(ii) $q \in A_1$. Since t_1 is unbranched, $q^\nu | t_1$ and, by (14.11),

$$g_\epsilon(t_1)|T_q = \sum C(d)\chi_2(q, t_2 d)\eta_d(q) f(\eta_d; t_1/d)$$
$$= \lambda_2(q; t_1, \epsilon, \chi) g_\epsilon(t_1),$$

by Lemma 16.2.

(iii) $q \in B_1 \cup C_{10} \cup C_{11}$.

$$g_\epsilon(t_1)|T_q = \sum_{q \nmid d} C(d) q_1 f(\eta_{dq}; t_1/dq)$$
$$+ q^{k-1} \sum_{q|d} C(d)\chi_1(q, t_1/d)\overline{\eta}_d(q) f(\eta_d; t_1/d)$$
$$= \sum_{q|d} f(\eta_d; t_1/d)\{q_1 C(d/q) + q^{k-1}\chi_1(q, t_1/d)\overline{\eta}_d(q) C(d)\}.$$

The term in curly brackets is zero by (16.17) and so

$$g_\epsilon(t_1)|T_q = 0. \tag{16.21}$$

(iv) $q \in B_{\nu-1}$. Since $t_1|q$ is branched, we have (16.21) again.

(v) $q \in B_\nu$.

$$g_\epsilon(t_1)|T_q = \sum_{q \nmid d} C(d)\{\chi_2(q, t_2 d)\eta_d(q) f(\eta_d; t_1/d) + f(\eta_{dq}; t_1/dq)\}.$$

There is no contribution for d divisible by q, since t_1/q is branched in this case; see (14.10). Thus we have on the right

$$\sum_{q \nmid d} C(d)\chi_2(q, t_2 d)\eta_d(q) f(\eta_d; t_1/d) + \sum_{q|d} C(d/q) f(\eta_d; t_1/d)$$
$$= \lambda_2(q; t_1, \epsilon, \chi) g_\epsilon(t_1).$$

(vi) $q \in D_1$.

$$g_\epsilon(t_1)|T_q = \sum_{q\nmid d} C(d)\{\chi_2(q, t_2 d)\eta_d(q) f(\eta_d; t_1/d) + q_1 f(\eta_d; t_1/qd)\}$$

$$+ \sum_{q|d} C(d) q^{k-1} \chi_1(q, t_1/d)\overline{\eta}_d(q) f(\eta_d; t_1/d)$$

$$= \sum_{q\nmid d} C(d)\lambda_2(q; t_1, \epsilon, \chi) f(\eta_d; t_1/d)$$

$$+ \sum_{q|d} f(\eta_d; t_1/d)\{q_1 C(d/q) + q^{k-1}\chi_1(q, t_1/q)\overline{\eta}_q(q) C(d)\},$$

on applying Lemma 16.2 to t_1/q in place of t_1. This gives

$$g_\epsilon(t_1)|T_q = \lambda_2(q; t_1, \epsilon, \chi) g_\epsilon(t_1).$$

(vii) $q \in C_{20} \cup C_{21}$.

$$g_\epsilon(t_1)|T_q = \sum_{q\nmid d} C(d)\{\chi_2(q, t_2 d)\eta_d(q) f(\eta_d; t_1/d) + f(\eta_{dq}; t_1/qd)\}$$

$$+ \sum_{q\|d} q_1 C(d) f(\eta_{dq}; t_1/dq)$$

$$+ \sum_{q^2|d} C(d) q^{k-1}\chi_1(q, t_1/d)\overline{\eta}_d(q) f(\eta_d; t_1/d).$$

When $q \in C_{21}$, the last two sums vanish, since then η_{dq} and η_d are not characters to the moduli $h(dq)$ and $h(d)$, respectively. Thus in this case, as in (v) we have

$$g_\epsilon(t_1)|T_q = \lambda_2(q; t_1, \epsilon, \chi) g_\epsilon(t_1).$$

It remains to consider $q \in C_{20}$. We have

$$g_\epsilon(t_1)|T_q = \sum_{q\nmid d} C(d)\lambda_2(q; t_1, \epsilon, \chi) f(\eta_d; t_1/d)$$

$$+ \sum_{q\|d} C(d/q) f(\eta_d; t_1/d)$$

$$+ \sum_{q^2|d} f(\eta_d; t_1/d)\{q_1 C(d/q) + C(d) q^{k-1}\chi_1(q, t_1/d)\overline{\eta}_d(q)\}.$$

By (16.17) and (16.18) it follows that

$$g_\epsilon(t_1)|T_q = \lambda_2(q; t_1, \epsilon, \chi) g_\epsilon(t_1).$$

Note that the values of the eigenvalues agree with Theorem 14.4.

It follows that the functions $g_\epsilon(t_1)$, for unbranched $t_1|N$ and ϵ a character modulo h, form a basis of eigenforms for all operators T_q, where q is a prime divisor of N. However, by (13.4) and Lemma 16.2 we see that they are also eigenforms for the operators T_n with $(n, N) = 1$. Hence Theorem 16.1 is proved.

17. An example.

We illustrate the results obtained by taking $N = p^2q^2$, where p and q are different primes and assume that

$$N(\epsilon_p) = p, \qquad N(\epsilon_q) = q^2.$$

The space is then of dimension $2p + 2$. The table lists the eigenforms and their eigenvalues (EV) under the action of the operators T_p and T_q.

t_1	$D(t_1)$	$g_\epsilon(t_1)$	EV for T_p	EV for T_q
1	1	$f(1;1)$	p^{k-1}	q^{k-1}
p	p	$f(\epsilon;p) \quad (\epsilon \neq \epsilon_p)$	0	$q^{k-1}\chi_1(q)\overline{\epsilon}(q)$
p	p	$f(\epsilon_p;p) - p^{-k}(p-1)f(1;1)$	0	q^{k-1}
p^2	p	$f(1;p^2)$	$\chi_2(p)$	$q^{k-1}\chi_1(q)$
q^2	1	$f(1;q^2)$	$p^{k-1}\chi_1(p)$	$\chi_2(q)$
pq^2	p	$f(\epsilon;pq^2) \quad (\epsilon \neq \epsilon_p)$	0	$\epsilon(q)$
pq^2	p	$f(\epsilon_p;pq^2) - p^{-k}(p-1)\chi_1(p)f(1;q^2)$	0	$\epsilon_p(q)$
p^2q^2	p	$f(1;p^2q^2) + f(1;pq^2)$	1	1

18. Concluding remarks.

In §1 of [3] it was stated that Hecke was aware that not every space of Eisenstein series was diagonalizable, and illustrated this fact in a particular case in Satz 45a of his paper [1], where he took $n = q^3$. This statement requires clarification.

Hecke was concerned with Eisenstein series belonging to the principal congruence group $\Gamma(N)$, and not to $\Gamma_0(N)$. The level N of the present paper he called Q, and he took a character ϵ modulo Q in place of our character χ. He put

$$Q = N = q^3, \qquad t_1 = t_2 = q^2,$$

and

$$\epsilon = \chi = \chi_1\chi_2,$$

where χ_1 and χ_2 were characters modulo q. Thus, in our notation,

$$\nu = 3, \qquad \delta = 0 \text{ or } 1\,.$$

Accordingly, t_1 and t_2 are not the divisors studied in the present paper, since their product is not N.

In Satz 45a Hecke assumed that the relevant space was diagonalizable and selected the eigenform corresponding to the Dirichlet series

$$(t_1, t_2)^{-s}L(s, \chi_1)L(s - k + 1, \chi_2),$$

that he had constructed in a previous theorem, namely Satz 44. This is a modular form F, which, since $t_1 t_2 = q^4$, is an infinite series in powers of

$$\exp(2\pi i\, t_1 t_2 nz/N) = \exp(2\pi i\, qnz) \quad (n \geq 0).$$

Thus $F(z + (1/q)) = F(z)$, and a contradiction is drawn from this fact. See also pp. 39-43 of Chapter IV of [2].

In conclusion, the following misprints in [3] may be noted.
In equation (7.1) q^{-1} should be replaced twice by $q - 1$.
On the right side of the equation following (8.7) $f_\epsilon(z, \tau_1/q)$ should be replaced by $f_\eta(z, \tau_1/q)$.

References

[1] E. HECKE. Über Modulfunktionen und die Dirichletschen Reihen mit Eulerscher Produktentwicklung. II. *Math. Ann.* **114** (1937) 316-351.

[2] ANDREW OGG. *Modular Forms and Dirichlet Series.* (New York, 1969).

[3] R. A. RANKIN. Diagonalizing Eisenstein Series, I, in *Analytical Number Theory,* Proceedings of a Conference in Honor of Paul T. Bateman edited by B. C. BERNDT, H. G. DIAMOND, H. HALBERSTAM, A. HILDEBRAND. (Birkhäuser, 1990).

[4] R. A. RANKIN. Diagonalizing Eisenstein Series, II, in *A Tribute to Emil Grosswald: Number Theory and Related Analysis. Contemporary Mathematics* Series. (American Mathematical Society, Providence, Rhode Island, 1991).

Department of Mathematics
University of Glasgow
Glasgow G12 8QJ

Some remarks on 2-generator hyperbolic 3-manifolds

Alan W. Reid

To Murray Macbeath on the occasion of his retirement

1. Introduction

By a *hyperbolic 3-manifold* we shall always mean a complete orientable hyperbolic 3-manifold of finite volume. A hyperbolic 3-manifold M is said to be *n-generator* if the minimal number of elements required to generate $\pi_1(M)$ is n. The focus of this paper is 2-generator hyperbolic 3-manifolds, the main aim being to give a construction of infinitely many closed hyperbolic 3-manifolds which are not 2-generator, but have a proper finite cover which is. Our interest in such examples was motivated by the deep results contained in [4] and [10] which relate questions on 2-generator subgroups of hyperbolic 3-manifold groups to estimates on the lower bound for the smallest volume of a closed hyperbolic 3-manifold. We also construct certain 2-generator Haken hyperbolic 3-manifolds whose existence helps to explain why more recent methods of Culler and Shalen (in preparation) seem to be necessary for estimating volumes of closed Haken hyperbolic 3-manifolds.

To describe the connection between the articles referred to above and the contents of this article we need to recall the definition of a Margulis number of a hyperbolic 3-manifold.

Let $M = \mathbf{H}^3/\Gamma$ be a closed hyperbolic 3-manifold and $\epsilon > 0$. Then ϵ is a *Margulis number* for M if for every point z of $\mathbf{H}^3$ and every pair of non-commuting elements γ and δ of Γ we have $\max\{\rho(\mathrm{z}, \gamma(\mathrm{z})), \rho(\mathrm{z}, \delta(\mathrm{z}))\} \geq \epsilon$, where ρ denotes the hyperbolic metric. Now it can be shown that if ϵ is a Margulis number for M, then M contains an embedded ball of radius $\epsilon/2$, see for example [4]. In particular by computing the hyperbolic volume of this embedded ball, an estimate of the volume of M can be made.

With M as above, in ([10], Corollary 5.3) it is shown that if every 2-generator subgroup of $\pi_1(M)$ is free, then $\frac{1}{2}\log 3$ is a Margulis number for M. This gives a volume estimate of vol(M) ≥ 0.1124, which is of the same order of magnitude as the smallest known (and conjectured smallest) volume, namely that of the Weeks manifold (which is 2-generator), which is obtained by $(5,1),(5,2)$-Dehn surgery on the Whitehead link, and has volume approximately $0.9427\ldots$; see [13]. It is also shown in ([10], Corollary 1.9) that if the $\mathbf{Z}_p$-rank of the first homology of M is at least 4 then every 2-generator subgroup of $\pi_1(M)$ is free. Recall, that if M is a closed hyperbolic 3-manifold and F a 2-generator subgroup of $\pi_1(M)$, then F has finite index or F is free, see ([8], Theorem 6.4.1).

The natural question that arises from [10] is whether there exist closed hyperbolic 3-manifolds which are not 2-generator, but have a finite cover that is? If the answer were no, then the results of [10] would give a volume estimate of the same order of magnitude as the Weeks manifold for all closed hyperbolic 3-manifolds which are not 2-generator.

Furthermore, it is shown in [4], that if the rank of the first homology of a closed hyperbolic 3-manifold M is at least 3, then the volume of M, is greater than $0.92\ldots$. The proof of this result proceeds by analysis of 2-generator subgroups of $\pi_1(M)$ and their action on the sphere-at-infinity to show that under this hypothesis on the first betti number, $\log 3$ is a Margulis number for M. Actually much more is shown but we will not refer to it here.

Of course, the condition on the first betti number is a restrictive one. For example most known manifolds of small volume have zero first betti number; in particular, the Weeks manifold has finite first homology. However, if the answer to the question raised above on 2-generator subgroups of $\pi_1(M)$ were no, the methods of [4] would apply to show that $\log 3$ is a Margulis number for all closed hyperbolic 3-manifolds that are not 2-generator. Hence morally at least, small volume manifolds should arise from the class of 2-generator manifolds, as experimental evidence suggests.

In more recent work, Culler and Shalen have shown how to obtain volume estimates for certain Haken hyperbolic 3-manifolds without the restriction on the first betti number given above. This restriction is replaced by a technical condition on embedded surfaces in M, see §5 for a further discussion. Again, the proof proceeds by analysing 2-generator subgroups of $\pi_1(M)$. In §5, which is an addition to an earlier version of this paper, we construct closed Haken hyperbolic 3-manifolds whose existence has implications for this more recent work of Culler and Shalen.

2. Preliminaries

In this section we recall some relevent facts that will be required in our constructions.

2.1 Let T denote the torus with an open disc removed. We can construct a hyperbolic 3-manifold M_ϕ by taking the mapping torus of a pseudo-Anosov homeomorphism ϕ of T which is the identity on ∂T [12]. M_ϕ is a fiber bundle over the circle with fiber a once punctured torus, and in the sequel will be referred to as a *hyperbolic punctured torus bundle.* The monodromy of such a fiber bundle can be represented by a hyperbolic element of $\mathrm{SL}(2, \mathbf{Z})$, i.e., an element with two real distinct eigenvalues. Given two hyperbolic punctured torus bundles with monodromies ϕ and ψ then M_ϕ and M_ψ are homeomorphic if and only if ϕ is $\mathrm{GL}(2, \mathbf{Z})$ conjugate to ψ or ψ^{-1}, see ([2], §1.3).

Let $R = \left(\begin{smallmatrix} 1 & 1 \\ 0 & 1 \end{smallmatrix}\right)$ and $L = \left(\begin{smallmatrix} 1 & 0 \\ 1 & 1 \end{smallmatrix}\right)$. R and L generate $\mathrm{SL}(2, \mathbf{Z})$ and so each hyperbolic element can be written as a word in R and L. In particular we can associate to a hyperbolic punctured torus bundle its *RL*-factorization. That is, any hyperbolic element ϕ of $\mathrm{SL}(2, \mathbf{Z})$ is conjugate to a word of the form $\pm R^{a_1} L^{a_2} \ldots L^{a_n}$ for integers $a_1, a_2, \ldots a_n$. This word is the *RL-factorization* of M_ϕ.

2.2 Evidently, from the construction of M_ϕ, the fundamental group of M_ϕ is an HNN-extension of a free group on 2-generators. By being careful we can choose a meridian/longitude pair for M_ϕ, i.e. a pair of generators for $H_1(\partial M_\phi; \mathbf{Z})$, for which a meridian x is the stable letter of the HNN-extension and the longitude ℓ is the boundary of a fiber. Referring to Figure 1 below, the meridian x is $t \times [0, 1]/\phi$ and ℓ is as shown.

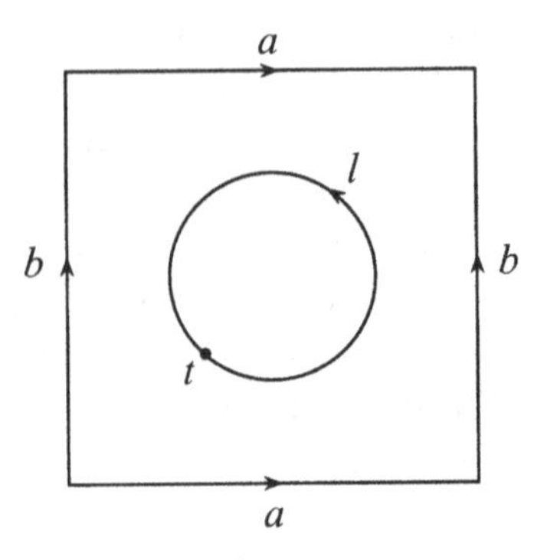

Figure 1

From the fibration of M_ϕ over the circle we have the following short exact sequence. Here we have specified a pair of generators a and b for the fundamental group of the fiber.

$$1 \to < a, b > \to \pi_1(M_\phi) = < a, b, x > \to \mathbf{Z} \to 1.$$

Of course by construction ℓ maps trivially, under the map to $\mathbf{Z}$ above. Also observe that every hyperbolic punctured torus bundle is at "worst" 3-generator.

2.3 For coprime integers p and q, by (p,q)-Dehn filling on M_ϕ we mean cutting off ∂M_ϕ and gluing in a solid torus V so that a meridian of V is identified with $x^p \ell^q$. The result of (p,q)-Dehn filling will be denoted by $M_\phi(p,q)$. By Thurston's hyperbolic Dehn surgery theorem (see [11]), this is a hyperbolic manifold for all but a finite number of p and q.

3. Results

Here we shall state the main theorem and prove some easy corollaries. The proof of Theorem 1 occupies the next section.

THEOREM 1. *Let M_ϕ be a hyperbolic punctured torus bundle. Then for $|p| > 1$, $M_\phi(p,q)$ has a proper 2-generator subgroup of finite index.*

COROLLARY 1. *There exist infinitely many closed 2-generator hyperbolic 3-manifolds which have a proper finite sheeted cover which is also 2-generator.*

PROOF. Simply take the punctured torus bundle itself to be 2-generator. For example all $(p,1)$ surgeries, for $|p| > 4$, on one component of the Whitehead link, see [7]. ∎

COROLLARY 2. *There exist infinitely many closed 3-generator hyperbolic 3-manifolds which have a proper finite sheeted cover that is 2-generator.*

PROOF. Consider the hyperbolic punctured torus bundle M whose monodromy as a matrix in $SL(2, \mathbf{Z})$ is $\begin{pmatrix} 5 & -2 \\ -2 & 1 \end{pmatrix}$. Then M is 3-generator as $H_1(M)$ is $\mathbf{Z} \oplus \mathbf{Z}_2 \oplus \mathbf{Z}_2$. Doing any $(2p, q)$-Dehn filling on M continues to be 3-generator — the $\mathbf{Z}_2$-rank of the first homology is 3. The result then follows by choosing p and q large enough so that the manifold obtained by the appropriate Dehn filling is hyperbolic. ∎

REMARK. As remarked in the Introduction, in [10] it is shown that if, for some prime p, the $\mathbf{Z}_p$-rank of the first homology of the closed orientable 3-manifold M is at least 4, then every 2-generator subgroup of $\pi_1(M)$ is free. The examples given in the proof of Corollary 2 show that this result fails for $\mathbf{Z}_p$-rank 3.

4. Proof of Theorem 1

Using the exact sequence of §2.2 we can construct a "canonical" family of p-fold cyclic covers of M_ϕ. Namely, compose the map to $\mathbf{Z}$ with reduction modulo p for any positive integer $p \geq 2$ and take the cover corresponding to the kernel. Denote these covering manifolds by $M_{\phi,p}$. These manifolds are also hyperbolic punctured torus bundles; they are simply the mapping torus of ϕ^p. Also a meridian for $M_{\phi,p}$ is x^p, with longitude the boundary of the fiber and this defines the fundamental group of $M_{\phi,p}$ as an HNN-extension of $< a, b >$ by x^p.

Now choose a pair of coprime integers (p, q), and consider the manifold $M_\phi(p, q)$ obtained by (p, q)-Dehn filling. The fundamental group of this manifold is obtained from that of M_ϕ by adjoining the relation $x^p \ell^q = 1$. Together with the exact sequence of §2.2 we get a homomorphism

$$\alpha_{p,q} : \pi_1(M_\phi(p, q)) \to \mathbf{Z}_p,$$

which is compatible with the exact sequence. That is, the cover of M_ϕ corresponding to the kernel of the homomorphism to $\mathbf{Z}_p$ obtained by factoring through $\pi_1(M_\phi(p, q))$ composed with $\alpha_{p,q}$ is $M_{\phi,p}$. However, this p-fold cover of $M_\phi(p, q)$ is simply that given by $(1, q)$-Dehn filling on $M_{\phi,p}$. Now the chosen meridian for $M_{\phi,p}$ is x^p, and since the corresponding longitude is a word in a and b (recall it is the boundary of a fiber) it follows that in $Ker(\alpha_{p,q})$, x^p is a word in a and b. Thus $Ker(\alpha_{p,q})$ is 2-generator, which proves the theorem. ∎

5. 2-generator hyperbolic Haken 3-manifolds

As stated in the Introduction, in more recent work in preparation, Culler and Shalen have obtained volume estimates for a "large" class (see below for a description) of closed, Haken hyperbolic 3-manifolds. In particular the assumption that the first betti number is at least 3 is replaced by restrictions on the embedded incompressible surfaces in the 3-manifold. This leads to a question about existence of certain Haken hyperbolic 3-manifold groups and the structure of their 2-generator subgroups (see below). Here we give a partial answer to the question.

Let us describe this restriction in more detail. We need to make a definition.

DEFINITION. A compact 3-manifold with boundary N is called a *relative graph manifold* if N is the union of two submanifolds E and T where E is an I-bundle over a (possibly disconnected and non-orientable) surface B, the intersection $A = T \cap E$ is the induced I-bundle over the boundary of B, and each component of of A is an annulus in the boundary of T which is homotopically non-trivial in T.

A closed embedded incompressible surface S in a 3-manifold M is called a *fibroid* if the (possibly disconnected) manifold obtained by splitting M along S is a relative graph manifold. A closed Haken hyperbolic 3-manifold M will be called *fibroid* (resp. *non-fibroid*) if it contains an embedded incompressible surface which is a fibroid (resp. if no embedded incompressible surface is a fibroid).

The result of Culler and Shalen then says, roughly (there is a further condition in the case where the surface separates), if M is a closed, orientable hyperbolic 3-manifold which contains an embedded incompressible suface which is not a fibroid, then vol(M) is at least 0.35. Again, the study of 2-generator subgroups of $\pi_1(M)$ is involved in obtaining the estimate, and as a consequence of these methods Culler and Shalen posed the following question:

QUESTION. Does there exist a closed Haken hyperbolic 3-manifold which is non-fibroid and whose fundamental group contains a 2-generator subgroup of finite index?

The relevance of the question is that if every 2-generator subgroup of a non-fibroid hyperbolic 3-manifold were of infinite index and had non-empty set of discontinuity on the sphere-at-infinity, then the previous work

[4], would apply to get a volume estimate of 0.92, superceding the volume estimate of 0.35 stated above.

We point out here that special cases of a fibroid are, the fiber in a fibration over the circle, or the common boundary of the union of two twisted I-bundles over a closed surface. Moreover, in this latter case the manifold is double covered by something which is fibered, see [6]. In the remainder of this section we give the following partial answer to the Question.

THEOREM 2. *There exist closed Haken hyperbolic 3-manifolds which are not fibered, not double covered by something which is fibered and which have a finite cover which is 2-generator.*

The proof is based on a refinement of the methods of the proof of Theorem 1 together with an analysis of surgery on boundary slopes of punctured torus bundles.

For convenience we remind the reader of some salient points. Let M be a compact, connected, irreducible, orientable 3-manifold such that ∂M is a torus. The unoriented isotopy class of a simple closed curve in ∂M will be called its *slope*. Now let S be a surface with non-empty boundary, properly embedded and incompressible in M; in what follows S will be called *essential* in M. The boundary components of S are disjoint non-trivial simple closed curves in ∂M, all having the same slope r, called the *boundary slope* of S. More generally a slope r on ∂M is called a *boundary slope* if it is the boundary slope of some embedded, properly embedded incompressible surface in M.

By specifying a framing, i.e. a basis for the first homology of the torus the slopes can be parametrized by $\mathbf{Q} \cup \infty$. If we fix a framing, and p/q (possibly ∞) is a boundary slope of M, by Dehn filling this boundary slope we mean performing (p, q)-Dehn filling on M, again with respect to the chosen framing. It is a theorem of Hatcher, [5] that M as above has only finitely many boundary slopes.

PROOF OF THEOREM 2. We first make a few necessary remarks on boundary slopes of hyperbolic punctured torus bundles. As stated above to make precise statements about boundary slopes, one has to fix a framing for ∂M_ϕ. We will be using the results of [2]. Their description of essential surfaces in hyperbolic punctured torus bundles and their boundary slopes is given in terms of a particular framing, which need not coincide with the choice made in §2.2. However, the description of how to transform between framings and the effect on the boundary slope is well described in §6.2 of [2]. It will be implicit from here on that in any reference to boundary slope, the framing is that described in §2.2, unless otherwise indicated.

LEMMA 3. *Let M_ϕ be a hyperbolic punctured torus bundle and assume that p/q is a boundary slope for which $M_\phi(p,q)$ is hyperbolic. Then $M_\phi(p,q)$ is Haken.*

PROOF. We will apply Theorem 2.0.3 of [3]. This describes exactly what happens when one does Dehn filling on a boundary slope of M_ϕ. The four possibilities are a connected sum of Lens Spaces, a Haken manifold, M_ϕ contains a closed embedded non-boundary parallel surface which may be compressible upon Dehn filling or the slope corresponds to the boundary slope of a surface which is a fiber.

By assumption, $M_\phi(p,q)$ is hyperbolic so the first possibility does not apply. From [2], M_ϕ has no closed embedded incompressible surfaces other than tori which are boundary parallel. This rules out the third possibility and the last case is ruled out as M_ϕ is fibered in a unique way, fibers being punctured tori of slope $0/1$. Dehn filling this slope gives the torus bundle over the circle with monodromy ϕ, and in particular is non-hyperbolic. Thus $M_\phi(p,q)$ is Haken. ∎

As noted in the proof of Lemma 3 there is always at least one Dehn filling on a hyperbolic punctured torus bundle that gives a non-hyperbolic manifold. The next result, due to Bleiler and Hodgson ([6], Theorem 10) shows that under conditions on the monodromy this is the only one. Notation used in the statement of Theorem 4 is that given in §2.1.

THEOREM 4 (BLEILER-HODGSON). *Let M_ϕ be a hyperbolic punctured torus bundle whose monodromy has an an RL-factorization* $\mathrm{SL}(2,\mathbf{Z})$ *conjugate to an element of the form, $\pm R^{a_1}L^{a_2}\dots L^{a_n}$, where $a_i > 0$ for each i and n is even. Then there is an integer $N > 0$ such that for all $n > N$ every Dehn filling but one on M_ϕ yields a hyperbolic 3-manifold, the exception being the case described above.*

We can now complete the proof of Theorem 2. This will entail a choice of ϕ in order to apply [2] and to restrict $\mathrm{Tor}(\mathrm{H}_1(\mathrm{M}_\phi,\mathbf{Z}))$ which will simplify calculations. We shall assume ϕ has an RL factorization of the form $-(R^pL^p)^q$ where p is an odd positive integer, $q = 2^n$ and n is a positive integer such that 2^n is greater than N as given in Theorem 4.

An elementary argument shows that $|\mathrm{Tor}(\mathrm{H}_1(\mathrm{M}_\phi,\mathbf{Z}))| = |\mathrm{trace}(\phi) - 2|$.

LEMMA 5. *Let ϕ be an element of* $\mathrm{SL}(2,\mathbf{Z})$ *of the form,* $\pm(R^pL^p)^q$ *where p is an odd positive integer and $q = 2^n$, for n a positive integer. Then trace(ϕ) is odd.*

PROOF. Let $X_p = R^pL^p$, then $X_p = \begin{pmatrix} 1+p^2 & p \\ p & 1 \end{pmatrix}$. It suffices to show that modulo 2, X_p^q has non-zero trace. Since p is odd, reducing coefficients of X_p modulo 2 gives $\begin{pmatrix} 0 & 1 \\ 1 & 1 \end{pmatrix}$. This is an element of order 3 in $\mathrm{SL}(2,\mathbf{F}_2)$, where $\mathbf{F}_2$ is the field of two elements. By choice of q, X_p^q will continue to have trace 1, as required. ∎

Therefore from Lemma 5 and the remark preceeding it we can conclude that $|\mathrm{Tor}(\mathrm{H}_1(\mathrm{M}_\phi,\mathbf{Z}))|$ is odd. Clearly the elements $-(R^pL^p)^q$ as described above are hyperbolic, so that the associated punctured torus bundles are hyperbolic.

From the classification of essential surfaces and their boundary slopes given in [2], we can read off that the hyperbolic punctured torus bundle with monodromy given by $-(L^pR^p)^q$ has an essential surface with two boundary components of boundary slope $-2/1$, with respect to the "standard framing" given in [2]. By conjugating with R^p we get ϕ as in Lemma 5. Now we can use the contents of §6.2 of [2] to describe the boundary slope in terms of the framing given in §2.2; a direct calculation shows that the boundary slope has the form $-2/r$ for some integer r. The important point to note is that we just require the numerator to be different from ± 1 so as to apply the method of proof of Theorem 1 (see below).

To conclude, $M = M_\phi(-2,r)$ is hyperbolic by Theorem 4 and Haken by Lemma 3. Indeed, by choice of ϕ, $H_1(M,\mathbf{Z})$ has the form, $\mathbf{Z}_2 \times A$ where $|A|$ is finite of odd order by Lemma 5. Thus M has a unique double cover. This double cover corresponds, as in the proof of Theorem 1, to a $(-1,r)$-Dehn filling on the hyperbolic punctured torus bundle that is the unique double cover of M_ϕ. In particular its homology is also finite, so it is not fibered. Finally the proof of Theorem 1 shows that the double cover of M is 2- generator. ∎

REMARKS.

(1) In the closed hyperbolic 3-manifolds constructed above the embedded incompressible surfaces are always geometrically finite. This follows since it is known that if an embedded incompressible surface in a hyperbolic 3-manifold is geometrically infinite and separating then the manifold is the union of two twisted I-bundles and, as pointed out above, it follows

from [6] that such a manifold is double covered by something which is fibered. Of course the construction shows that this cannot be the case here. In addition, by [9], Corollary 3, it follows that the surface cannot be totally geodesic.

(2) As defined above, the class of surfaces which are fibroid contains properly those which are geometrically infinite. Unfortunately there seems to be little at present that one can say geometrically that would distinguish a fibroid which is quasi-Fuchsian from other quasi-Fuchsian surfaces.

Acknowledgements

Much of the content of the paper was written whilst the author was at the University of Aberdeen. This work was supported by the SERC, and NSF grant numbers DMS-8505550 administered through M.S.R.I and DMS-9108050 which the author gratefully acknowledges. The author also wishes to thank C. Maclachlan and W. Neumann for many useful discussions, and P. Shalen for correspondence on his work with M. Culler on volumes of hyperbolic manifolds.

References

[1] S. A. Bleiler and C. D. Hodgson, Spherical space forms and Dehn filling. Preprint.

[2] M. Culler, W. Jaco and H. Rubinstein, Incompressible surfaces in once-punctured torus bundles. *Proc. London Math. Soc.* (3), **45** (1982), 385-419.

[3] M. Culler, C. McA. Gordon, J. Luecke and P. B. Shalen, Dehn surgery on knots. *Ann. of Math.* **125** (1987), 237-300.

[4] M. Culler and P. B. Shalen, Paradoxical decompositions, 2-generator Kleinian groups and volumes of hyperbolic 3-manifolds. Preprint.

[5] A. Hatcher, On the boundary curves of incompressible surfaces. *Pacific J. Math.* **99** (1982), 373-377.

[6] J. Hempel, 3-Manifolds. *Annals of Math. Studies*, **86** (1976), Princeton University Press.

[7] C. D. Hodgson, G. R. Meyerhoff and J. R. Weeks, Surgeries on the Whitehead link yield geometrically similar manifolds. To appear in TOPOLOGY '90, *Proceedings of the Research Semester in Low Dimensional Topology at Ohio State Univ.*

[8] W. JACO AND P. B. SHALEN, Seifert fibered spaces in 3-manifolds. *Memoirs of the A.M.S.*, **220** (1979).

[9] W. MENASCO AND A. W. REID, Totally geodesic surfaces in hyperbolic link complements. To appear in TOPOLOGY '90, *Proceedings of the Research Semester in Low Dimensional Topology at Ohio State Univ.*

[10] P. B. SHALEN AND P. WAGREICH, Growth rates, $\mathbf{Z}_p$-homology, and volumes of hyperbolic 3-manifolds. Preprint.

[11] W. THURSTON, The Geometry and Topology of 3- manifolds. Mimeographed lecture notes, Princeton Univ., (1978).

[12] W. THURSTON, Hyperbolic structures on 3- manifolds II: Surface groups and 3-manifolds which fiber over the circle. Preprint.

[13] J. R. WEEKS, Ph. D Thesis, Princeton Univ., (1985).

Mathematical Sciences Research Institute
1000 Centennial Drive
Berkeley, CA 94720
U. S. A.
and
Department of Mathematics
The Ohio State University
231 W. 18th Ave.
Columbus, Ohio 43210
U. S. A

Email: reid@msri.org

Uniformization, graded Riemann surfaces and supersymmetry

P. Teofilatto

Dedicated to A. M. Macbeath on his retirement

Abstract. A projective structure on a Riemann surface M is achieved by lifting to a universal covering space of M the ratio of solutions of a linear differential equation defined on the surface. We show that these solutions, and vector fields preserving the projective structure induced on M, define a sheaf of graded Lie algebras. At the same time, we construct a sheaf of graded Lie algebras over the Teichmüller family of Riemann surfaces, which is related to the process of simultaneous uniformization for the family. This structure constitutes a modular supersymmetry, an extended notion of local sl_2-symmetry, which is motivated by Grand Unification Theory in theoretical physics.

Introduction

A projective structure on a Riemann surface M is a reduction of its pseudo-group of holomorphic coordinate transformations to the group of fractional linear transformations PL(2, C). On a Riemann surface of genus $g \geq 2$, this can be achieved by choosing a *projective connection*, an element of an affine space over the vector space of globally defined holomorphic quadratic differentials [7]. A new complex atlas on M with projective coordinate transformations is obtained by composing the old coordinates with the ratio of two linearly independent solutions of a differential equation defined on the surface.

A significant fact, proved in [9], is that solutions of such a uniformizing equation must be local spinors, sections of a square root of the tangent bundle of M. Starting from this point, we prove in this paper that the process of projective uniformization of a Riemann surface M is related to a sheaf of *graded Lie algebras* defined on M (Theorem 1).

This result extends to moduli: let us consider the bundle $\mathbf{P}_g$ of all projective connections over all marked Riemann surfaces, defined in [10]. Then a generalization of the uniformization theorem, due to Bers [3], selects a holomorphic section of this bundle. By means of this section and a (local) holomorphic section of the family of spin bundles over the Teichmüller space, we obtain a sheaf of graded Lie algebras over the Teichmüller space (theorem 2), intrinsically related to the process of *simultaneous* uniformization as the previous sheaf of graded Lie algebras is related to the projective uniformization of a *single* Riemann surface. This result motivates consideration of such structures for general holomorphic families of surfaces; we indicate how to identify the bundle of relative $\frac{3}{2}$-differentials as the space of infinitesimal deformations. Our constructions thereby establish a link between classical uniformization theory of Riemann surfaces and the graded extensions of Teichmüller theory for Riemann surfaces introduced recently in superstring theory ([1],[11],[12]). After defining graded Riemann surface, we prove in Section 2 that such an object can be constructed from the graded Lie algebra sheaf arising in the projective uniformization. Then we show in Section 3 that, after a suitable grading of the sheaves (given via the exterior algebra functor Λ between bundles and graded manifolds), there is coincidence between the space of infinitesimal deformations of the objects constructed here and the analogues of Teichmüller space for graded/super Riemann surfaces discussed by other authors.

We belive that this connection between classical uniformization theory, graded complex structures on surfaces and super-symmetry is of some intrinsic interest. Moreover it is important in understanding the action of the modular group on this super-moduli space (a problem treated in [8]), and has a bearing on the definition of a measure of integration for superstring theory purposes [15].

2. Graded Riemann surfaces

Physical theories aiming to provide a unified description of all fundamental interactions among particles are usually described by perturbation series, which are currently affected by divergences due to gravitational interactions. One such model is *String Theory*, where the primary objects are not particles but strings, one-dimensional objects whose space-time evolution describes surfaces, compact orientable surfaces in the case of closed strings. This relates string theory to many interesting aspects of the theory of Riemann surfaces. It is hoped that divergences will be avoided by introducing "supersymmetry", leading to a so-called *superstring* theory. To represent supersymmetry, many mathematical theories have to be enlarged by adding their corresponding "odd-counterparts". The infinitesimal version of such symmetry is a graded Lie algebra.

DEFINITION 1. A *graded Lie algebra* is a $\mathbb{Z}_2$-graded algebra $A = A_0 \bigoplus A_1$ such that its $\mathbb{Z}_2$ graded product, which is symmetric (respectively skew symmetric) on the 0 (respectively 1) graded part, verifies the generalized Jacobi identity

$$(-1)^{\|X\|\|Z\|}\langle X,\langle Y,Z\rangle\rangle+(-1)^{\|X\|\|Y\|}\langle Y,\langle Z,X\rangle\rangle+(-1)^{\|Z\|\|Y\|}\langle Z,\langle X,Y\rangle\rangle = 0$$

where $\| * \| = i$ if $* \in A_i$, $i = 0,1$. We call A_0, A_1 the *even* and *odd* sectors of the graded Lie algebra A.

EXAMPLE: The algebra osp(2|1). Take three even (L_{-1},L_0, L_1) and two odd ($G_{\frac{1}{2}}, G_{-\frac{1}{2}}$) generators with defining relations:

$$\begin{aligned}
[L_i, L_j] &= (j-i)L_{i+j} \\
\{G_\alpha, G_\beta\} &= 2L_{\alpha+\beta} \\
[L_i, G_\alpha] &= (\alpha + i/2)G_{\alpha+i} = -[G_\alpha, L_i]
\end{aligned}$$

where $i,j = -1,0,1$, $\alpha,\beta = \frac{1}{2},-\frac{1}{2}$, and sums over indices are modulo 2. Then one checks that this defines a graded Lie algebra, whose even part is the Lie algebra $\mathrm{sl}(2,\mathbb{C})$. This algebra is central in our work.

Just as derivations (vector fields) on manifolds define sheaves of Lie algebras, sheaves of graded Lie algebras are defined by derivations on graded manifolds. For our purposes, it is enough to define graded manifolds of dimension $(1|1)^\dagger$.

DEFINITION 2. A holomorphic (1|1) *graded manifold* is a pair $(M, \mathcal{A})$ where:
i) M is a complex manifold of dimension 1.
ii) $\mathcal{A}$ is the graded sheaf of sections of the exterior algebra bundle $\bigwedge E$, with E some holomorphic line bundle on M.

---

† a general $(m|n)$ graded manifold is defined by a sheaf of Z_2 graded algebras locally isomorphic to the sheaf of sections of the bundle $\wedge E$ of exterior algebras, with E a bundle of rank n over the m dimensional manifold M.

By definition, if U is an open set of M where E trivializes, then the local sections of $\mathcal{A}$ over U can be written as $f_0(z)+f_1(z)\theta$ where f_0, f_1 belong to $\mathcal{O}_M(U)$ and $\theta \in \mathcal{O}(E)(U)$, so that $\theta^2 = 0$ in $\mathcal{A}(U)$. Thus $\mathcal{A}(U)$ is a direct sum $\mathcal{A}(U)_0 \bigoplus \mathcal{A}(U)_1 = \mathcal{O}(U) \bigoplus \mathcal{O}(U)\theta$.

The grading on $\mathcal{A}$ induces a grading on End$(\mathcal{A})$ and the notion of graded derivation respects this grading: if U is an open set of M, define a *graded derivation* of $\mathcal{A}(U)$ to be an endomorphism $X : \mathcal{A}(U) \to \mathcal{A}(U)$, such that

$$X(f\ g) = X(f)g + (-1)^{\|X\|\ \|f\|} f X(g)$$

where $\|f\| = 0$ if $f \in \mathcal{A}(U)_0$, $\|f\| = 1$ if $f \in \mathcal{A}(U)_1$ and $\|X\| = 0$ or 1 as X preserves or changes parity. The set of graded derivations of $\mathcal{A}(U)$ is denoted Der$(U, \mathcal{A})$. We need the following elementary results (for a proof,see [2]):

LEMMA 1. *The presheaf $U \to$ Der$(U, \mathcal{A})$ defines a sheaf Der$(M, \mathcal{A})$ over M; it is the sheaf of graded derivations of $(M, \mathcal{A})$.*

LEMMA 2. *If X, Y are graded derivations, define the Lie bracket of X with Y*

$$\langle X, Y\rangle = XY - (-1)^{\|X\|\|Y\|}\langle Y, X\rangle$$

Then $\langle$-, -$\rangle$ makes Der$(M, \mathcal{A})$ a sheaf of graded Lie algebras.

LEMMA 3. *Define the elements $\dfrac{\partial}{\partial z}, \dfrac{\partial}{\partial \theta}$ of Der$(M, \mathcal{A})$ as follows:*

$$\frac{\partial}{\partial z}(f_0(z) + f_1(z)\theta) = f_0'(z) + f_1'(z)\theta$$

$$\frac{\partial}{\partial \theta}(f_0(z) + f_1(z)\theta) = f_1(z)$$

Then Der$(U, \mathcal{A})$ is a free $\mathcal{A}(U)$ module with basis $\dfrac{\partial}{\partial z}, \dfrac{\partial}{\partial \theta}$.

To define a *graded Riemann surface* a condition has to be imposed on the graded Lie algebra sheaf Der (c.f. [1],[12]):

DEFINITION 3. A *graded Riemann surface* is a holomorphic $(1|1)$ graded manifold $(M, \mathcal{A})$ endowed with a locally free subsheaf $\mathcal{T}_1 \subset$ Der$(M, \mathcal{A})$ of rank $(0|1)$ such that, if D is a local base (generator) of $\mathcal{T}_1$, then $(D, \frac{1}{2}\langle D, D\rangle)$ is a local basis of Der$(M, \mathcal{A})$.

To construct a graded structure on a Riemann surface M, it is enough to choose one of the 2^{2g} square roots of the canonical bundle K. Namely, let $L \to M$ be such a bundle, then apply the functor $\bigwedge$ from the category of vector bundles to the category of graded manifolds, sending L to $(M, \mathcal{A})$, where $\mathcal{A}$ is the sheaf of sections of $\bigwedge L$. Then the graded derivation defined by $D = \dfrac{\partial}{\partial \theta} + \theta \dfrac{\partial}{\partial z}$, where z is a local coordinate of M and θ is a local holomorphic section of $L^{\ddagger}$, generates over $\mathcal{A}$ a subsheaf of $\mathrm{Der}(M, \mathcal{A})$, and it verifies: $\frac{1}{2}\langle D, D\rangle = \dfrac{\partial}{\partial z}$; therefore $(D, \langle D, D\rangle)$ generates $\mathrm{Der}(M, \mathcal{A})$, and a graded Riemann surface is defined.

Conversely, let $(M, \mathcal{A})$ be a graded Riemann surface, and $\mathcal{T}_1$ its subsheaf of Der of rank $(0|1)$. Then it follows that $\mathrm{Der}(M, \mathcal{A})/\mathcal{T}_1$ is isomorphic to the tangent bundle TM, and the isomorphism $\mathcal{T}_1 \bigotimes \mathcal{T}_1 \to \mathrm{Der}(M, \mathcal{A})/\mathcal{T}_1$ induces an isomorphism $\mathcal{T}_1^2 \cong TM$. Thus a spin bundle over M is defined, and we have:

PROPOSITION 1. *$\bigwedge$ is an isomorphism from the category of spin structures on Riemann surfaces (and morphisms preserving such structures) to the category of graded Riemann surfaces (with their morphisms).*

EXAMPLE. Take the Riemann sphere $\mathbb{C}\mathrm{P}^1$ with its tautological line bundle L, then the $(1|1)$ graded manifold $\mathbf{GCP}^1 = (\mathbb{C}\mathrm{P}^1, \mathcal{A})$, with $\mathcal{A} \cong \bigwedge L$ is a graded Riemann surface.

2. Projective uniformization and sheaves of graded Lie algebras

Next we shall develop a deeper connection between graded Riemann surfaces and an important construction in Riemann surface theory. Namely we can associate a graded Riemann surface $(M, \mathcal{A})$ to the process of projective uniformization of the Riemann surface M. The sheaf of graded derivations of $(M, \mathcal{A})$ will correspond to a sheaf of graded Lie algebras over M arising naturally from the projective uniformization.

Recall that a *projective structure* on a Riemann surface is given by an equivalence class of projective atlas on M, so that the transition functions for local coordinate transformations lie in the group $\mathrm{PL}(2, \mathbb{C})$.

A projective structure is obtained by deformation of local coordinates of M, $\{U_\alpha, z_\alpha\}$ via a solution of the equation

$$\{w_\alpha, z_\alpha\} = h_\alpha. \tag{1}$$

‡ Alternatively, with [2] we can regard $\wedge L$ as $\mathcal{O}_M \oplus \Pi L$, where Π is the parity changing functor; then θ is actually a section of ΠL, and $\frac{\partial}{\partial \theta}$ is the dual section.

where {-,-} is the Schwarzian derivative and (h_α) belongs to the space of *projective connections*, an affine space over the space of global quadratic differential forms on M. We describe this process briefly; for a detailed account the reader should refer to [7], [9].

From the transformation properties of the Schwarzian derivative and of (h_α) it follows that the new atlas on M, $\{U_\alpha, w_\alpha \circ z_\alpha\}$, has projective transition functions. Moreover the w_α can be patched together to define a local homeomorphism $f : \mathcal{U} \longrightarrow \mathbf{C}P^1$ from the universal covering space $\mathcal{U}$ of M to $\mathbf{C}P^1$ (the *developing map*) and a (monodromy) homomorphism ρ is defined from the group $\Gamma \cong \pi_1(M)$ of covering transformations to PL(2, C), such that $f(\gamma z) = \rho(\gamma) f(z)$. If $\rho(\Gamma)$ acts discontinously on $f(\mathcal{U})$, then $f(U)/\rho(\Gamma)$ is isomorphic to $M \cong \mathcal{U}/\Gamma$, and this is called a *projective uniformization* of M. Hence, given a projective connection h, we can vary the way of regarding M via patching of open sets on $\mathbf{C}P^1$, or in other words we can deform the action of the group Γ by means of the pair (f, ρ).

It is well known that w_α is a solution of (1) iff w_α is a ratio of two linearly independent solutions of the linear equation:

$$u''_\alpha + \frac{1}{2} h_\alpha u_\alpha = 0. \tag{2}$$

Hawley and Schiffer and, independently, Gunning proved the following significant fact [9].

PROPOSITION 2. *Solutions of (2) are (local) sections of L^{-1}, the dual of a square root of the canonical bundle K of M.*

Namely it follows from the transformation properties of the projective connection h that if u_α,v_β are solutions on overlapping charts then, on the overlap, the relation $u_\alpha(f_{\alpha\beta}(z_\beta)) = v_\beta(z_\beta) f'_{\alpha\beta}(z_\beta)^{\frac{1}{2}}$ must hold, and so the solutions of the differential equation $u'' + \frac{1}{2}hu = 0$, defined on the Riemann surface M, transform as differential forms of order $-\frac{1}{2}$.

Now to construct a projective atlas on M a choice of a spin structure L must be made. Solutions of the equation (2) are sections of L^{-1} defining a sheaf over M; this sheaf will form the odd sector of our sheaf of graded Lie algebras, which is in fact close at hand: the even sector is generated by the odd one and it is the sheaf of tangent vectors preserving the projective structure on M. These vectors are defined by an equation arising in the theory of deformations of projective structure (see [4]), which is now briefly recalled.

An *infinitesimal deformation* of the projective structure or, equivalently, of the pair (f, ρ), can be achieved by finding solutions $\dot{f}_\alpha$ of the equation

$$\{f_\alpha + \epsilon \dot{f}_\alpha, z_\alpha\} = h_\alpha + \epsilon \phi_\alpha. \tag{3}$$

where f_α is a solution of (1) and ϕ_α is in the tangent space to the projective connections, that is a quadratic differential. It follows that $\dot{f}_\alpha$ is a solution of (3) if and only if

$$(f_\alpha)^* = \frac{\dot{f}_\alpha}{f'_\alpha}$$

is a holomorphic vector field (section of K^{-1}) satisfying the equation

$$({f_\alpha}^*)''' + 2h_\alpha({f_\alpha}^*)' + (h_\alpha)'{f_\alpha}^* = \phi_\alpha.$$

Furthermore the homogeneous form of this equation

$$({f_\alpha}^*)''' + 2h_\alpha({f_\alpha}^*)' + (h_\alpha)'{f_\alpha}^* = 0 \tag{4}$$

determines vector fields which are infinitesimal automorphisms of the projective structure.

Now we can state the core result of this paper, connecting the projective uniformization of Riemann surfaces with the graded structures of §1.

THEOREM 1. *Solutions of equations (2) and (4) define a sheaf $\mathcal{A}$ of graded Lie algebras over M, with stalks isomorphic to $osp(2,1)$.*

PROOF. If u^1_α , u^2_α are linearly independent solutions of (2), define the h-projective coordinates $w^*_\alpha = w_\alpha \circ z_\alpha$, where $w_\alpha = u^1_\alpha / u^2_\alpha$. In terms of these coordinates $\{h_\alpha\}$ is represented by the zero connection. Thus the sheaf of solutions of (2) is isomorphic to the sheaf of solutions of $d^2 u^*_\alpha / dw^{*2}_\alpha = 0$; therefore u^*_α is a section of L^{-1} of the form $a_0 + a_1 w^*_\alpha$, i.e. a polynomial of degree one. We denote by $\mathcal{P}_1(L^{-1})$ the subsheaf of $\mathcal{O}(L^{-1})$ consisting of such degree 1 polynomials.

Solutions of (4) for $h = 0$ satisfy $d^3 f^*_\alpha / dw^{*3}_\alpha = 0$; hence f^*_α are sections of $\mathcal{O}(K^{-1})$ which are polynomials of degree 2 , defining a subsheaf $\mathcal{P}_2(L^{-2})$ of $\mathcal{O}(L^{-2})$. If $\left(\frac{\partial}{\partial\theta}, z\frac{\partial}{\partial\theta}\right)$ are generators of the sheaf $\mathcal{P}_1(L^{-1})$, then tensor products of such sections give a basis $\left(\frac{\partial}{\partial z}, z\frac{\partial}{\partial z}, z^2\frac{\partial}{\partial z}\right)$ of $\mathcal{P}_2(L^{-2})$. Now

$\mathcal{A} = \mathcal{A}_0 \bigoplus \mathcal{A}_1 = \mathcal{P}_2(L^{-2}) \bigoplus \mathcal{P}_1(L^{-1})$ has a structure of Z_2-graded algebra with the product $\langle$-,-$\rangle$, where

$$\langle -,- \rangle : \mathcal{P}_2(L^{-2}) \times \mathcal{P}_2(L^{-2}) \longrightarrow \mathcal{P}_2(L^{-2})$$

is the commutator on fields,

$$\langle -,- \rangle : \mathcal{P}_1(L^{-1}) \times \mathcal{P}_1(L^{-1}) \longrightarrow \mathcal{P}_2(L^{-2})$$

is twice the tensor product of sections of L^{-1}, and

$$\langle -,- \rangle : \mathcal{P}_2(L^{-2}) \times \mathcal{P}_1(L^{-1}) \longrightarrow \mathcal{P}_1(L^{-1})$$

is the Lie derivative of vector fields on sections of L^{-1}, given by

$$\langle f\frac{d}{dz}, \eta\frac{\partial}{\partial\theta}\rangle = (f\frac{d\eta}{dz} - \frac{1}{2}\frac{df}{dz}\eta)\frac{\partial}{\partial\theta}.$$

It is now a simple calculation to verify that $\mathcal{A}$ is a graded Lie algebra sheaf on M whose stalks are isomorphic to osp(2|1). ∎

This theorem holds for any Riemann surface of genus $g \geq 2$, and it applies also to CP^1, on which there is a natural (and unique) projective structure and a unique spin structure defined by the tautological line bundle $L \to \mathsf{CP}^1$. The subsheaves $\mathcal{P}_1(L^{-1})$, $\mathcal{P}_2(L^{-2})$ of L^{-1}, $T\mathsf{CP}^1$ consisting of polynomials of degree one and two respectively, give the global sections of the bundles L^{-1} and $T\mathsf{CP}^1$. Therefore $H^0(\mathsf{CP}^1, T\mathsf{CP}^1) \bigoplus H^0(\mathsf{CP}^1, L^{-1})$, with the product $\langle$- ,-$\rangle$ defined in Theorem 1, is a graded Lie algebra isomorphic to osp(2|1). It follows that the graded Lie algebra sheaf $\mathcal{A}(M)$ on M can be also regarded as the pull-back by the developing map f of the graded Lie algebra $H^0(\mathsf{CP}^1, T\mathsf{CP}^1) \bigoplus H^0(\mathsf{CP}^1, L^{-1})$. This latter is easily seen to be equivalent to the (global) graded derivations of the graded Riemann sphere :

LEMMA 4. *The graded Lie algebra* $H^0(\mathsf{CP}^1, T\mathsf{CP}^1) \bigoplus H^0(\mathsf{CP}^1, L^{-1})$ *is isomorphic to the globally defined graded derivations of* GCP^1 .

PROOF. We define an isomorphism of graded algebras as follows.

$$G_{-\frac{1}{2}} = \frac{\partial}{\partial\theta} \longmapsto \mathcal{G}_{-\frac{1}{2}} := \frac{\partial}{\partial\theta} + \theta\frac{\partial}{\partial z} = D$$

This map carries the sections $\frac{\partial}{\partial\theta}$, $z\frac{\partial}{\partial\theta}$ to the globally defined graded derivations D, zD of $\mathbf{GCP}^1$. Now we generate the globally defined even derivations using the anticommutator:

$$\frac{1}{2}\{\mathcal{G}_{-\frac{1}{2}},\mathcal{G}_{-\frac{1}{2}}\} = \mathcal{L}_{-1} = \frac{\partial}{\partial z}$$
$$\frac{1}{2}\{\mathcal{G}_{-\frac{1}{2}},\mathcal{G}_{\frac{1}{2}}\} = \mathcal{L}_0 = z\frac{\partial}{\partial z} + \frac{1}{2}\theta\frac{\partial}{\partial\theta}$$
$$\frac{1}{2}\{\mathcal{G}_{\frac{1}{2}},\mathcal{G}_{\frac{1}{2}}\} = \mathcal{L}_1 = z^2\frac{\partial}{\partial z} + z\theta\frac{\partial}{\partial\theta}.$$

One checks that $(\mathcal{L}_{-1},\mathcal{L}_0,\mathcal{L}_1,\mathcal{G}_{-\frac{1}{2}},\mathcal{G}_{\frac{1}{2}})$ define an osp(2|1) algebra.

3. Families of graded Riemann surfaces and sheaves of graded Lie algebras over the Teichmüller family

Graded Riemann surfaces are motivated by the supersymmetric extension of string theory. The physical content (namely the multiloop contribution) of string theory can be reduced in each genus to the calculation of integrals defined on the space of moduli of conformally inequivalent Riemann surfaces, and it is known that local coordinates (moduli) for this space belong to a $3g-3$ dimensional complex open set isomorphic to the Teichmüller space T_g.

In superstring theory, the so called "supermoduli" have to be taken in account together with moduli, so that one needs to compute integrals defined on a $(3g-3|2g-2)$ dimensional graded manifold, the *supermoduli space*. This space has been introduced in sheaf theoretical terms by Manin:

DEFINITION 4. A *super-family of graded Riemann surfaces* is given by a smooth proper morphism

$$\pi : (Y, \mathcal{A}_Y) \longrightarrow (W, \mathcal{A}_W)$$

of graded manifolds, where $(Y, \mathcal{A}_Y)$ has dimension $(m+1|n+1)$ and $(W, \mathcal{A}_W)$ has dimension $(m|n)$, together with a subsheaf $\mathcal{D}$ of rank $(0|1)$ of the sheaf of relative graded derivations, Der_{rel}, such that $\mathcal{D}\bigotimes\mathcal{D}$ is isomorphic to $Der_{rel}\,/\,\mathcal{D}$.

We have used the term super-family of graded Riemann surfaces since it is true that a graded Riemann surface can be regarded as a super-family over a point: $(Y, \mathcal{A}_Y) \to (s, *)$: its structure is induced by the sheaf $\mathcal{T}_1$ of odd derivations over the Riemann surface Y_s. The family of Definition 4 is called SUSY-family in [12] and super Riemann surface in [11]; an appropriate graded extension of Teichmüller theory was also developed by these authors.

PROPOSITION 3. *The space of local deformations of super Riemann surfaces is the graded manifold* $(T_g, \bigwedge \Psi_g)$, *where* T_g *is the Teichmüller space and* Ψ_g *is the bundle of global relative* $\frac{3}{2}$*-differentials on* T_g.

For the proof see [11] or [12]. The bundle Ψ_g exists, for instance, by results of [5].

Here we outline a different approach to supermoduli, leading to an equivalent construction, based on §2.

DEFINITION 5. Given a holomorphic family of Riemann surfaces $W \to S$, a *family of* $osp(2|1)$ *algebras* $\mathcal{B}$ *over* $W \to S$ is a holomorphic sheaf $\mathcal{B}$ over W, whose restriction to any fiber W_s of W, is a sheaf of graded Lie algebras over W_s with stalks isomorphic to osp(2|1).

A typical example is the "Teichmüller family of osp(2|1) algebras" to be constructed below in Theorem 2. This will show that as the uniformization of a single Riemann surface is related to a sheaf of osp(2|1) algebras over it, in the same way the process of simultaneous uniformization of a family of Riemann surfaces leads to a sheaf of osp(2|1) algebras defined over the family.

THEOREM 2. *The process of simultaneous uniformization of Riemann surfaces, holomorphic in moduli, determines a sheaf* $\mathcal{A}$ *of graded* $osp(2|1)$ *Lie algebras over the Teichmüller family.*

PROOF. First we summarise the Bers construction of the Teichmüller family $\pi : V_g \to T_g$. For more details, see [3], [13].

Fix a reference Riemann surface M with its Fuchsian uniformization: $M = \mathcal{U}/\Gamma$ where $\mathcal{U}$ is the upper half plane and $\Gamma \subset \mathrm{PSL}(2, \mathbb{R})$. Then let

$$T_g = \{f_\mu : \mathbb{CP}^1 \longrightarrow \mathbb{CP}^1, \mu\text{-quasiconformal}\} /\!\sim$$

be the Teichmüller space, where the μ's are measurable complex functions on $\mathcal{U}$ with $\sup_{z \in \mathcal{U}} |\mu(z)| < 1$ and satisfying

$$\mu(z) = \mu(\gamma z) \frac{\overline{\gamma'(z)}}{\gamma'(z)} \quad \text{for} \quad \gamma \in \Gamma.$$

Each μ is extended to $\mathbf{CP}^1$ by $\mu \equiv 0$ on the lower half plane $\mathcal{L}$; for any μ, f_μ is the unique normalised solution of a Beltrami equation $\bar{\partial} f_\mu = \mu \partial f_\mu$ and the equivalence relation in the definition of T_g reads as follows: $f_\mu \sim f_\nu$ iff $f_\mu = f_\nu$ on $\mathcal{L}$.

To describe the family V_g, we note first that the domain $U_\tau = f_\mu(\mathcal{U})$ depends only on the equivalence class $\tau = [f_\mu]$, and that its boundary $\mathcal{C}_\tau = \{f_\mu(x),\ x \in \mathbf{R} \cup \infty\}$ is a holomorphic function of τ. Moreover the uniformizing group Γ is transformed to a *quasi-fuchsian* group $\Gamma_\tau = f_\mu \Gamma f_\mu{}^{-1}$ with $\mathcal{C}_\tau$ as limit set. Now $F_g = \{(z, \tau) \in \mathcal{U}_\tau \times T_g\}$ is a fiber space over T_g, and the Γ action on $F_g : (\gamma, z, \tau) \mapsto (\gamma_\tau(z), \tau)$, where $\gamma_\tau = f_\mu \gamma f_\mu{}^{-1}$, produces a complex fiber space $\pi : V_g \to T_g$, whose fibers are the Riemann surfaces $M_\tau = \mathcal{U}_\tau / \Gamma_\tau$. In other words we have an assignment of projective structures on M_τ which is holomorphic in τ, the modular parameter. Since projective structures on M_τ are in one to one correspondence with projective connections, we can say that the Bers construction of the family $V_g \to T_g$ singles out a holomorphic section σ of the bundle $\mathbf{P}_g$ of all projective connections over all marked Riemann surfaces [10]. Namely, the value of σ on $\tau \in T_g$ is the projective connection on M_τ characterised by the Schwarzian derivative of the function f_τ, which is univalent on $\mathcal{L}$.

$$\{f_\tau, z\} = \phi_\tau = \sigma(\tau).$$

Now we choose a (local) holomorphic section of the family of spin bundles $\mathbf{L}_g$ over the Teichmüller family [14] and denote by L_τ the fibers. Then, for each $\tau \in T_g$ we can define the sheaf $\mathcal{A}(\tau)$ over the surface M_τ, as in §2, by the data $(\sigma(\tau),\ L_\tau{}^{-1})$. Since everything varies holomorphically with τ,

$$\boldsymbol{\mathcal{A}} = \bigcup_{\tau \in T_g} \mathcal{A}(\tau)$$

is a holomorphic sheaf of graded Lie algebras over the family $V_g \to T_g$. ∎

The sheaf $\boldsymbol{\mathcal{A}}$ is intrinsically related to the process of simultaneous uniformization of Riemann surfaces: any stalk of $\boldsymbol{\mathcal{A}}$ gives solutions of equation (2) and vector fields preserving that projective structure for which the uniformization process is holomorphic with respect to τ in T_g.

To conclude, we outline the deformation theory of these super-families, and comment briefly on further developments. The deformations of the sheaf of graded Lie algebras $\mathcal{A}$ over a fixed Riemann surface M are described by the *Eichler cohomology sets*:

$$H^1(M, \mathcal{A}_0) \bigoplus H^1(M, \mathcal{A}_1)$$

where $H^1(M, \mathcal{A}_0)$ represents the infinitesimal variation of the complex and projective structure of M, while $H^1(M, \mathcal{A}_1)$ represents:

i) variations of the chosen spin structure L^{-1}, that is non-trivial affine bundles over L^{-1} (which are parametrized by $H^1(M, \mathcal{O}(L^{-1}))$;

ii) infinitesimal displacements along the fiber $\pi^{-1}(0)$ of the bundle $\Psi_g \to V_g$ of relative $\frac{3}{2}$-differentials.

These two variations are obtained from monodromy of solutions of non-homogeneous equations paralleling (2):

$$u''_\alpha + \tfrac{1}{2} h_\alpha u_\alpha = \psi_\alpha$$

where ψ_α are $\frac{3}{2}$-differential forms on M.

REMARK 1. By Kodaira-Spencer theory, deformations of families of osp(2|1) algebras $\mathcal{B}$ over $W \to S$ are given by a sheaf $\mathcal{F}$ over $W \to S$ such that for a distinguished point $s_0 \in S$, we have $\mathcal{F}(s_0) \simeq \mathcal{B}$. For marked families, the universal property of the Teichmüller family V_g enables us to identify $\mathcal{F}$ in terms of the sheaf of sections of the bundle Ψ_g over $V_g \to T_g$ of relative $\frac{3}{2}$-differentials, since holomorphicity on moduli and the choice of a preferred spin structure L_g allows deformations of type ii) only. Applying the functor $\bigwedge$ we recover the graded manifold $(T_g, \mathcal{A})$ with $\mathcal{A}$ the sheaf of sections of $\bigwedge \Psi_g$, that is, the space of infinitesimal deformations for families of graded Riemann surfaces mentioned earlier in Proposition 3.

This is not surprising because of the correspondence between super-families of graded Riemann surfaces and osp(2|1) families. Namely, given a family of osp(2|1) algebras $\mathcal{B}$, we have a system of generators for the sheaf $\mathcal{B}$:

$$(L_{-1}, L_0, L_1, G_{-\frac{1}{2}}, G_{\frac{1}{2}})(z, \tau)$$

holomorphic in z and τ. We can identify these with derivations of the graded manifolds M_τ by extending the rule $G_{\frac{1}{2}} \mapsto D$ as in Lemma 4. It then follows easily that a family of (1|1) graded manifolds is determined and that the relative derivations $D(z, \tau)$ generate a sheaf $\mathcal{D}$ with the properties required to make this into a super-family of graded Riemann surfaces.

Conversely, given such a super-family, a family of osp(2|1) algebras may be constructed in terms of the kernel of the operators $D_\tau^5 = \partial_\tau^2 D_\tau$, which vary holomorphically with τ and act on sections of the sheaf $\mathcal{A}$.

REMARK 2. To achieve the desired supermoduli spaces it remains to work out the action of the modular groups on our spaces of local deformations; for some progress on this, see [8].

REMARK 3. For superstring applications, one has to find the correct measure of integration on the space of supermoduli. This requires a generalization of the Weil-Petersson inner product used in constructing the bosonic string measure. Such an inner product is regarded in [6] as the integral of the exterior product of sl(2, C)-valued one forms, the product being taken with respect to the Killing form on sl(2, C). Using the construction described here, we give in [15] a natural generalisation of this to osp(2|1) valued forms inducing a metric on supermoduli, by defining the appropriate graded exterior product of osp(2|1)-valued one forms.

REMARK 4. The algebraic relation between solutions of (2) and (4) is a particular case of a more general relation: solutions of (2) generate, by $(r-1)$-symmetric tensor powers, solutions of differential equations of order r defined over the bundles L^{1-r}. The sheaves $\mathcal{P}_{r-1}(L^{1-r})$ of solutions to such equations belong to exact sequences

$$0 \longrightarrow \mathcal{P}_{r-1}(L^{1-r}) \longrightarrow \mathcal{O}(L^{1-r}) \xrightarrow{D_\phi^r} \mathcal{O}(L^{r+1}) \longrightarrow 0$$

where the D_ϕ^r resemble the differential operators occuring in the study of the KdV hierarchy.

Acknowledgment. The author is grateful to W. Harvey for his guidance and constant encouragement.

References

[1] M. BATCHELOR and P. BRYANT. *Comm. Math. Phys.* **114** (1988) 243-255.

[2] F. BEREZIN. *Introduction to superanalysis* (Reidel 1987).

[3] L. BERS. *Acta Math.* **130** (1973) 89-126.

[4] C. J. EARLE. in *Annals of Math. Studies* **97** (1981) 87-99.

[5] C. J. EARLE, I. KRA. Half-canonical divisors on variable Riemann surfaces. *J. Math Kyoto U.* **26** (1986) 39-64.

[6] W. GOLDMAN. *Adv. in Math.* **54** (1984) 200-225.
[7] R. C. GUNNING. *Lectures on Riemann surfaces* Math. Notes n.2 Princeton (1965).
[8] W. HARVEY. in *The Interface of Mathematics and Particle Physics*, ed. Segal, Quillen and Tsou (Oxford Univ. Press 1990) 187-191.
[9] N. HAWLEY AND M. SHIFFER. in *Acta Math.* **115** (1966) 119-236.
[10] J. H. HUBBARD. in *Annals of Math. Studies* **97** (1981) 257-275.
[11] C. LE BRUN and M. ROTHSTEIN. *Comm. Math. Phys.* **117** (1988) 159-178.
[12] YU I. MANIN. in *Functional Anal. and Appl.* **0** (1986) 244-246.
[13] S. NAG. *Complex Analytic Theory of Teichmüller Spaces.* (John Wiley & Sons, New York, 1988).
[14] P. SIPE. *Math. Ann.* **260** (1982) 67-92.
[15] P. TEOFILATTO. in *Lect. Notes in Phys.* **375** (Springer Verlag 1991).

Scuola di Ingegneria Aerospaziale
Universitá di Roma "La Sapienza"
via Eudossiana 16, 00100 Roma
Email:TEOFILATTO@ROMA2.INFN.IT

Generating sets for finite groups

Richard M. Thomas

To Murray Macbeath on the occasion of his retirement

1. Introduction

Given positive integers m, n and k, we are interested in finite groups generated by two elements a and b satisfying the relations $a^m = b^n = (ab)^k = 1$. To put this another way, if (m, n, k) denotes the *triangle group* defined by the presentation

$$< a, b : a^m = b^n = (ab)^k = 1 >,$$

then we are looking for finite homomorphic images of (m, n, k). Let

$$\alpha = \frac{1}{m} + \frac{1}{n} + \frac{1}{k}.$$

If $\alpha > 1$, then (m, n, k) is finite, and, if $\alpha = 1$, (m, n, k) is solvable; we are interested in the case where $\alpha < 1$. The largest possible such value of α is $\frac{41}{42}$, in which case we have a *Hurwitz group*, that is a finite group generated by two elements a and b satisfying the relations $a^2 = b^3 = (ab)^7 = 1$; for a general survey of Hurwitz groups, see [6].

One intriguing question is the following: given particular (usually small) values of m, n and k, how can we construct interesting finite homomorphic images of (m, n, k)? In particular, how can we construct interesting Hurwitz groups? Since any Hurwitz group G (and, in fact, any image of (m, n, k)

with m, n and k pair-wise co-prime) must have a maximal normal subgroup M with G/M a non-abelian simple group, it is reasonable to start by considering which of the finite simple groups are Hurwitz groups. The smallest Hurwitz group is $PSL(2,7)$, which has the presentation

$$< a, b : a^2 = b^3 = (ab)^7 = [a,b]^4 = 1 > .$$

This leads naturally to the following question: if $(2,3,7;q)$ is the group defined by the presentation

$$< a, b : a^2 = b^3 = (ab)^7 = [a,b]^q = 1 >,$$

when is $(2,3,7;q)$ finite? One may readily check that $(2,3,7;q)$ is trivial for $q \leq 3$; for the remaining cases, we have:

(i) $(2,3,7;4) \cong PSL(2,7)$ [2];
(ii) $(2,3,7;5)$ is trivial [2];
(iii) $(2,3,7;6) \cong PSL(2,13)$ [2];
(iv) $(2,3,7;7) \cong PSL(2,13)$ [24];
(v) $(2,3,7;8) \cong E_{64}.PSL(2,7)$ [19];
(vi) $(2,3,7;9)$ is infinite [23, 18];
(vii) $(2,3,7;10)$ is infinite [14, 12];
(viii) $(2,3,7;11)$ is infinite [10];
(ix) $(2,3,7;q)$ is infinite for $q \geq 12$ [16, 15].

Of course, we can have finite factor groups of $(2,3,7;q)$ even when $q \geq 9$; for example, the first Janko group J_1 is a homomorphic image of $(2,3,7;q)$ for $q = 10, 11, 15$ and 19 (see [22] for example). We may extend the question to ask when the group $(2,3,p;q)$ defined by the presentation

$$< a, b : a^2 = b^3 = (ab)^p = [a,b]^q = 1 >$$

is finite, where $p \geq 7$. Since $(2,3,p;2)$ is isomorphic to A_4 if $p \equiv 3 \pmod 6$, $A_4 \times C_2$ if $p \equiv 0 \pmod 6$, and is trivial otherwise, and the group $(2,3,p;3)$ has order $\frac{3p^2}{2}$ if p is even, and is trivial otherwise [25], we will assume that $q \geq 4$. We have the following:

(i) $(2,3,8;4) \cong PGL(2,7)$ [7];
(ii) $(2,3,8;5) \cong C_3.A_6.C_2$ [7];
(iii) $(2,3,9;4) \cong A_4$ [7];
(iv) $(2,3,9;5) \cong C_3.PSL(2,19)$ [26];
(v) $(2,3,10;4) \cong C_3.A_6.C_2$ [7];
(vi) $(2,3,11;4) \cong PSL(2,23)$ [7].

Also, it was shown in [8, 9] that, if p is even, then

$$(2,3,p;q) \text{ is finite} \Leftrightarrow \cos(\tfrac{4\pi}{p}) + \cos(\tfrac{2\pi}{q}) < \tfrac{1}{2},$$

and it was conjectured that this criterion should hold even without the assumption that p is even. Combined with the results mentioned above, and with the possible exception of $(p,q) = (13,4)$, this was shown to be the case in [16], giving:

With the possible exception of $(p,q) = (13,4)$, *the group* $(2,3,p;q)$ *is infinite if and only if* p *and* q *satisfy one of the following conditions:*

(i) $p = 7$, $q \geq 9$; *(ii)* $p = 8$ *or* 9, $q \geq 6$;
(iii) $p = 10$ *or* 11, $q \geq 5$; *(iv)* $p \geq 12$, $q \geq 4$.

Returning to the question as to which finite simple groups are Hurwitz groups, Macbeath showed [21] that $PSL(2,p^n)$ is a Hurwitz group (where p is prime and $n \geq 1$) if and only if one of the following three conditions holds:

(i) $n = 1, p = 7$; (ii) $n = 1, p \equiv 1 \pmod 7$;
(iii) $n = 3, p \equiv \pm 2, \pm 3 \pmod 7$.

It was shown in [14] that all but finitely many of the alternating groups A_n are Hurwitz groups, and then, in [4], that A_n is a Hurwitz group for $n \geq 168$ and for all but 64 integers in the range $3 \leq n \leq 167$; for further details, and for a summary of which finite simple groups are known to be Hurwitz groups, see [6], and, for a classification of all the Hurwitz groups of order less that 10^6, see [5].

Having determined that a particular group K is a homomorphic image of (m,n,k), we would like to construct new such groups G with a normal subgroup N such that G/N is isomorphic to K. An elegant approach is that introduced by Macbeath in [20]: if Δ is the group $(2,3,7)$, Λ is a normal subgroup of finite index in Δ, K is the Hurwitz group Δ/Λ, and m is an arbitrary positive integer, then $\Lambda/\Lambda'\Lambda^m$ is a finite abelian group and $\Delta/\Lambda'\Lambda^m$ is a Hurwitz group G with normal abelian subgroup $N = \Lambda/\Lambda'\Lambda^m$ and G/N isomorphic to K.

We present here another method of forming certain new finite homomorphic images of triangle groups from old. If K is a triangle group with a faithful irreducible representation over F_p with the elements of order m and n acting fixed-point-freely, we show how to construct a larger such group by taking a semi-direct product of a p-group with K. To be precise, we will prove the following:

THEOREM A. *Let K be a finite group generated by non-trivial elements u and v satisfying the relations $u^m = v^n = (uv)^k = 1$, and suppose that p is a prime with $O_{p'}(K) \neq 1$. Let G be a semi-direct product NK, where N is an normal p-subgroup of G, $\Phi(N)$ is central in G, and where K acts faithfully and irreducibly on $N/\Phi(N)$ with u and v acting fixed-point-freely on $N/\Phi(N)^\sharp$. Then G is a homomorphic image of (m, n, k).*

Before we prove this result, a few comments are appropriate.

Clearly a necessary condition that $G = NK$ be a homomorphic image of a triangle group is that G can be generated by two elements. It was shown in [27] that, if K is a d-generator finite group with $d > 1$ and $O_{p'}(K) \neq 1$, and if K acts faithfully and irreducibly on an elementary abelian p-group N, then the semi-direct product NK can be generated by d elements. This was generalized in [1], where it was shown that the hypothesis that $O_{p'}(K)$ is non-trivial could be omitted. In our case, K is a 2-generator group, and so G is generated by xc and yd for some $x, y \in N$, where c and d generate K. By a result of Gaschütz [11], we may choose c and d to be u and v.

If N is elementary abelian, and if an element of a minimal generating set $\{u, v\}$ for K is a p'-element centralizing a non-trivial element of N, say v has order n and centralizes $x \in N^\sharp$, then we would have $< (vx)^n > = < x^n > = < x >$ and $< (vx)^p > = < v^p > = < v >$, and so $G = < u, vx >$. Unfortunately, this does not suffice, as we are interested in showing that the new pair of elements satisfies the appropriate relations, *as well as* generating the group, and we do not have $(vx)^n = 1$. In fact, while NK is always 2-generated for a Hurwitz group K acting faithfully and irreducibly on N, NK need not be a Hurwitz group; for example, any extension of an elementary abelian group N by $PSL(2, 7)$ acting faithfully and irreducibly on N is 2-generated, but not all such extensions are Hurwitz groups; see [3] for details.

We can say a little more than is explicitly stated in Theorem A. The essence of the result is that, if we have a homomorphism θ of (m, n, k) onto K mapping the generators of (m, n, k) of orders m and n onto u and v respectively, then there is a homomorphism ϕ of (m, n, k) onto G. If u, v and uv have orders *precisely* m, n and k in K, then we have generators u' and v' for G such that u', v' and $u'v'$ have orders m, n and k respectively (where $u' \in Nu$ and $v' \in Nv$). So the kernel of ϕ is torsion-free, and hence a surface group. Moreover, ϕ is compatible with θ, i.e. if we let $\psi : G \to K$ be the natural homomorphism defined by $(nk)\psi = k$ for $n \in N$ and $k \in K$, then $\theta = \phi \circ \psi$.

2. Notation and preliminary results

In this section, we outline the notation we use, and state some standard results. All groups from now on are finite. (This should be thought of as a convention, as opposed to a theorem!).

We hope the notation is reasonably standard, and so confine ourselves to a few remarks. We let F_p denote the field of p elements for any prime p. For any group G, $G^\sharp$ is the set of non-trivial elements of G, $\Phi(G)$ is the Frattini subgroup of G, and $O_{p'}(G)$ the largest normal subgroup of G whose order is not divisible by the prime p. With regard to conjugacy and commutator conventions, a^x stands for $x^{-1}ax$ and $[a,x]$ for $a^{-1}x^{-1}ax$. For any prime power n, E_n denotes the elementary abelian group of order n.

For the convenience of the reader, we also list some of the standard results we shall be using. We start with the Schur-Zassenhaus theorem:

2.1. Let G be a group, π be a set of primes and H be a normal Hall π-subgroup of G (i.e. a normal π-subgroup such that $[G:H]$ is a π'-number). Then there is a complement C to H in G, and all such complements are conjugate in G.

PROOF. See (6.2.1) of [13] for example.

2.2. Let G be a group, $x \in G$ and ϕ be a fixed-point-free automorphism of G of order n. Let S be the semi-direct product $G < c >$, where $< c >$ is cyclic of order n and c acts as ϕ on G. Then xc has order n in S.

PROOF. By (10.1.1) of [13], we have that $x.x^{c^{-1}}.x^{c^{-2}}\ldots x^{c^{-(n-1)}} = 1$, i.e. $(xc)^n = 1$. Since xcG has order n in S/G, xc has order n in G as required.

2.3. Let P be a normal p-subgroup of G and C be a subgroup of G such that $G = PC$, and suppose that F is a characteristic subgroup of P such that $P \cap C \leq F$ and $[F,C] = 1$. Then $O_{p'}(G/P) = O_{p'}(C)P/P$.

PROOF. Since $O_{p'}(C)$ is normal in C, $O_{p'}(C)P$ is normal in $CP = G$, so that $O_{p'}(C)P/P$ is a normal p'-subgroup of G/P, and hence $O_{p'}(C)P/P \leq O_{p'}(G/P)$.

Let L be the subgroup of G containing P such that $O_{p'}(G/P) = L/P$. Using (2.1), we may write L as PD where D is a p'-subgroup of C; in particular, $F \cap D = 1$. Since $[F,D] = 1$, we have that $FD = F \times D$, so that D is the unique Hall p'-subgroup of FD.

Since L is normal in G, we have that $[L,C] \leq L$, and so $[D,C] \leq L$. But $[D,C] \leq C$, so that $[D,C] \leq L \cap C = PD \cap C = (P \cap C)D \leq FD$.

Since C normalizes F, we have that $[FD, C] \leq FD$. Now C normalizes FD and D is characteristic in FD, so that C must normalize D. We see that $D \leq O_{p'}(C)$, and then $O_{p'}(G/P) = DP/P \leq O_{p'}(C)P/P$ as required.

2.4. If G is a group, N is a normal subgroup of G, $U \leq G$ and $N \leq \Phi(U)$, then $N \leq \Phi(G)$.

PROOF. See (9.3.7) of [17] for example.

3. Proofs of results

Our approach is to modify the argument of [27] to show that, under the hypotheses of Theorem A, we can find a generating pair for G satisfying the appropriate relations. Rather than confuse the reader (and the author) by trying to describe how the argument in [27] should be modified, we give the new proof in full. Essentially, the work in proving Theorem A is summed up in the following result:

THEOREM B. *Let G be a finite group with a normal p-subgroup N such that G/N acts faithfully and irreducibly on $N/\Phi(N)$. Suppose that u and v are elements of G such that G/N is generated by uN and vN, that $\Phi(N)$ is central in G, and that $O_{p'}(G/N) \neq 1$. Then for some $x \in N$*

$$G = < ux, x^{-1}v > .$$

PROOF OF THEOREM B. Let $K = < u, v >$, so that $G = NK$, and let $Q = O_{p'}(K)$ and $V = N/\Phi(N)$. Let Σ be a set of coset representatives for $\Phi(N)$ in N. For any $x \in \Sigma$, we let K_x denote $< ux, x^{-1}v >$, so that $G = NK_x$. If $K_x = G$ for some $x \in \Sigma$, we have finished; so suppose that $K_x < G$ for all $x \in \Sigma$.

Now $V \cap K_x\Phi(N)/\Phi(N)$ is a K-invariant subgroup of V. If

$$V \cap K_x\Phi(N)/\Phi(N) = V,$$

then we have that $N = N \cap K_x\Phi(N) = (K_x \cap N)\Phi(N)$ and

$$G = K_xN = K_x\Phi(N).$$

Since $\Phi(N) \leq \Phi(G)$ by (2.4), we have that $G = K_x$, a contradiction. So we must have that $V \cap K_x\Phi(N)/\Phi(N)$ is a proper K-invariant subgroup of V, and so $V \cap K_x\Phi(N)/\Phi(N) = 1$, i.e. $K_x \cap N \leq \Phi(N)$ for all $x \in \Sigma$.

Let Q_x denote $O_{p'}(K_x)$; in particular, uv normalizes Q_x for all $x \in \Sigma$. Since $NQ_x/N = O_{p'}(G/N) = NQ/N$ by (2.3), we have that $Q \neq 1$ (by hypothesis) and that $NQ_x = NQ$ for all $x \in \Sigma$, where $Q_x \cap N = 1$. Since G/N acts faithfully on V, we have $C_V(Q_x) < V$; since K_x normalizes $C_V(Q_x)$ and K_x acts irreducibly on V, we have that $C_V(Q_x) = 1$.

If $Q_x\Phi(N) = Q_y\Phi(N)$ for $x, y \in \Sigma$, then $Q_x\Phi(N)$ is normalized by ux and uy, and hence by $x^{-1}y$. Since $x^{-1}y \in N$, we have

$$[Q_x\Phi(N), x^{-1}y] \leq Q_x\Phi(N) \cap N = \Phi(N),$$

so that $x^{-1}y\Phi(N) \in C_V(Q_x) = 1$, i.e. $x\Phi(N) = y\Phi(N)$, and then $x = y$. So, if $x \neq y$, then $Q_x\Phi(N) \neq Q_y\Phi(N)$; thus

$$|\{Q_x\Phi(N) : x \in \Sigma\}| = |\Sigma|.$$

Now, since $Q_x \cap N = 1$ for all $x \in \Sigma$, each $Q_x\Phi(N)/\Phi(N)$ is a p'-complement to the normal p-subgroup $N/\Phi(N)$ of $NQ/\Phi(N)$; so, given $x \in \Sigma$, we have $Q_x\Phi(N)/\Phi(N) = (Q\Phi(N)/\Phi(N))^{z\Phi(N)}$ for some $z \in N$ by (2.1). Writing z as sf with $s \in \Sigma$ and $f \in \Phi(N)$, we have $Q_x\Phi(N)/\Phi(N) = (Q\Phi(N)/\Phi(N))^{s\Phi(N)}$ so that

$$\{Q_x\Phi(N) : x \in \Sigma\} \subseteq \{Q^s\Phi(N) : s \in \Sigma\}.$$

But $|\{Q_x\Phi(N) : x \in \Sigma\}| = |\Sigma|$, and so

$$\{Q_x\Phi(N) : x \in \Sigma\} = \{Q^s\Phi(N) : s \in \Sigma\}.$$

Let t be any element of Σ, so that $Q^{t^{-1}}\Phi(N) = Q_x\Phi(N)$ for some $x \in \Sigma$. Since uv normalizes Q_x, uv normalizes $Q^{t^{-1}}\Phi(N)$, i.e. $Q^{t^{-1}uv}\Phi(N) = Q^{t^{-1}}\Phi(N)$, and hence $Q^{t^{-1}uvt}\Phi(N) = Q\Phi(N)$. Since $Q^{(uv)^{-1}} = Q$, we have that $Q^{(uv)^{-1}t^{-1}(uv)t}\Phi(N) = Q\Phi(N)$, i.e. $[uv, t]$ normalizes $Q\Phi(N)$. Since $[uv, t] \in [uv, N] \leq N$, we have that

$$[\,[uv, t], Q\Phi(N)\,] \leq N \cap Q\Phi(N) = \Phi(N);$$

hence $[uv, t]\Phi(N) \in C_V(Q) = 1$, and so $[uv, t] \in \Phi(N)$. So we must have that $[uv, t] \in \Phi(N)$ for all $t \in \Sigma$, so that uv centralizes V, contradicting the fact that K acts faithfully on V. This completes the proof of Theorem B.

Having proved Theorem B, we can now deduce Theorem A:

PROOF OF THEOREM A. By Theorem B, there exists $x \in N$ with

$$< ux, x^{-1}v > = G.$$

Since u and v act fixed-point-freely on N, $u_1 = ux$ and $v_1 = x^{-1}v$ generate G and satisfy $u_1^m = v_1^n = (u_1v_1)^k = 1$ by (2.2), and the result follows.

Acknowledgements

The main motivation for writing this paper was to pay tribute to the work of Murray Macbeath, and, in particular, the lovely work on Hurwitz groups in [20] and [21]. This interest was stimulated by Marston Conder's excellent survey [6], and we have drawn on that paper heavily in preparing the present work. It is a pleasure to acknowledge the influence of both these authors, and to thank the referee for his/her comments, which improved the presentation of this paper, and also Hilary Craig for all her help and encouragement.

References

[1] M. ASCHBACHER AND R. GURALNICK. Some applications of the first cohomology group. *J. Algebra* **90** (1984) 446-460.

[2] H. R. BRAHANA. Certain perfect groups generated by two operators of orders two and three. *Amer. J. Math.* **50** (1928) 345-356.

[3] J. M. COHEN. On Hurwitz extensions by $PSL_2(7)$. *Math. Proc. Cambridge Phil. Soc.* **86** (1979) 395-400.

[4] M. D. E. CONDER. Generators for alternating and symmetric groups. *J. London Math. Soc.* **22** (1980) 75-86.

[5] M. D. E. CONDER. The genus of compact Riemann surfaces with maximal automorphism group. *J. Algebra* **108** (1987) 204-247.

[6] M. D. E. CONDER. Hurwitz groups - a brief survey. *Bull. Amer. Math. Soc.* **23** (1990) 359-370.

[7] H. S. M. COXETER. The abstract groups $G^{m,n,p}$. *Trans. Amer. Math. Soc.* **45** (1939) 73-150.

[8] H. S. M. COXETER. Groups generated by unitary reflections of period two. *Canadian J. Math.* **9** (1957) 243-272.

[9] H. S. M. COXETER. The abstract group $G^{3,7,16}$. *Proc. Edinburgh Math. Soc.* **13** (1962) 47-61 and 189.

[10] M. EDJVET. An example of an infinite group. *This volume.*

[11] W. GASCHÜTZ. Zu einem von B. H. und H. Neumann gestellten Problem. *Math. Nachr.* **14** (1955) 249-252.

[12] L. C. GROVE AND J. M. MCSHANE. On Coxeter's groups $G^{p,q,r}$, in *Groups St. Andrews 1989, Volume 1* (eds. Campbell, Robertson). London Math. Soc. Lecture Note Series **159** (Cambridge University Press, Cambridge 1991) 211-213.

[13] D. GORENSTEIN. *Finite Groups.* (Harper and Row, New York, 1968).

[14] G. HIGMAN *The groups* $G^{3,7,n}$. (Unpublished notes, Mathematical Institute, University of Oxford.)

[15] D. F. HOLT AND W. PLESKEN. A cohomological criterion for a finitely presented group to be infinite. *Preprint.*

[16] J. HOWIE AND R. M. THOMAS. The groups $(2,3,p;q)$; asphericity and a conjecture of Coxeter. *J. Algebra*, to appear.

[17] R. KOCHENDÖRFFER. *Group Theory.* (McGraw-Hill, London, 1966).

[18] J. LEECH. Note on the abstract group $(2,3,7;9)$. *Proc. Cambridge Phil. Soc.* **62** (1966) 7-10.

[19] J. LEECH AND J. MENNICKE. Note on a conjecture of Coxeter. *Proc. Glasgow Math. Assoc.* **5** (1969) 25-29.

[20] A. M. MACBEATH. On a theorem of Hurwitz. *Proc. Glasgow Math. Assoc.* **5** (1961) 90-96.

[21] A. M. MACBEATH. Generators of the linear fractional groups. In *Number Theory* (Proc. Sympos. Pure Math. **XII**, Amer. Math. Soc., Providence, R.I., 1969) 14-32.

[22] J. MCKAY AND K-C. YOUNG. The non-abelian simple groups G, $|G| < 10^6$ - minimal generating pairs. *Math. Comp.* **33** (1979) 812-814.

[23] C. C. SIMS. On the group $(2,3,7;9)$. *Notices Amer. Math. Soc.* **11N** (1964) 687-688.

[24] A. SINKOV. A set of defining relations for the simple group of order 1092. *Bull. Amer. Math. Soc.* **41** (1935) 237-240.

[25] A. SINKOV. The group defined by the relations $S^l = T^m = (S^{-1}T^{-1}ST)^p = 1$. *Duke Math. J.* **2** (1936) 74-83.

[26] A. SINKOV. The number of abstract definitions of $LF(2,p)$ as a quotient group of $(2,3,n)$. *J. Algebra* **12** (1969) 525-532.

[27] R. M. THOMAS. On the number of generators for certain finite groups. *J. Algebra* **71** (1981) 576-582.

Department of Computing Studies
University of Leicester
University Road
Leicester LE1 7RH
Email: rmt@uk.ac.le

Group actions on trees with and without fixed points

David L. Wilkens

To Murray Macbeath on the occasion of his retirement

In [4; 3.1] it is shown that an **R**-tree T, on which a group G acts, has a unique minimal invariant subtree if there are elements of G with no fixed points. The situation where each element of G fixes some point of T splits into two cases depending on whether the action is bounded or not. In the unbounded case, when T has no invariant subtree, it is shown in Theorem 1 that G is given by an infinite tower of subgroups. In [3] Chiswell constructs an action on a tree that corresponds to a given Lyndon length function defined on G. This construction is used in Theorem 2 to establish a necessary and sufficient condition for two length functions to arise from the same action of G on some tree T, again in the case where each element of G fixes some point of T.

An **R**-*tree* T is a non-empty metric space, with metric d, such that there is no subspace homeomorphic to a circle, and for any two points u, $v \in T$ there is a unique isometry $\alpha : [0, r] \to T$, with $\alpha(0) = u$, $\alpha(r) = v$, where $r = d(u, v)$. It is shown in section 4 of [5] that the completion of an **R**-tree is again an **R**-tree. The definition is originally due to Tits [8], where completeness is assumed. In this paper all **R**-trees will be assumed to be complete. This allows the results of Theorem 1 to be economically expressed. Basic properties of **R**-trees are established in [1] and [4], where reference may be made for the following.

Let a group G act as a group of isometries on an $\mathbf{R}$-tree T. For $x \in G$ let $T^x = \{u \in T \mid xu = u\}$, which is either empty or is a subtree of T. Let $N = \{x \in G \mid T^x \neq \emptyset\}$. Then N is a normal subset of G. An element $x \notin N$ has a unique axis A_x in T, that is an isometric image of $\mathbf{R}$ on which x acts as a translation. A subset H of G is said to have *bounded action* if for each $u \in T$ the set $\{d(u, xu) \mid x \in H\}$ is bounded.

A *Lyndon length function* on G is a function $l : G \to \mathbf{R}$ such that for all $x, y, z \in G$

$$\begin{aligned} &\text{A1}'. \quad l(1) = 0 \\ &\text{A2}. \quad l(x) = l(x^{-1}) \\ &\text{A4}. \quad c(x,y) < c(x,z) \text{ implies } c(x,y) = c(y,z), \\ &\qquad\quad \text{where } 2c(x,y) = l(x) + l(y) - l(xy^{-1}). \end{aligned}$$

The definition is due to Lyndon [7], and it is an easy consequence of the axioms that $l(x)$, $c(x,y) \geq 0$.

In [3] Chiswell showed that if G acts on T then for $u \in T$ a length function l_u is given by $l_u(x) = d(u, xu)$. He also gave a construction of a tree T, and an action of G, that corresponds to a given length function l. That the construction yields an $\mathbf{R}$-tree is proved in [2] and [5]. Lemma 3 of [6] states that $x \in N$ if and only if $l(x^2) \leq l(x)$. We will make frequent use of Chiswell's construction, which is given below, in the notation of this paper. T is the completion of the following space.

Points are equivalence classes $[x, m]$, for $x \in G$, where $0 \leq m \leq l(x)$, under the equivalence relation $(x, m) \sim (y, n)$ if and only if $m = n \leq c(x^{-1}, y^{-1})$. The base-point $u = [x, 0]$, for any $x \in G$.

The metric d is given by

$$d([x,m],[y,n]) = \begin{cases} |m-n| & \text{if } \min(m,n) \leq c(x^{-1},y^{-1}) \\ m+n-2c(x^{-1},y^{-1}) & \text{if } \min(m,n) > c(x^{-1},y^{-1}) \end{cases}$$

The action of G is defined by

$$y[x,m] = \begin{cases} [y,\ l(y) - m] & \text{if } m \leq c(x^{-1},y) \\ [yx,\ l(y) + m - 2c(x^{-1},y)] & \text{if } m > c(x^{-1},y) \end{cases}$$

The following is Proposition 3.3 of [10].

LEMMA 1. *If $x, y, xy \in N$ then two of $l(x)$, $l(y)$, $l(xy)$ are equal with the third no greater.*

LEMMA 2. *Let l be a length function with $N = G$ having corresponding action on the Chiswell tree T. If $x, y \in G$, with $0 \leq 2b \leq l(x)$ and $w = [x, b] \in T$, then*

$$yw = \begin{cases} w & \text{if } l(y) \leq 2b \\ [y, l(y) - b] & \text{if } l(y) > 2b \end{cases}$$

and $l_w(y) = \max(l(y) - 2b, 0)$.

PROOF. If $l(y) < 2b \leq l(x)$ then by Lemma 1, $l(yx) = l(x)$, and so $2c(x^{-1}, y) = l(y) < 2b$. Hence $y[x, b] = [yx, l(y) + b - 2c(x^{-1}, y)] = [yx, b]$. By Lemma 1, $l(x^{-1}yx) \leq l(yx) = l(x)$, and thus $2c(x^{-1}, (yx)^{-1}) = 2l(x) - l(x^{-1}yx) \geq l(x) \geq 2b$. Hence $[yx, b] = [x, b] = w$.

If $l(y) \geq 2b$ then by Lemma 1, $l(yx) \leq \max(l(x), l(y))$ and so $2c(x^{-1}, y) \geq \min(l(x), l(y)) \geq 2b$. Also by replacing y by y^{-1}, $c(x^{-1}, y^{-1}) \geq b$. Hence $y[x, b] = [y, l(y) - b]$. For $l(y) = 2b$ then $y[x, b] = [y, b] = [x, b]$.

Now $l_w(y) = d(w, yw)$ and so for $l(y) \leq 2b$, $l_w(y) = 0$. If $l(y) > 2b$ then by Lemma 1, $l(x^{-1}y) = l(y)$ and so $2c(x^{-1}, y^{-1}) = l(x) \geq 2b = 2\min(b, l(y) - b)$. Thus $l_w(y) = d([x, b], [y, l(y) - b]) = (l(y) - b) - b = l(y) - 2b$. ∎

An action of a group on an **R**-tree is said to be *minimal* if there is no proper subtree on which G acts. If G acts on T and $u \in T$ then the completion of the union of the geodesics from u to xu , for all $x \in G$, is a subtree T_u of T, which is a copy of Chiswell's tree corresponding to the length function $l = l_u$. It follows that any minimal action must be on a Chiswell tree arising from some length function.

THEOREM 1. *Let G act on an **R**-tree T.*

(i) If $N \neq G$ then T has a unique minimal invariant subtree, namely the completion of the union of all axes in T.

(ii) If $N = G$ has bounded action then T contains minimal invariant subtrees, which are single points fixed by G.

(iii) If $N = G$ has unbounded action then T contains no minimal invariant subtree. This case can occur if and only if G has a tower of subgroups

$$H_0 \subset H_1 \subset H_2 \subset \dots \quad \text{with} \quad \bigcup_{i \geq 0} H_i = G.$$

Proof. Case (i) is Proposition 3.1 of [4], except that we require our trees to be complete.

If $G = N$ has bounded action then by Theorem 3.2 of [11], G fixes some point of T. A minimal invariant subtree is therefore given by any such point fixed by G.

Suppose $G = N$ has unbounded action. To show that any such action cannot be minimal we can assume that T is given by Chiswell's construction from an unbounded length function l. Take $x \in G$ with $2b = l(x) > 0$ and $w = [x, b] \in T$. For $y \in G$, yw is given by Lemma 2. If $l(y) \leq l(x)$ then $yw = w$ and if $l(y) > l(x)$ then $yw = [y, l(y) - b]$. Now by Lemma 1, if $l(y) > l(x)$ then $l(x^{-1}y) = l(y)$ and so $2c(x^{-1}, y^{-1}) = l(x) = 2b$. Hence $[y, b] = [x, b] = w$. It follows that the geodesic from w to yw contains only points $[y, m]$ with $m \geq b$. The completion of the union of all such geodesics, which is the Chiswell tree arising from l_w, is thus a proper invariant subtree of T.

If l is unbounded then there exist $0 = r_0 < r_1 < r_2 < \ldots$ with $r_n \to \infty$ and elements $x_i \in G$ with $l(x_i) = r_i$. Let $H_i = \{x \in G \mid l(x) \leq r_i\}$, then by Lemma 1, H_i is a subgroup of G, and

$$H_0 \subset H_1 \subset H_2 \subset \ldots \quad \text{with} \bigcup_{i \geq 0} H_i = G.$$

Conversely if there is such a tower of subgroups of G then proposition 4 of [9] shows that an unbounded length function l with $N = G$ is given by $l(x) = \inf\{i \, ; \, x \in H_i\}$. ∎

It is possible for a bounded l to be given by an infinite tower of subgroups of G, as described in [9]. Completeness is required here for the corresponding Chiswell tree to have a fixed point of G (see the proof of Theorem 3.2 of [11]).

For two length functions $l,\, l' : G \to \mathbf{R}$, write $l \sim l'$ if there is an action of G on some tree T, and points $u,\, v \in T$ with $l = l_u,\ l' = l_v$. A length function l is said to be *non-Archimedean* if $N = G$.

Theorem 2. *Let $l,\, l' : G \to \mathbf{R}$ be non-Archimedean length functions. Then $l \sim l'$ if and only if there exists $r,\, s \geq 0$ such that for all $y \in G$*

$$\max(l(y) - r, 0) = \max(l'(y) - s, 0).$$

Proof. The condition can only be satisfied if both l and l' are bounded or both are unbounded. If both are bounded then for suitable $r,\, s \geq 0$ the lengths can be reduced to zero and the condition is satisfied. By Theorem 3.2 of [11] the corresponding actions of G on Chiswell's trees have fixed points. Identifying the two trees at a fixed point gives a tree from which both l and l' are defined.

Suppose that l and l' arise from an unbounded action of G on a tree T, with $l = l_u$ and $l' = l_v$. Then T contains invariant subtrees T_u and T_v which are given by Chiswell's construction from l and l' respectively. If T_u and T_v are disjoint then by [4; 1.1] there is a unique spanning geodesic from a point $u' \in T_u$ to a point $v' \in T_v$. For $x \in G$, $xu' \in T_u$ and $xv' \in T_v$ and since $d(xu', xv') = d(u', v')$ it follows that $xu' = u'$ and $xv' = v'$. Hence G fixes each point of the spanning geodesic. But this is not possible for an unbounded action by Theorem 3.2 of [11]. The subtrees T_u and T_v must therefore contain a common point w.

Since $w \in T_u$, $w = [x, b]$ for some $x \in G$ with $0 \le b \le l(x)$. If $2b \le l(x)$ then by applying Lemma 2 to the action of G on the Chiswell tree T_u, $l_w(y) = \max(l(y) - 2b, 0)$. If $l(x) < 2b$ then $2b' = 2l(x) - 2b < l(x)$. By Lemma 1, $2c(x^{-1}, x) \ge l(x) > 2b'$ and so $x[x, b'] = [x, l(x) - b'] = [x, b]$. So $w = xw'$ where $w' = [x, b']$. Hence $l_w(y) = l_{xw'}(y) = d(yxw', xw') = d(x^{-1}yxw', w') = l_{w'}(x^{-1}yx) = \max(l(x^{-1}yx) - 2b', 0)$ by Lemma 2. Since $w \in T_v$ the same analysis can be carried out replacing l by l'. Moreover we can assume that $w = [x, b]$ in T_u with $2b \le l(x)$, or otherwise w could be replaced by xw. Hence for some r, $s \ge 0$ and $x \in G$, $l_w(y) = \max(l(y) - r, 0)$, $l'_w(y) = \max(l'(x^{-1}yx) - s, 0)$. Equating $l_w(y)$ and $l'_w(y)$ gives the condition $\max(l(y) - r, 0) = \max(l'(x^{-1}yx) - s, 0)$. The condition still holds if r, s are increased by the same amount; and so r, s can be taken as large as required. By applying Lemma 1 twice $l'(x^{-1}y) = l'(y)$ and $l'(x^{-1}yx) = l'(y)$ if $l'(y) > l'(x)$. Also $l'(x^{-1}y) \le l'(x)$ and $l'(x^{-1}yx) \le l'(x)$ if $l'(y) \le l'(x)$. Thus for $s \ge l'(x)$, $\max(l'(x^{-1}yx) - s, 0) = \max(l'(y) - s, 0)$, giving the required condition.

Conversely suppose that length functions l, l' are such that $\max(l(y) - r, 0) = \max(l'(y) - s, 0)$. Chiswell's construction gives a tree T_u with $l_u = l$, and a tree T_v with $l_v = l'$. By Lemma 2 each of these trees contains a subtree on which G acts arising from Chiswell's construction associated with the length function $l''(y) = \max(l(y) - r, 0) = \max(l'(y) - s, 0)$. Identifying these subtrees in T_u and T_v gives a tree from which both l and l' are defined. In the first paragraph of this proof, and also here, two trees are identified by a common subtree. That this results in a tree follows from section 3 of [5] or section 2 of [11] concerning the four-point condition for trees.

References

[1] R. C. ALPERIN and H. BASS. Length functions of group actions on Λ-trees, in *Combinatorial group theory and topology*, S. Gersten and J. Stallings, eds. (Princeton, N.J., Princeton Univ. Press, 1986) 265-378.

[2] R. C. ALPERIN and K. N. MOSS. Complete trees for groups with a real-valued length function. *J. London Math. Soc.* (2) **31** (1985) 55-68.

[3] I. M. CHISWELL. Abstract length functions in groups. *Math. Proc. Cambridge Phil. Soc.* **80** (1976) 451-463.

[4] M. CULLER and J. W. MORGAN, Group actions on **R**-trees. *Proc. London Math. Soc.* (3) **55** (1987), 571-604.

[5] W. IMRICH. On metric properties of tree-like spaces, *Beiträge zur Graphentheorie und deren Anwendungen*, (ed by Sektion Marök der Technischen Hoshschule Ilmenau, Oberhof, DDR 1977) 129-156

[6] W. IMRICH and G. SCHWARZ. Trees and length functions on groups. *Combinatorial Mathematics* (Marseille-Luminy, 1981) 347-359, (Amsterdam, 1983)

[7] R. C. LYNDON. Length functions in groups. *Math. Scand.* **12** (1963) 209-234

[8] J. TITS. A 'Theorem of Lie-Kolchin' for trees. *Contributions to algebra* (Academic Press, New York, 1977) 377-388

[9] D. L. WILKENS. On non-Archimedean lengths in groups. *Mathematika* **23** (1976) 57-61

[10] D. L. WILKENS. Length functions and normal subgroups. *J. London Math. Soc.* (2) **22** (1980) 439-448

[11] D. L. WILKENS. Group actions on trees and length functions. *Michigan Math. J.* **35** (1988) 141-150

School of Mathematics and Statistics
University of Birmingham
Birmingham B15 2TT

Printed in the United States
By Bookmasters